Water, Peace, and War

Water, Peace, and War

Confronting the
Global Water Crisis

Brahma Chellaney

ROWMAN & LITTLEFIELD PUBLISHERS, INC.
Lanham • Boulder • New York • Toronto • Plymouth, UK

Published by Rowman & Littlefield Publishers, Inc.
A wholly owned subsidiary of The Rowman & Littlefield Publishing Group, Inc.
4501 Forbes Boulevard, Suite 200, Lanham, Maryland 20706
www.rowman.com

10 Thornbury Road, Plymouth PL6 7PP, United Kingdom

British Library Cataloguing in Publication Information Available

Library of Congress Cataloging-in-Publication Data
Water, peace, and war : confronting the global water crisis / Brahma Chellaney.
    pages cm
    Includes bibliographical references and index.
    ISBN 978-1-4422-2139-0 (hbk. : alk. paper) — ISBN 978-1-4422-2140-6 (electronic)
1. Water resources development. 2. Water-supply—Political aspects. 3. Globalization—Economic aspects. I. Title.
    HD1691.C44 2013
    333.91—dc23

                                                                              2012050091

Printed in the United States of America

Contents

List of Illustrations   ix

Introduction: Our Most Precious Resource under Threat   xi

1   The Specter of Water Wars   **1**
   Water, the New Oil   5
     Falling Water Tables and Declining River Flows   8
     Implications of the Widening Water Gap   11
     A Basic Right Denied to Many   17
   The Fundamentals of Resource Pressures   19
     Our Changing World   21
     Resource Acquisition as a Geopolitical Driver   24
     Key Factors behind Spiraling Resource Consumption   28
   The Increased Risk of Water Conflicts   35
     Weak Water Accords and Institutions   39
     Waging Water War by Another Name   46
     Water as an Object, Target, or Tool of Warfare   55

2   The Power of Water   **59**
   The Green/Blue Water Equation   62
   Water Stress and Rising Food Demand   64
     The Role of Irrigation in Resource Depletion   73
     Changing Diets Accentuate Water Scarcity   77
   The Intersection of Water and Energy   84
     Water Appropriations for Energy Production   86
     Hydropower: Strengths, Impacts, and Vulnerabilities   92
     Nuclear Power's Water Risks   95
     Addressing Water-Related Energy Challenges   101

Can "Virtual Water" Trade Be the Answer?   104
   A Nascent, Evolving Field   106
   Striking Realities   107
   The Limits of Virtual-Water Trade Benefits   111

3 The Future of Water   **119**
New Challenges, New Beginnings   120
The Commoditization and Marketing of Water   125
   The Rise of the Bottled-Water Industry   128
   Water Trading within and between Nations   133
New Threats to Water: Environmental and Climate Change   141
   Assessing the Growing Anthropogenic Impacts   143
   The Hydrological Impact of Environmental Change   147
   Water-Related Challenges in an Era of Global Warming 151
Forestalling a Thirsty Future   160
   Water Refugees and Water Warriors   161
   The Dangers of a Parched Future   165

4 Changing Water Cooperation, Competition, and Conflict   **175**
Water Control as Potential Political Weapon   176
Water as a Driver of Territorial Disputes   181
   Redrawing Political Maps Compounds Water Challenges 183
   The Nexus between Territorial and Water Feuds   187
Lessons from Past Water Cooperation   202
   The Danube and Rhine Commissions   204
   The U.S.-Canada and U.S.-Mexico Treaties   207
   Building Transnational Water Institutions:
     An Evolutionary Process   213
Today's Grating Hydropolitics   217
   The Intersection of Water Scarcity, Overpopulation,
     and Terrorism   218
   Fashioning Water Pipelines for Peace   225
A Water Hegemon with No Modern Historical Parallel   229
   China's Establishment of Hydro-Hegemony   230
   The Wages of Dam Frenzy   236

5 Shaping Water for Peace and Profit   **243**
Building International Water Norms and Rules   246
   Strengthening International Rules   248
   The UN Convention, Helsinki Rules, and Berlin Rules 252
   The Role of the International Court of Justice   257
Building Bilateral or Basinwide Institutions   259
   Transnational "No Harm" Principle in Practice   259
   Moderating the Problematic Doctrine of Prior Appropriation 261
   Broader Parameters to Alleviate Conflict   263

Pathways to Help Contain the Risks 268
   Stemming the Security Risks 270
   Mitigating the Economic Risks 277
Why Cooperative Politics Holds the Key 291
   Law versus Politics 293
   Investing in Water Peace 296

Appendixes
   A: Web Links to International Water Norms 301
   B: Genuine Intercountry Water-Sharing Agreements Currently
      in Effect 303

Notes 307

Glossary 357

Index 365

About the Author 399

Illustrations

TABLES

1.1 World Population Milestones   29

2.1 Global per Capita Cropland and Irrigated Land Area   69

2.2 Average Water Requirements of Main Food Products   78

2.3 World's Top Ten Rice Exporting and Importing Countries   116

3.1 Accelerated Thawing of Snow and Ice Layers and Its
Significance for Freshwater Supply   156

3.2 Dependence of the Major Rivers on Glacial and
Snow Meltwaters That Originate in or around the
Great Himalayan Range   158

4.1 Internal Freshwater Resources in Different Continents
and Subregions   219

4.2 Major Rivers Flowing Out of Chinese Territory to
Other Countries   232

MAPS

1.1 Israel's War-Secured Throttlehold on Subregional
Water Resources   51

3.1 The Murray-Darling Basin   122

4.1 Rivers of the Indus System, and the Treaty Line
       Partitioning Rivers   194

4.2 The Euphrates-Tigris Basin and Turkey's GAP Program   198

4.3 The Kurdish-Majority Region   202

4.4 China's Major Rivers and the Three Routes of Its Great
       South–North Water Diversion Project   240

FIGURES

1.1 Comparison of Freshwater Availability in Different Continents,
       per Capita   9

1.2 Average Water Use per Person per Day in Selected Countries   15

1.3 The Sharp Rise in Global Commodity Prices since the 1970s   36

2.1 Global Water Use by Different Sectors   65

2.2 World Demand for Cereals, 1965–2030   65

2.3 Water Lost in Irrigation   76

2.4 The Impact of Growing Meat Consumption on Global
       Water Resources   80

2.5 Water Consumption in Thermoelectric-Power Generation
       Using Different Cooling Technologies   87

4.1 World's Poorest States in Aggregate Water Resources   220

4.2 The Arc of Islam Faces Vicious Circle   221

5.1 Per Capita Freshwater Availability in the World's Fifteen
       Leading Economies   269

5.2 Water-Storage Capacity of Selected Countries   287

Introduction

Our Most Precious Resource under Threat

When the present international institutions were set up after World War II, few anticipated that freshwater shortages would emerge as a serious economic and security challenge. Many at that time thought water resources were so abundant that they would never run out, just as many believed there were plentiful quantities of precious metals and energy resources to sustain global needs. Yet the golden age of safe, cheap, and easily available water has come to an end in most parts of the world, replaced by a new era of increasing supply and quality constraints. Even some regions that traditionally boasted a wealth of freshwater resources are now beginning to seriously fret about adequate water availability in the coming years. Indeed, the availability of all natural resources has started coming under pressure because of the world's phenomenal economic and demographic expansion and soaring consumption levels.

Water, as a life-creating, life-supporting, and life-enhancing resource, poses the biggest challenge of all natural resources because it has no replacement and is not internationally traded like oil, gas, and mineral ores. This makes adaptation to water scarcity more onerous.[1] Inadequate access to clean water has emerged, for instance, as one of the world's biggest public-health problems by contributing to the spread of deadly bacteria and microorganisms. In fact, as the world enters an unfamiliar era of serious water stress, compounded by degradation and depletion of many watercourses (rivers, lakes, and underground aquifers), the specter of wars over freshwater looms on the horizon.[2]

Politicians and media often debate geopolitical prospects that are *not* going to happen while ignoring significant developments with long-term ramifications unfolding before their eyes. The growing freshwater stress in large parts of the world is one such geopolitical development with grim, long-term implications. Global demand is outstripping the earth's water-renewable capacity, forcing communities to tap deep aquifers where water has slowly accumulated over thousands, maybe even millions, of years.

Yet this situation—made worse by water pollution, profligate practices, uneven distribution of water resources, and festering water disputes arising from political boundaries crisscrossing natural water flows—barely rates high-level international consideration. Just reflect on one sobering fact: the retail price of bottled water is already higher than the international spot price of crude oil. Is it thus any surprise that many investors are beginning to view water as the new oil of the twenty-first century? But unlike oil, water has no known substitutes, making it more valuable from a long-term perspective.

Two-thirds of our planet is covered with water (much of it seawater), yet the portion of the world population living in water-stressed conditions is set to rise from slightly more than half today to two-thirds in the 2020s. These conditions will be worse in some regions than others, with the United Nations projecting that 47 percent of the global population is likely to be living in areas reeling under "high" water stress by the end of the next decade.[3] Water shortages already are bringing the battle between water conservation and economic development into sharp relief.

Against this background, shared water has increasingly become a divisive issue in the intercountry and intracountry context. Intracountry water disputes are rife. They trouble many countries, including the world's most powerful economy, the United States, where water wrangles were once restricted to the arid West but have inexorably spread eastward in response to urban and population growth, which has created water shortages and conflicting uses. Some long-standing internal conflicts in the developing world are rooted in water issues. The bloody Darfur conflict in Sudan's far west, for example, arose from clashes over water and grazing rights.[4] In subregions with ethnically overlapping political boundaries, intracountry water conflicts can even assume intercountry dimensions.

Although intracountry water conflicts are more frequent and violent, intercountry feuds raise greater concerns because they threaten wider peace and security and impede regional collaboration and integration. In fact, the sources of interstate disputes—ranging from water storage and diversion projects to the absence of basin-level rules on harnessing shared resources—tend to foster unregulated competition and political tensions. Water feuds are especially intense in Asia, the Middle East, and Africa. Examples include the acrimony and charged competition among river-basin states over the waters of Africa's Nile, the Amu Darya in Central Asia, and the Tigris-Euphrates in the Middle East. Even where formal water arrangements are in force, as between India and Pakistan, between Malaysia and Singapore, or between Israel and Jordan, tensions over water pose a security risk. But as the periodically strained interriparian relations between the United States and Mexico show, water can lurk as an underlying factor in intercountry frictions.

Of all the environmental-security challenges confronting the world, ensuring an adequate supply of clean water has become the most pressing. This is a challenge that holds wide socioeconomic and security ramifications. Internationally, meeting the increasing water needs of agriculture, industry, and households is becoming more difficult than ever. As hundreds of millions of citizens climb up the economic ladder and attain middle-class status, major increases in water demand will be driven not merely by population and economic growth but by energy, manufacturing, and food-production needs to meet rising consumption levels.

Growing prosperity, population size, and economic development are not the only factors behind the soaring consumption: the global population is also getting fatter, especially in wealthier countries.[5] This promises to have a big impact on water demand, as fat people consume more water-intensive resources like food and energy than those who are fit, thus indirectly driving overexploitation of natural resources, deforestation, and the release of greenhouse gases. Yet the obesity threat to water resources and the environment has received little attention.

Several of the world's fast-growing economies are already in or near water-stressed conditions, underscoring the potential danger to their continued rapid economic growth from an unmitigated water crisis. Water paucity is also becoming an important challenge in some developed countries, such as Spain, South Korea, and Australia. For investors, water shortages carry risks that potentially are as damaging as nonperforming loans, industrial overcapacity, real-estate bubbles, and political corruption. In some respects, the long-term economic risks related to scarce water resources may actually rank above other problems.

In fact, at a time when nontraditional threats are weighing heavily in the national-security calculus of many states and becoming internationally prominent, there are growing risks of unconventional water conflicts, waged with the aid of economic or riparian leverage, terrorist proxies, or other covert means. A U.S. intelligence assessment publicly released in 2012 warned that "the use of water as a weapon will become more common during the next 10 years, with more powerful upstream nations impeding or cutting off downstream flow." The report—prepared at the request of the State Department, and based on a classified National Intelligence Estimate—said the use of water to "further terrorist objectives" will also become more likely, with "extremists, terrorists, and rogue states" threatening physical infrastructure, including dams, as "convenient and high-publicity targets."[6]

Ominously, with the easily available surface-water and groundwater resources getting depleted or degraded, many nations are turning to nontraditional sources of water. These sources range from deep, "fossil" (ancient) aquifers cut off from any significant recharge to wastewater reclamation. Interbasin water transfer (IBWT) projects also symbolize the pursuit of unconventional sources of water.

In this light, it is significant that the global rise of nontraditional security threats has coincided with the increasing resort of countries and communities to nontraditional sources of water—a phenomenon likely to be accelerated by efforts to alleviate water stress. Just as the new nontraditional security challenges often transcend national frontiers, several of the nontraditional sources of water being tapped are "not restricted by the confines of watershed boundaries."[7] They include desalination, wastewater reclamation, the tapping of fossil aquifers, and IBWT.

The juxtaposition of the changing nature of threats and conflicts with the changing sources of water is portentous in the context of spreading water scarcity. Whereas water recycling and desalination are positive trends, IBWT projects and fossil-aquifer exploitation carry serious environmental costs. They may also hold transboundary implications and thus become an additional source of interriparian friction and conflict.

The changing nature of conflict and growing water shortages actually constitute a potent mix. Water scarcity's fateful link with religious extremism and political turmoil has been highlighted by the popular uprisings and unrest since early 2011 in a number of countries that fall below the international water-poverty threshold—defined as per capita availability of less than 1,000 cubic meters of water annually. The Arab Spring movements—which have helped to empower Islamists while simultaneously unleashing broader sectarian forces that have brought the fragile state structure of the Middle East under strain—were triggered not just by political repression and injustice but also by rising food prices, linked to the worsening regional freshwater crisis.[8] Water scarcity, by promising to engender greater political and social upheaval, poses an existential threat to the Arab world, where water already seems a more valuable resource than oil or gas in several states.

More broadly, given the critical role of water resources in economic production and social progress, it may be no accident that many of the nations widely regarded as failed or failing states are water famished. In fact, some armed water conflicts rage unnoticed within weak states. For example, violent skirmishes over water and grazing rights still recur between the pastoral populations along the border between the Republic of South Sudan and the Central African Republic, and between Ethiopia and Kenya. As water sources recede, pastoralists have to range more widely in search of water and grazing land, bringing them in conflict with other nomadic herdsmen, with rival herders at times ready to fight for and die over access to scarce water and grass.

THE PAST, THE PRESENT, AND THE FUTURE

Water helps to sustain and expand the modern economies and preserve the ecosystems on which human civilization depends. Water powered the industrial revolution—including through the steam engine, the water turbine, and water-intensive manufacturing technologies—and helped disprove the "Malthusian catastrophe" thesis, which contended that population growth would swamp the earth's agricultural production capacity, leading to famine and a return to subsistence-level conditions.[9] Thomas Malthus, an English economist and demographer familiar with England's history of plagues, famines, and resource scarcities, contended at the end of the eighteenth century that available food supply would dictate human population size because food production can only increase arithmetically ($3 + 3 + 3 = 9$) while population would grow geometrically ($3 \times 3 \times 3 = 27$).

His dark thesis was turned on its head by the scientific advances ushered in by the industrial revolution, the transportation revolution, the green revolution, and the biotech revolution. With a declining percentage of human society engaged in agriculture, the world has managed to produce increasing quantities of food, thanks to intense use of water and fertilizers and new farm varieties and techniques.

Yet, with water use having grown at more than twice the rate of population increase in the past one hundred years, water shortages threaten to crimp future economic growth, even as they sharpen current regional hydropolitics and challenge human ability to innovate and live in harmony with nature. Food security, for example, is emerging as a major challenge by itself, largely due to water constraints. High levels of fertilizer application have actually caused water pollution and eutrophication—a process where waterways receive a heavy concentration of nutrients like phosphates and nitrates, resulting in excessive algae growth. Crop-yield growth has slowed globally since the late 1980s, although rising yields are critical to food security.[10] According to the United Nations Food and Agriculture Organization, yield growth accounted for more than two-thirds of the increase in crop production in the developing world in the past half century.

Despite having averted the Malthusian catastrophe, the world faces varied resource constraints that have resurrected the warning issued by a nongovernmental think tank, the Club of Rome, in its contentious 1972 book, *The Limits of Growth*. The book concluded—on the basis of computer models developed at MIT in an era where, with the advent of the first microprocessor, computers commanded reverential respect—that if high rates of consumption, population growth, and resource degradation and depletion persisted, "the limits of growth on this planet will be reached sometime within the next 100 years. The most probable result will be a rather sudden and uncontrolled decline in both population and industrial capacity."[11] Although decades later that conclusion still sounds alarmist, the world admittedly faces the imperative to gradually shift to more sustainable development.

The conventional wisdom that market forces would come to the rescue by finding a way out from the scarcity of any natural resource proved true in much of the twentieth century as discovery of new sources of supply and technological innovations helped to largely stabilize resource prices, even though international commodity prices began climbing from the early 1970s. The high price volatility in the twenty-first century, however, points to the emerging strains on resource systems due to rapidly rising consumption levels, environmental degradation, and other factors, thus increasing the risks of unprecedented resource-related shocks in the future.[12] Demand for many resources is rising faster than the available supply, reflected in the geographical spread of water shortages and the increasing prices of commodities like energy, grains, steel, and metals.

More than one-fifth of the global population still lacks ready access to potable water. No less shocking is the fact that more people today own or use a mobile phone than have access to water-sanitation services.[13]

Because the degradation of groundwater, unlike surface water, is not visible to the human eye, groundwater resources tend to be more recklessly exploited, in keeping with the "out of sight, out of mind" tendency. With water being pumped out faster than nature's replenishment capacity, aquifer depletion is drying up groundwater-fed wetlands and lakes and causing the land surface to sink owing to the emptying of the underlying water. In over-tapped coastal aquifers, saline seawater is flowing in to replace the freshwater that has been pumped out. Saltwater intrusion and the formation of sinkholes serve as a striking testament to the overexploitation of groundwater resources. River depletion is also a problem, with increasing quantities of river waters being rerouted for irrigation, energy production, mining, and manufacturing. Many hydropower plants impound river waters in massive reservoirs.

Meanwhile, a consequence of the current water shortages is that local resistance is growing within some countries to government or corporate decisions to build water-intensive energy or manufacturing plants in areas already water stressed. Nongovernmental organizations and citizens' groups have led grassroots movements against the setting up of water-guzzling industries in areas where industrial and food production demands have constrained local water supply. Such protests, especially in democratic states, have even halted or considerably delayed major projects, driving up their costs.[14] But with electricity demand in warm, water-scarce regions increasing at a faster pace than elsewhere, resistance to new power plants tends to only exacerbate the energy shortfall.

The greater the seasonal variability of freshwater availability in an economy, the greater that economy's risk of experiencing water-induced shocks. To moderate such risks, major investments are necessary, for example, in building a large national capacity to store water in the wet season for release in low-availability periods.

Indeed, with the existing global water-supply infrastructure coming under strain and water quality deteriorating, considerable financial costs must be incurred nationally to tackle rising demand, including the building of new water facilities, networks, and services; investing in clean-water technologies; tapping nonconventional water sources; and introducing smart water-pricing policies. Otherwise, constraints on water quality and quantity will exact growing socioeconomic costs while encouraging users to drain surface water and dig even deeper wells. Without corrective measures, acute water stress, coupled with environmental and climate change, is bound to cause lifestyle changes and socioeconomic disruptions.

Wealthy and emerging economies will seek to buffer the impacts of water scarcity by investing in storage infrastructure, clean-water technologies, and other engineering and technological measures that can help mitigate the distress. It is the underdeveloped economies, many of them already reeling under water and environmental stresses, that will likely bear the brunt of the global water crisis. Poor, drought-prone countries will also remain highly vulnerable to water-related economic shocks because they lack financial, institutional, and human resources to build improved storage, invest in new technologies, and set up plants to recycle water or treat brackish or contaminated water.

The water crisis is also raising intercountry security risks, highlighted by increasingly grating hydropolitics. Just as the scramble for energy resources has long shaped geopolitical rivalries, the struggle for water is becoming a defining fulcrum of regional politics and security, roiling interriparian relations. The State Department announced in 2010 that it was upgrading water scarcity to "a central U.S. foreign policy concern." After all, intercountry water discord impinges on American interests, including impeding collaboration between U.S. allies and friends and creating regional instability. With the source of a water feud often serving as the obstacle to regional cooperation, the escalating competition over transboundary water resources threatens to become a trigger for larger geopolitical conflict.

Water war as a concept may not mesh with the conventional construct of warfare, especially for those who visualize war with tanks, combat planes, and attack submarines. Yet armies don't necessarily have to march to battle to seize or defend water resources. Water wars can be waged and won by nonmilitary means, such as by reengineering transboundary flows. But when military force has been employed by any country in the post–World War II period to change the regional water map—irrespective of whether its intent was apparent or hidden—the plan was conceived at the highest political level and the target was headwaters-controlling territory.

Conquest of territory is no longer practicable in a world in which sanctity of borders has become a powerful norm to help underpin peace and stability.[15] Yet water can be fashioned into a weapon by various means, including commandeering shared resources. A dominant riparian, even if located downstream, can enjoy the lion's share of river-basin resources by resorting to coercive diplomacy to deter the upstream construction of dams, barrages, and other water diversions. In the case of a transboundary aquifer, a regional power seeking to retain its pumping advantage can deter a co-basin state from setting up a competing extraction project.

In the small number of cases where a river and an aquifer flow in opposite directions across a political boundary, one riparian may have the ability to undergird a mutually dependent relationship with another, more-advantageously-placed riparian by manipulating the alternative river or aquifer. But in situations where riparian ascendancy is unhindered—and unreceptive to international norms and principles and to the interests of a subjacent state—a low-intensity water conflict, at a minimum, is likely to arise and may even prompt the aggrieved nation to build countervailing military power.

Averting water wars demands rules-based cooperation, water sharing, uninterrupted data flow, and dispute-settlement mechanisms. Transparency, collaboration, and sharing are the most essential elements. Despite the ubiquity of transnational basins, the number of genuine water-sharing treaties in the world, unfortunately, remains remarkably small. Most transboundary river, lake, and aquifer basins, in fact, lack any treaty-based arrangements to promote cooperation between co-riparian states on harnessing and managing the common resources.

This century has yet to witness a single new water-sharing treaty covering a major river, aquifer, or lake. This suggests that in an era of growing water stress, it will be more challenging to conclude intercountry water-sharing treaties. And when they do materialize, they are unlikely to mirror the generous terms of some of the treaties that were signed in the past when water shortages were not known or significant. Because the scope of water collaboration is set to alter with the changing circumstances, the past examples hold only a limited utility, even though they remain important models that illustrate what cooperation can yield or how tensions can be minimized.

If there is good news, it is that the spotlight on potentially serious climate-change impacts has helped to move water challenges into the international mainstream. There is increasing appreciation that water security needs to be an important component of national, regional, and international security policies. The water "securitization" trends, however, have also highlighted the increased risks that feuds over shared water resources hold for peace and stability.

For corporations, the costs of doing business in water-stressed economies are set to mount, even as water-intensive industries come under increasing pressure to reduce their water footprints through greater efficiency. The potential risks for investors will likely rise. But the business opportunities from scarcity, including greater privatization and commoditization of water resources, are expected to more than balance the risks.

OUTLINE OF THE BOOK

This book is a study of the global linkages between water and peace for all those interested in the issues that relate to the world's most vital yet most undervalued natural resource. By synthesizing research materials from the scientific fields (including hydrology, geology, agriculture, medicine, sustainable science, and engineering) and policy fields (environment, security, international law, and international relations), the book brings out the basic interconnections between different academic disciplines for a holistic understanding of the challenges relating to water—a cross-sectoral resource whose management involves diverse specialists, such as hydrologists, engineers, economists, lawyers, political scientists, sociologists, geographers, biologists, and geologists.

As a truly multidisciplinary study, the book presents an integrated picture of the implications for international peace and security of spreading water stress and sharpening water competition. The technical material is explained in simple terms and fully blended with the book's broader geopolitical focus. The book is thematically structured—a more challenging yet more illuminating way of examining the relationship between water and peace. It keeps the thematic exposition of its arguments global in scope throughout, using specific regional examples as illustrations.

In fact, as the first comprehensive, policy-oriented study of the geostrategic dimensions of the global water crisis, this volume fills a conspicuous gap in the burgeoning literature on water. It can thus claim to be a pioneering study.

By placing complex, multifaceted issues in a larger strategic framework and explaining them in ways that even a general reader can comprehend, the book aims to become the recognized and readable authority on a central challenge the world confronts.

The book presents the big picture in chapter 1, including the factors that have left humanity to contend with growing water shortages. Water scarcity now haunts not just the arid lands but also the fertile plains in several regions, including the cradles of human civilization. The factors bringing water resources under strain include rapid economic and population growth, soaring consumption levels, fast-paced urbanization, and an obesity epidemic. The human population, having expanded sixfold in numbers in the past two centuries and having gained extra body weight especially in recent decades, is at risk of becoming too large and fat for earth to manage.

One consequence is that groundwater has become the world's most extracted natural resource. The sharpening struggle for freshwater resources threatens to overshadow the current struggle for energy resources. The chapter also weighs the peace and security implications of the increasing mistrust and divisiveness over transboundary water issues, the weakness of the existing conflict-prevention mechanisms, past cases where water wars were waged by another name, and the growing risk of water systems becoming targets of attack.

In chapter 2, the book turns to the central role of water in socioeconomic advances and the specific challenges water shortages pose for achieving greater food and energy production and manufacturing. There is also the imperative to maintain a natural balance between "blue water"—the resources in rivers, lakes, and aquifers used for economic activity and direct human consumption—and "green water," or the water from precipitation that naturally supports rainfed agriculture and other plants and trees. At a time when international food prices are spiraling and malnutrition remains rampant in large parts of the world, food-related challenges are being compounded by unsustainably heavy water abstractions for irrigation as well as by changing diets, especially the greater intake of meat, whose production is notoriously water intensive.

To meet the rapidly rising consumption levels, water stress dictates that the world grow more food with less water. Yet ill-advised biofuel subsidies and mandates, as the chapter shows, have resulted in food production competing with biofuels for scarce water and land resources, as well as in the diversion of vast amounts of farm produce away from the food chain, thus aiding food-price inflation. Paucity of water resources, meanwhile, is driving some cash-rich countries to produce food for their home markets on farmland leased in impoverished nations where many citizens remain hungry or undernourished. The sharp increase in global meat production, for its part, means more demand for animal feed, which in turn is speeding up water-consumption rates. Even more than human population growth, the fast-expanding livestock population is bringing water resources under strain. The livestock population is many times bigger than the human population and its ecological footprint thus is also larger. Almost 10 billion animals are slaughtered every year in the United States alone.

The chapter investigates two key questions: To what extent can international trade in "virtual water"—the freshwater used in the production processes and thus embedded in traded goods and services—help alleviate water distress? And given the fast-rising demand for energy to support economic growth and the continuing lack of access to electricity by nearly one-fifth of the global population, how can the world produce more power without exacerbating water shortages?

The bulk of the world's water abstractions already go to meet the thirsty nature of food and energy production, with higher energy prices actually translating into higher food prices. Adequate availability of local water resources, for example, is a must for setting up a thermoelectric power plant fueled by, say, coal, uranium, natural gas, or oil. The cooling technology such plants employ greatly determines the extent of their water use. The building of water-guzzling nuclear power plants has increasingly been pushed to the coastlines so that they rely on seawater. Yet natural disasters like storms, hurricanes, and tsunamis are becoming more common, owing to climate change, which will also cause a rise in ocean levels, making seaside reactors more vulnerable—a danger highlighted by the Fukushima disaster.

Chapter 3 explores what the future may bring by examining new water-related issues, as exemplified by the dramatic rise of the bottled-water industry since the 1990s and the threats from environmental and climate change. The growing trend toward the commoditization of the world's underpriced, increasingly scarce water seems unstoppable as businesses seek to capitalize on shortages by convincing politicians that only a free marketplace can meet the challenges. Yet this trend is widening the global rich-poor gap, with the prosperous largely relying on bottled drinking water even as many poor people lack access to a basic water supply.

Some water markets, where water rights are tradable like property rights, have developed, including in Australia, Chile, and the American West. This trend suggests that as water scarcity intensifies, moves to internationally trade bulk water could gain commercial traction. The chapter examines the constraints that must be overcome for that to become an everyday reality, not just a temporary arrangement to alleviate a searing drought, as when Spain's Catalonia region imported freshwater from France by tankers in 2008.

The chapter also analyzes threats to water resources resulting from environmental change and climate change, drawing a distinction between the two. The greatest impact of human-induced alterations to natural systems is on freshwater ecosystems. Climate and freshwater systems are interlinked, and changes in one system tend to bring about changes in the other. Weather-modification technologies, meanwhile, are opening a new frontier for human intervention, including for potentially nonpeaceful uses. The weather-modi-fication activities currently being pursued, such as artificially inducing rain or snow, carry unpredictable effects for existing climatic patterns that are not confined by national boundaries. The chapter additionally looks at the emergence of water warlords in some parched regions and the potential for large-scale flows of refugees displaced by severe water scarcity.

Chapter 4 probes the changing patterns of cooperation, competition, and conflict over water and their broader, long-range implications for peace and security. It highlights the lessons to be drawn from past water cooperation. However, historical examples may hold little value for guiding future coop-eration because water shortages in the past were geographically confined to arid areas and were relatively unknown in the rest of the world. Tellingly, efforts to frame an international convention defining shared-water principles and norms did not gain momentum until the 1990s—that is, until rapid eco-nomic, demographic, and consumption growth had placed the world on the cusp of a new era of serious water stress.

The chapter details how water can be fashioned into a political weapon, wieldable without any resort to military force yet able to keep a co-riparian state under sustained pressure. It also analyzes the potent nexus between wa-ter and territorial disputes. The rejiggering of political frontiers as part of the process of decolonization led to a number of intractable water disputes and other still-festering conflicts in the developing world. Political boundaries have changed even after decolonization due to internal or regional conflicts or national dysfunction, resulting in more divided basins and more water dis-putes. Water remains a major driver of several territorial disputes and separat-ist problems. It thus holds the key to resolving such political feuds and fights.

Another perilous nexus examined is between water scarcity, overpopula-tion, and terrorism. The arc of Islam—which has most of the world's poorest states in aggregate water resources, the principal springboards of interna-tional terrorism, and exploding populations—risks getting irredeemably trapped in a circle in which demographic and economic pressures exacerbate water scarcity, which in turn spurs greater unemployment, food insecurity, extremism, and political instability, forcing governments to step up water and other resource-depleting subsidies.

This veritable vicious cycle might generate extreme water scarcity, wider political turmoil, a deepening jihad culture, and more transnational extremist violence. In fact, many of the states in the arc of Islam are modern constructs without historical identities that were set up by departing colonial powers. But after being held together for decades by monarchs, military rulers, and other despots, these artificial states have come under increasing internal and regional pressure, making their future uncertain.

Power equations will remain central to interriparian relations across the world, determining the extent of cooperation or competition and the potential for conflict. In this context, the chapter, while examining the unique cases of Turkey and China, assesses the factors that help build a regional hydro-supremacy. China, seeking to extend its power and control to transnational water resources after having forcibly absorbed the sprawling, water-rich Tibetan Plateau and Xinjiang, has emerged as a hydro-hegemon with no peer in the world. It is the source of river flows to more countries than any other riparian power, as well as the world's undisputed dam-building leader. It is assembling significant leverage over downriver countries through projects to reengineer cross-border outflows, even as it has become increasingly assertive in pressing its territorial claims with its smaller neighbors.

The final chapter focuses on the international norms, policies, and measures necessary to avert a parched future and water wars. The policy-oriented proposals outlined in detail to mitigate growing economic and security risks are both technical and political in nature. The ambiguities and open issues in international law relating to shared waters must be addressed, or else no effective regime can possibly emerge to help forestall conflicts. Ways must also be found to rein in the *doctrine of prior appropriation*, which has set off "dam racing" between riparian neighbors. Under this doctrine—which is based on the maxim "First in time, first in right"—the first appropriator of water (whether an upstream or downstream state) acquires the priority right to its future use against any subsequent user or appropriator.

The chapter, while acknowledging the complexities, power play, and other political constraints involved in managing the global water crisis, defines a range of relevant ideas and concrete proposals for securing our water future, even at the risk of some of them seeming visionary in the present divisiveness of international geopolitics.

Appendix A lists Web links to the key texts, conventions, and draft articles aimed at establishing global norms for transboundary water resources. The most important of the texts—the 1997 United Nations Convention on the Law of the Non-Navigational Uses of International Watercourses, which lays down rules for transnational water resources and represents the first accord to establish an international water law—has yet to take effect. Appendix B gives a tabular rundown on genuine freshwater-sharing agreements between countries. What it shows is that, in a world with 276 transboundary river and lake basins and an even greater number of cross-border aquifers, there currently are only eighteen mutually binding agreements with a specific formula dividing shared waters. Several of these accords, despite their water-sharing provisions, provide for no means for resolving disputes or for promoting larger water collaboration.

Issues relating to transnational water resources admittedly pose intricate challenges, yet there are ways to manage intercountry water competition and to shield the ecosystems central to human progress. Intercountry water institutions, if founded on the principles of fair resource allocation and of doing no harm to any co-riparian, and if underpinned by conflict-resolution mechanisms, are likely to facilitate structured cooperation and to help stem the risk of disputes flaring into confrontation or armed conflict. But if such institutions do not include a key riparian, no meaningful basin community will emerge. The task is thus to persuade the stronger riparians that institutionalized water cooperation would serve their interests by promoting broader, mutually beneficial regional collaboration and integration.

The most basic contribution of this book, as one of its prepublication anonymous peer reviewers pointed out, is that it makes evident that water is emerging as a more important issue for the fate of humankind than the food crisis, the population and urban explosion, climate change, deforestation, economic slowdown, peak oil, and other oft-cited challenges. It does so by detailing how water is connected with many of the problems people and pundits fret about. Water issues are intimately related to global warming, energy shortages, stresses on food supply, population pressures, pollution, environmental degradation, global epidemics, and natural disasters. Reducing the global hunger gap or meeting the rapidly growing electricity demand, for example, is fundamentally a water-resource challenge, although other factors and inputs are also important.

"Whiskey is for drinking, water is for fighting over." This tongue-in-cheek comment attributed to Mark Twain in an era when the early European settlers in California were digging for gold may sound less facetious in an increasingly water-stressed world, especially with the growing risks of water wars in which water sources and flows become national-security assets to be fought over or controlled through regular armies or surrogate militias, including terrorist organizations.[16] The link between water scarcity and violent conflict is deep rooted.

Yet the political risks come with important opportunities for building peace, cooperation, and rules-based joint management, just as the risks for investors are tied with significant business opportunities. The water crisis, while bringing governments under pressure to change their business-as-usual approach, is opening important new areas for investment, innovation, and profit.

Better water management is integral to building a more harmonious and sustainable world. Water security, after all, is essential for economic, food, and environmental security; public health; national well-being; regional peace; and international stability. It is past time the international community recognized the centrality of water among global challenges and focused attention on the prime problem that holds the key to dealing with other pressing issues.

Chapter One

The Specter of Water Wars

Water wars are no longer just the stuff of Hollywood melodramas. With water stress spreading across much of the world, the next flash point could well be water. The battles of yesterday were fought over land, including empire-building colonies. Those of today are over energy. But the battles of tomorrow are likely to be over the most precious of all natural resources—water—which some investment strategists envisage potentially becoming the single most important physical-commodity asset class in a water-scarce world. Water—the world's most essential yet today's most underappreciated commodity—could eventually, as an asset class, overtake other resources that remain in great demand, such as oil, copper, iron, and agricultural commodities.[1]

International shifts in the basic-commodity, energy, and metal markets, with growth in demand moving from West to East, reflect larger geopolitical shifts. The market shifts have promoted resource competition in distant lands, such as Africa, between the old and new economic giants. A new resource-related Great Game has unfolded, centered on building new geopolitical partnerships, gaining a larger share of strategic resources, and influencing the direction of energy pipelines and other resource exports. The impact of such geopolitics upon markets is already tangible. But if water were to emerge as a physical-commodity asset class, it would open the path to the West meeting some of the water demand in the East. Water-rich Canada, the Nordic countries, and Russia are well positioned to cash in on their "blue gold" wealth by emerging as the Saudi Arabias of the freshwater world.

Water is a sustainer of life and livelihoods and an enabler of development. Yet it has already become an overexploited resource, with the resultant shortages, degradation, and competition triggering intercountry and intrastate sharing disputes. This, in turn, is straining environmental sustainability,

1

sharpening territorial feuds over or separatism in water-rich regions, and threatening to slow overall global economic growth. What is common between Tibet; the Golan Heights; the traditional Kurdish homeland, which straddles the Tigris-Euphrates River Basin; Kashmir; and the Fergana Valley of Central Asia? They are all strategically located water-rich regions racked by separatist unrest or territorial disputes. Experience has shown that water scarcity occurring in combination with other sources of tensions—including territorial disputes, environmental degradation, poverty, and weak or absent regional institutions—easily stokes conflict.

Water wars—in a political, diplomatic, or economic sense—are already being waged between riparian neighbors in several regions, fueling a cycle of bitter recrimination, exacerbating water challenges, and fostering mistrust that impedes broader regional cooperation and integration. Without any shots being fired, rising costs continue to be exacted. This shows that water wars are not just a future peril but a little-publicized reality already confronting the international community. In fact, in a silent hydrological war, the resources of transnational rivers, aquifers, and lakes have become the targets of rival appropriation, with the tools of increasing competition ranging from hydro-engineering works to cross-border support for proxies.

Driving the rival appropriation plans and water nationalism is the notion that sharing waters is a zero-sum game. The danger that the current or emerging riparian battles may slide into armed conflict looms large on the international horizon, given the extent of the water crisis confronting humanity—a crisis that threatens to aggravate the already-grave food situation and slow down the rapid expansion of energy supplies. The international community is on the cusp of a new era in which serious water shortages, if unaddressed, will likely impinge on peace, social stability, and rapid economic modernization. A report reflecting the joint judgment of U.S. intelligence agencies has warned that the use of water as a weapon of war or a tool of terrorism will become more likely over the next decade in some regions, with some states using shared waters to exert leverage over their neighbors and to secure regional influence.[2] Water threatens to become the world's next major security threat.

Securing a larger portion of the shared water resources in a region has already become a flash point in intercountry relationships. There is often little incentive to conserve or protect supplies for users beyond national borders, unless there are specific water-sharing arrangements in place. Like arms racing, "dam racing" has emerged as a geopolitical concern, especially in Asia, where the world's fastest economic growth is being accompanied by the world's fastest increase in military spending and the world's fiercest competition for natural resources, especially water and energy. As riparian neighbors in several regions compete to appropriate the resources of shared rivers by building dams, reservoirs, barrages, irrigation networks, and other structures, the relationships between upstream and downstream states are often characterized by mutual distrust and discord.

Water scarcity and declining water quality, meanwhile, are reaching alarming proportions in several parts of the world, as illustrated by the drying up of the Colorado, Yellow, and Indus river deltas; the polluted waterways in South Africa; and water rationing in regions like California and northern China.

In history, too, societies faced water quality and quantity challenges owing to overuse or contamination of resources. The drying up of local water sources indeed led to profound moves, such as the abandonment of the Mughal Empire's new capital, Fatehpur Sikri ("the City of Victory") in India. This royal city of palaces and imposing public buildings, built with brilliant red sandstone in a blend of Hindu and Islamic architectural styles, was largely completed in 1573 with little regard for the sustained availability of water resources. Just before it was abandoned in 1585, it was described by English traveler Ralph Fitch as "considerably larger than London and more populous." Now a World Heritage site, Fatehpur Sikri represents one of the greatest accomplishments of Mughal architecture, along with the famed Taj Mahal, located barely forty-five kilometers away.[3]

In this century, Sanaa in Yemen may become the first capital city to run out of water, with its groundwater reserves officially projected to last only up to 2025 at the current rate of consumption—a prospect that could turn the now-bustling city of 2 million residents into a ghost town like Fatehpur Sikri. Sanaa, established in the sixth century BCE, relies entirely on subterranean water resources, but its population has quadrupled since the 1980s, even as subsidized diesel fuel has encouraged unregulated pumping of groundwater. Groundwater extraction rates in Sanaa are four times higher than natural replenishment.[4] Sanaa's impending water catastrophe could engulf large parts of Yemen, where water is already a key instigator of conflict.

Sanaa, however, is not the only city that faces the specter of using up all its water resources. Abu Dhabi, capital of the oil-rich United Arab Emirates, says it is likely to exhaust its groundwater reserves by midcentury, while the Pakistani city of Quetta is expected to run out of water even earlier unless additional supplies are diverted to it from elsewhere. Tripoli, the Libyan capital, and other northern Mediterranean cities such as Benghazi, Sirte, and Misrata in Libya now rely on "fossil" (ancient) groundwater transported from the Sahara Desert in the deep south by the so-called Great Manmade River Project (GMRP)—a showpiece of Colonel Muammar el-Qaddafi's forty-two-year rule that the now-slain dictator pompously described as "the eighth wonder of the world."

Qaddafi—a twenty-seven-year-old junior military officer who deposed King Idris in a 1969 coup just before international oil prices began to surge—tapped Libya's new oil wealth to launch major infrastructure projects like the GMRP. Nicknamed the "Mad Dog of the Middle East" by U.S. president Ronald Reagan, Qaddafi came to symbolize the rise of a new generation of dictators that spent their vast oil revenues to finance international terrorism and suppress their own people. The GMRP, however, helped transform Libya's seminomadic tribes into an overwhelmingly urban population concentrated along the northern coastline, where most of the Libyans now live, in less than 2 percent of the country's total area.

But water consumption is now increasing so fast in the northern cities that Libya risks fully depleting the fossil waters that had slowly accumulated in underground strata over many thousands of years.[5] These waters—trapped in separate subbasins, and originally almost equal in volume to the Black Sea—were accidentally discovered while exploring for new oil fields.

Even in the United States, several large cities risk running out of water, including Los Angeles, Las Vegas, and Phoenix.[6] Atlanta's use of Lake Lanier is the cause of a protracted water dispute between Georgia, Alabama, and Florida. In the towns along Europe's southern shores, from Greece to the tip of Portugal, water rationing has become the norm. These developments show that water stress, although more widespread and intense in the developing world, now afflicts parts of the West. Scarcity of freshwater is an increasingly nagging issue, for example, in the arid regions of Spain and in one-third of the United States, which in 2012 was gripped by a drought unparalleled in scale for more than half a century.

In California, the Southwest, and Texas, water withdrawals are already greater than the renewable water supply, which means that tomorrow's resources are being tapped to meet today's needs.[7] The arid West has remained the fastest-growing region in the United States for most of the period since 1970, and the pressures on water resources are "changing virtually every aspect of municipal, industrial, and agricultural water practice."[8] The unregulated exploitation of the freshwater resources of the Great Lakes, whose basin is home to more than 40 million Americans and Canadians, has also raised deep concerns.[9]

But when entire countries are racked by water distress—with the effects fueling increased food prices and spurring greater resource competition—political and social convulsions may become unavoidable. One common factor in the popular uprisings of 2011 that engulfed a series of North African and Middle Eastern states was the popular anger over an issue directly tied to the regional water crisis—rising food prices. These states share one ominous link: they all fall below the international water-poverty line, defined as annual availability of less than 1,000 cubic meters per head. Two of the states hit by political unrest—the petrodollar-rich Kuwait and Bahrain—actually rank among the world's five poorest states in aggregate internal water resources.

Historically, the depletion or degradation of water resources and the ensuing impacts on food production and the environment have caused eco-meltdown and the fall of some civilizations. Land and water degradation resulting from intensive agriculture can lead to salinization and a sharp decline in the productivity of soils. There are several historical examples of societies fatally undermining their ecological security. The early Sumerian civilization, which emerged in the lower reaches of the Tigris and Euphrates rivers about 4000 BCE, was brought down by the deterioration of water quality through soil erosion and salinization, leading to greatly diminished wheat and barley yields. In the case of Central America's Maya civilization, which thrived for a millennium or so before going into decline around 800 CE, reckless deforestation promoted land and water degradation and undermined agriculture, setting the stage for its downfall.[10]

As in history, future human migration patterns will be influenced by water availability, or the lack of it. A large exodus of people from water-scarce regions to water-sufficient areas will swamp the latter socially and economically, potentially triggering local backlashes and straining their internal security and environmental sustainability. Millions of thirsty Yemenis, for example, may be forced to abandon their country's capital and other water-starved cities, with many such "water refugees" seeking to move to the nearby wealthy but water-stressed Persian Gulf nations or even to Europe. Such possible scenarios are a reminder that resources like water and land must not be misused or utilized in excess of nature's capacity for their replenishment or regeneration.

WATER, THE NEW OIL

Water in the twenty-first century could easily become what oil was to the twentieth century—a source of both wealth and conflict. If one compares the price of mineral water in the supermarket with the international spot-market price of crude oil, water is already the new oil. Despite the rise in oil prices, the crude oil spot price is still lower than the retail price of mineral water, or even plain bottled water.

The rise of the bottled-water industry in the past two decades underscores how the era of safe, bountiful water has given way to the commercialization of a resource whose availability and quality have come under growing pressure. Lack of pure drinking water, by causing illness, is the greatest killer on the globe, especially of children. Compounding the situation is the fact that most people don't pay an economic price for water due to government subsidies.

Geopolitically, just as oil in the last century played a role in determining the ascent or decline of states, so the rise and fall of powers in this century could be influenced by water. In the way oil shaped international geopolitics in the twentieth century, the struggle for water is set to define many interstate relationships in this century and to increase the risk that today's silent water wars may heat up. In truth, water wars can be fought and won without firing a single shot—by quietly building an upstream hydroengineering infrastructure to commandeer shared resources.

At a time when territorial disputes and separatist struggles are often being driven by resource issues, water indeed is becoming the new oil. In fact, just as the search for new oil has extended to Canadian tar sands and "tight rock" formations like shale as well as to forbidding places such as the Arctic region, the deep waters of the South Atlantic, the icy Sea of Okhotsh in the North Pacific, and other resource frontiers, the competition for greater access to freshwater supplies has led nations to exploit the resources in deep, ancient fossil aquifers and to build dams in remote, high-altitude, ecologically sensitive areas like the Tibetan Plateau and the Himalayan belt.

Water is a renewable resource. But tapping fossil water, unlike other groundwater, amounts to extraction of a virtually nonrenewable resource that accumulated over thousands, or even millions, of years. Fossil aquifers—often very deep geologic formations—are practically nonreplenishable. Tapping fossil water is thus like pumping oil. Recycling treated wastewater and desalinating seawater or brackish water also represent efforts to develop new, unconventional sources of sustainable supply, on lines similar to what is happening in the energy sector, where oil extracted from the tar sands under the boreal forest in Alberta, Canada, and gas and oil hauled out from shale rock have proved a game changer.

But unlike oil—dependence on which can be reduced by either tapping other sources of energy or switching to other means of generating electricity—water has no known substitutes. The share of global electricity produced from oil has shrunk from 23 percent in 1977 to 5 percent in 2012. It is projected to decline further to 3 percent by 2030.[11] Oil now is predominantly used as transportation fuel, in the form of gasoline and diesel. Even so, the world is likely to move toward electric automobiles that employ energy stored in their batteries—recharged by common household electricity—or fuel-cell vehicles that convert onboard-stored hydrogen gas and oxygen from the air into electricity to drive the electric motor, with grid power being used to make the hydrogen.

So, while promising new technologies raise hope that oil may ultimately be replaced by other sources of energy, water is simply irreplaceable. As UN secretary general Kofi Annan warned at the advent of the twenty-first century, "If we are not careful, future wars are going to be about water and not about oil."[12] He later amplified this warning by saying, "By 2025, two-thirds of the world's population is likely to live in countries with moderate or severe water shortages. Fierce national competition over water resources has prompted fears that water issues contain the seeds of violent conflict."[13] With the UN seeking to spotlight the growing global water challenges, Annan's successor, Ban Ki-moon, has struck a similar note, cautioning that water shortages in the world could serve as a catalyst in "transforming peaceful competition into violence" and deploring the fact that "too often where we need water, we find guns."[14]

Water, as a central key to sustainable livelihoods and development, presents a unique challenge. Not only is water the most fundamental of finite resources, but it also has no substitutes for most uses and is expensive to transport over long distances. Whereas countries can scour the world for oil, liquefied natural gas, mineral ores, and metals to keep their economic machines humming, water cannot be secured through long-distance international trade deals. The only option water-poor countries have is to sustainably optimize their water resources and, when it involves transboundary basins, to collaborate with co-riparian states on holistic resource management. If they have sufficient, sustainable foreign-exchange reserves, they can, of course, try to alleviate their water distress by importing, rather than producing, water-intensive products, ranging from grains and meat to industrial goods like paper, fabric, and plastic. Yet few states have addressed their water situations by such means.

For one thing, many nations are reluctant to take on political and financial risks by becoming dependent on other states for basic products, including food, which often may be cheaper to grow domestically than to import. Paradoxically, some water-distressed nations are major international food exporters. For another, in terms of water use per dollar of economic output, many industries are very thirsty, underscoring the link between water resources and economic advancement. The true level of water consumption by any industry can be assessed only by examining the embedded water—the real value of how much water has been used to create a product or service. Although grain farming and power generation are the two most water-intensive sectors in terms of direct water withdrawals, a study of the American economy has found that most water use in the United States—60 percent—is indirect, via supply chains, with 96 percent of sectors utilizing more water indirectly than directly.[15]

In this light, the growing freshwater shortages across much of the world pose major socioeconomic and security challenges. Water stress and water in-security are set to become more widespread and acute. Water stress actually is a stepping-stone to water scarcity. Water stress is most commonly defined as per capita water availability below 1,700 cubic meters per year, while water scarcity signifies annual availability below 1,000 cubic meters.[16]

The global average per capita freshwater availability per year for all eco-nomic activity and household use, after declining from about 9,000 to 6,079 cubic meters between 1988 and 2012, is projected to fall to approximately 5,000 cubic meters by 2025.[17] When freshwater availability declines to less than 2,000 cubic meters per inhabitant per year, the situation can potentially act as a serious constraint on rapid socioeconomic development and environ-mental protection. Yet several dozen countries already fall in this category, ranging from South Korea and India to Belgium and South Africa. And when freshwater availability, as has already happened in Tunisia, Algeria, Jordan, Cyprus, Kenya, the Maldives, Rwanda, and a number of other nations, falls below 1,000 cubic meters per head annually—the international water-poverty threshold—the constraints can be severe. Because several factors, including climate and level of industrial and agricultural development, determine basic water needs, no single benchmark, however, can fully capture water-shortage conditions. Also, given the spatial imbalances in water distribution within most countries, water stress can occur at any level of supply, depending on demand and other circumstances.

The global water crisis has essentially turned into a dual supply-and-quality problem. Against this background, the idea of making bulk water an internationally tradable commodity looks appealing to many investors, in spite of the fact that water is essentially local and is difficult and expensive to transport. Opportunities for profit, of course, extend to investments in solving the problems tied to the decades-old overuse and mismanagement of water resources. One thing is certain: the price of water is set to rise.

Falling Water Tables and Declining River Flows

Natural resources central to human life, including freshwater, quality soil, and biodiversity, are being degraded and depleted at an alarming rate. Water, of course, is the single most important resource, without which there can be no food or industrial production. People can survive war and plague, but they cannot survive without water.

Groundwater has traditionally been an insurance against drought condi-tions, but now its rising importance in global freshwater supply has made it the world's most extracted natural resource. With users exploiting ground-

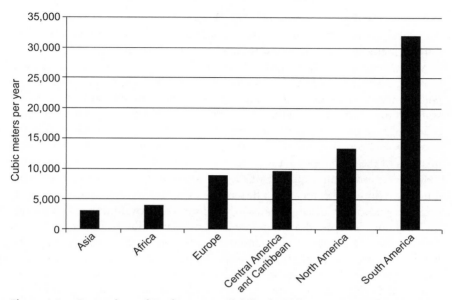

Figure 1.1. Comparison of Freshwater Availability in Different Continents, per Capita
Source: Data drawn from Aquastat, 2013.

water in several regions at rates surpassing nature's recharge capacity, falling water tables and other hydrologic changes are causing serious ecological effects, including a sharp drop-off in flows of many freshwater springs and rivers and a steady rise of algae in them. About 40 percent of irrigation globally—with India, China, and the United States in the lead—relies on groundwater, including fossil water, with depleting reserves threatening to significantly impact crop production.[18]

It is the countless millions of pump-operated wells, which have followed the dramatic, cost-effective improvements in drilling and pumping technologies in recent decades, that have turned groundwater into an overexploited resource.[19] The unbridled extraction—which more than doubled between 1960 and 2000 to 734 billion cubic meters per year, and now threatens to approach 1,000 billion cubic meters per year[20]—represents the greatest human-induced alteration of the hydrologic cycle, with groundwater depletion affecting natural streamflows, groundwater-fed wetlands, and related ecosystems.

Even in a country relatively well endowed with surface-water resources—the United States—groundwater abstractions aggregate to about 23 percent of all freshwater withdrawals daily.[21] For example, in Florida—a state that receives a lot of rain and historically boasted an abundance of water—the overpumping of groundwater to meet the demands from population, urban, and agricultural growths has contributed to the flow rate of many of the artesian springs and rivers in the northern and central regions dropping by up to a third, including the Silver and Rainbow springs and rivers.

More broadly, global withdrawals of surface and underground waters have more than tripled since 1960, with water demand increasing by an estimated 64 billion cubic meters a year.[22] Yet huge numbers of people still lack running water in their homes. More than one person in five in the world has no ready access to safe drinking water, and one in three lacks proper water-sanitation services. But access to mobile phone ownership and usage has reached three-quarters of the global population.[23]

In fact, more than a quarter of the world population does not even have a toilet, with open defecation allowing flies to spread diseases. Each year, diarrhea and other ailments caused by unsanitary conditions and poor hygiene claim the lives of an estimated 1.5 million children under the age of five. Increasing access to safe sanitation is thus critical to improving public health and reducing infant and child mortality. After all, sanitation and hygiene are the cheapest, most effective preventive medicines for the poor. This challenge is compounded by the broader depletion and contamination of water resources, which in turn contributes to increasing incidence of waterborne illnesses.

Rapid development, population growth, an obesity epidemic, soaring food and industrial-product needs, and fast-paced urbanization are inexorably increasing demand for the world's finite water resources. Any area's renewable water resources comprise the amount brought by rainfall and incoming river flows, minus the amount lost through evapotranspiration (the loss of plant-related water to the air through evaporation and transpiration). The evapotranspiration rate varies hugely across regions. In arid regions, like North Africa and the Middle East, barely 18 percent of the rainfall and incoming flows remains after evapotranspiration, but in humid coastal areas, such as southeastern China, retention can be as high as 50 percent.

The Water Exploitation Index (WEI) defines the share of long-term average renewable water resources that is annually abstracted, with a value of more than 40 percent signifying severe water shortages and unsustainable consumption. A sizable number of developing nations have a WEI value of over 40 percent. In fact, the index shows that water scarcity is largely concentrated in the developing world. For example, the majority of states in Asia and Africa, unlike Europe, face severe water paucity. Asia's per capita freshwater availability is almost three times lower than Europe's. Yet within Europe itself, Cyprus has a WEI value of 64 percent, and Belgium and Spain—with values slightly higher than 30 percent—are water stressed.[24]

There can also be major differences in water availability within a country, with some areas being water rich while others cope with water poverty, as in the United States and China. Within most developing countries, the level of access to safe drinking water often illustrates the divide between the rich and the poor. Water poverty can keep people trapped in extreme economic poverty.

At a time when the West has ceased to be the principal locomotive of global economic growth, one major question that arises is whether the major emerging economies have sufficient water resources to sustain their continued rapid growth in the coming decades and thereby prop up the world economy. The fact is that other than in Brazil and Russia, water demand is increasingly outstripping supply in the world's fast-growing economies, including Chile, China, India, South Africa, South Korea, and Vietnam. A host of other economies, from Mexico and Egypt to Israel and Iran, face difficult choices for meeting their water needs. Had their populations been much smaller, most of the emerging economies would have benefited from a better balance between population size and available natural resources.

More broadly, the water shortages in the world open up a litany of risks for economies and investors. These risks are potentially greater than nonperforming loans, real-estate bubbles, and political corruption. Of the 150 large international corporations that responded to a 2010 questionnaire sent out by the Carbon Disclosure Project—a group monitoring the climate-change-related strategies of the world's leading corporate bodies—39 percent reported that water-related problems were already serious enough to cause "detrimental impacts" to their businesses.[25]

Commercial or state decisions on where to set up new manufacturing or energy plants are increasingly being constrained by local availability of adequate water resources. In situations where competing users are seeking access to a scarce resource, a decision to erect a new water-intensive industry can meet grassroots resistance. International beverage giants, as the most visible users of local water resources, have faced grassroots protests and regulatory penalties for depleting groundwater in some developing parts of the world.

Implications of the Widening Water Gap

With water shortages in many parts of the world becoming an everyday reality, closing the gap between water supply and demand poses one of the most difficult and complex challenges today. Global water demand, according to one report, is likely to outstrip supply by as much as 40 percent by 2030, but one-third of the international population, living mostly in the developing world, is projected to confront a supply-demand deficit higher than 50 percent.[26] This will lead to a situation where even water essential for environmental needs is siphoned off for human use.

Such a "water gap" projection is based on an average global economic-growth scenario without water-efficiency gains. The estimated deficit will be smaller if measures to improve water productivity significantly lessen the water intensity of the world economy.

The water gap, however, will not be easy to bridge. For investors, the growing water crisis means that the costs of doing business are set to mount steadily. To help protect their bottom line, many companies are trying to reduce their heavy dependence on water by embracing new techniques, ranging from sanitizing plastic bottles with purified air instead of water to recycling wastewater through expensive, on-site treatment. Yet they face rising water-related risks of physical disruptions to operations and supply chains, declining availability and deteriorating quality of water, rising costs, changing regulatory regimes, and reputational damage from fines or litigation for overexploiting or contaminating a life-sustaining resource.

Water paucity and pollution challenge a host of industries—from beverage makers and high-tech companies to the metals industry and energy companies. Microchip makers, for example, use millions of gallons of water a day. So, more efficient water use by such manufacturers can translate into big savings for them while making greater water supply available to other local users. For instance, the giant multinational IBM has cut its water use by 27 percent at its plant in Vermont, saving some $3.5 million in company costs.[27] Industrial water demand is often intertwined with energy use, with energy-intensive industries also being water intensive. While some sectors of production, such as transportation equipment and food and beverages, have high water-consumption rates, other sectors demand high water flows for their manufacturing processes, including the chemical industry, which is also a major source of water pollution.

Faced with growing water shortages, a number of countries—ranging from China and Mexico to Australia and the United States—have been tapping the resources of the virtually nonrechargeable fossil aquifers. The possibility of fossil-water reserves running out is thus not limited to the riverless Arabian Peninsula and Libya. In the United States, one major fossil aquifer, the Ogallala—which extends northward from northwestern Texas to southern South Dakota and is the world's largest—is the source of 27 percent of the nation's irrigated agriculture, besides supplying drinking water to several of the Great Plains states. Named after a Native American tribe, this aquifer's resources, mostly accumulated over the past 13,000 years, are dwindling alarmingly, especially in its central and southern parts.[28] The global overexploitation of water resources, however, extends beyond aquifers, as exemplified by the creeping problem of river depletion.

To compound matters, the world now confronts an expanding problem of water pollution. With the environmental impacts of water contamination increasingly troubling most transboundary basins, water quality, not just water quantity, is set to become a major interstate issue.[29] Agriculture and industry both contribute to, and in turn are affected by, the degradation of water resources and their diminished availability. Agriculture accounts for the bulk of global water withdrawals, yet industry represents the fastest growth in water demand.

Given the close relationship between water and development, the concentration of water shortages in the developing world is ominous. Per capita availability of water in the rich world overall is far higher than in the developing world. It is also significant that many of the states that either rank very low in the United Nations Development Programme's annual Human Development Index or risk failing (according to the Failed States Index maintained by two Washington-based organizations) are water-stressed underdeveloped economies, including Burundi, Burkina Faso, Ethiopia, Malawi, Pakistan, Rwanda, Somalia, Sudan, and Yemen.[30]

The intersection between water distress and rising extremism in Yemen, Gaza, Afghanistan, Pakistan, and Somalia has significant transnational ramifications, with these lands turning into incubators of jihadist terror. In the Afghanistan-Pakistan ("AfPak") belt, the impacts of water scarcity, aggravated by explosive population growth, are "fueling dangerous tensions" that hold wider security implications, according to a U.S. congressional report.[31] Yemen, which already embodies the characteristics of a failed state, is in danger of becoming the first country in the world to run out of water, serving as a forewarning of the conflict and refugee exodus that are likely to follow when population growth overwhelms natural-resource availability.[32]

Wells in Yemen could run dry by midcentury. The battle lines over increasingly scarce water resources have engendered mounting civil unrest and three insurgencies in a country that has already become the seat of Al Qaeda in the Arabian Peninsula, or AQAP, a transnational terrorist organization. In parts of southern Yemen, a Qaeda affiliate, Ansar al-Sharia, functions as a de facto government. Yemen, the Arab world's most economically backward state, has one of the world's highest rates of population growth (2.9 percent per year, as compared with the global average of just under 1.2 percent) but one of its lowest levels of per capita freshwater availability—91.64 cubic meters a year.[33] Whereas ordinary citizens toil every day to get some water, Yemen's civilian and military elites have drilled wells in their residential backyards despite a government prohibition against such water wildcatting.

At the opposite end of the spectrum are Iceland and Guyana, which lead the list of countries endowed with bounteous freshwater availability per head. Other exceptionally water-rich states in per capita availability include Suriname, Gabon, Canada, and Norway.[34] In aggregate resources, however, Brazil is the virtual king of the freshwater world, although its semiarid, densely peopled northeast is water scarce. Brazil is the world's most water loaded nation (with 8,233 billion cubic meters of long-term average annual renewable resources), followed by the Russian Federation (4,508 billion cubic meters), the United States (3,069 billion cubic meters), and Canada (2,902 billion cubic meters).[35]

As people become prosperous, they tend to use more water—as well as more energy, the production of which demands huge volumes of water. Many wealthy nations are considered wasteful users of water (see figure 1.2), although some, like Finland, score high points for efficient water-use practices. The United States has the dubious distinction of having been graded the world's most water-profligate state by the Water Poverty Index, based on five components—resources, access, capacity, use, and environment.[36]

In fact, for many resources, including energy, America is the largest per capita consumer. It is not among the countries that top in per capita freshwater availability; yet its average daily water use per person ranks the highest in the world. The daily water use per American averages between 303 and 378 liters, according to the U.S. Geological Survey, or, if the United Nations Development Programme is to be believed, more than 550 liters.[37] In contrast, the lowest water use of 15 liters or less per head per day is in Ethiopia, Haiti, Mozambique, Rwanda, and Uganda.[38]

In poor communities in the developing world, girls and women are engaged in the daily chore of hauling water, often over long distances. Females indeed play a central part in the provision and management of water in such communities. Women's pivotal role as providers and users of water and as guardians of the living environment, according to the 1992 Dublin Statement, "has seldom been reflected in institutional arrangements for the development and management of water resources. Acceptance and implementation of this principle require positive policies to address women's specific needs and to equip and empower women to participate at all levels in water-resources programs, including decision-making and implementation, in ways defined by them."[39]

Females in the developing world continue to bear the brunt of water woes.[40] Lack of access to toilets at school means that menstruating girls cannot change sanitary napkins. As a result, girls after reaching puberty start missing school for one week each month before dropping out of school altogether. The daily task of fetching water from often distant wells and tankers

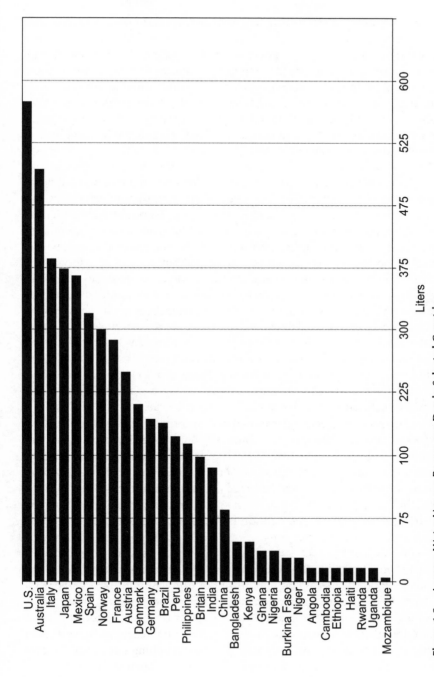

Figure 1.2.   Average Water Use per Person per Day in Selected Countries

Source: United Nations Development Program, *Human Development Report*, 2006.

also chains many females to lifelong illiteracy. And without education, the poverty cycle continues.

The core issue thus is water poverty which, as one author has noted, traps people "in a primitive day-to-day-struggle. Water poverty is, quite literally, de-civilizing."[41] In poorer societies, women, besides shouldering the burden of hauling water and managing households, make up the larger share of the agricultural workforce, including the majority of the 1.5 billion people in the world earning less than three dollars a day. They are thus on the front lines of the likely impact of climate change.

More fundamentally, a serious water crisis now confronts the world, and all the indicators are that it will get worse. The crisis centers on continuing mismanagement of water resources, even as decision makers gamely pay lip service to concepts like equity and sustainability. Inertia at the "leadership level and a world population not fully aware of the scale of the problem (and in many cases not sufficiently empowered to do much about it) means we fail to take the needed timely corrective actions and put the concepts to work."[42] In keeping with the asymmetrical distribution of water resources across the world, national water priorities vary considerably, with the least-developed countries giving priority to ensuring adequate water supply for food production and the water-sufficient advanced economies at the top end of the Human Development Index prioritizing the preservation of freshwater ecosystems.[43]

The potential water constraints on rapid economic growth raise several challenging issues, including how industrial and farm policies are likely to be reshaped in parched countries, ways to sustainably manage water resources, technological innovations to improve water productivity and to make water recycling economical, new public-private partnerships to channel greater investments into the water sector, and new water diplomacy to minimize conflict by promoting bilateral or basinwide cooperation on transboundary resources. Water has emerged as a source of increasing competition and discord within and between nations, spurring new tensions over shared basin resources and local resistance to setting up water-intensive industries. Such charged competition is likely to impinge on the economic-growth story in the coming years.

The incontrovertible fact is that improving water efficiency and productivity has become central to the task of lessening the supply-demand gap. Basin resources cannot be sustainably optimized or efficiently utilized without the involvement of all stakeholders. In regions where watercourses traverse international boundaries, the challenges to improve the quality, quantity, and distribution of water necessitate greater intercountry cooperation, lest interriparian conflicts emerge in a major way. Water challenges are at the heart of human development and regional peace and security.

A Basic Right Denied to Many

Water resources must be managed in a holistic way so that social and economic development is integrally linked with the protection of natural ecosystems. Yet, far from such management becoming the international norm, many people in the world are without the most basic human right that others take for granted: safe, clean, adequate water. Water is central to the realization of a fundamental right enshrined in the 1948 Universal Declaration of Human Rights: "Everyone has the right to a standard of living adequate for the health and well-being of himself and his family."[44] In fact, the United Nations General Assembly and the UN Human Rights Council in 2010 separately recognized the right to safe drinking water and sanitation as a human right central to a decent life.

At least half of the eight Millennium Development Goals adopted by world leaders in 2000 are linked to water, including eradicating poverty and hunger; reducing child mortality; combating HIV/AIDS, malaria, and other diseases; and ensuring environmental sustainability. For instance, poverty alleviation simply is not possible without the availability of minimally adequate water supplies. Nor can food shortages be remedied without local farmers being able to access sufficient water supplies to grow crops. Significantly, people with better access to water "tend to have lower levels of undernourishment. If water is a key ingredient to food security, lack of it can be a major cause of famine and undernourishment, especially in areas where people depend on local agriculture for food and income."[45] Containing disease and child mortality similarly demands improved access to clean drinking water and to sanitation.

Access to clean freshwater must be treated as a universal human right to ensure the health and well-being of citizens. Almost four-fifths of all countries actually recognize the right to water.[46] Yet, despite many countries defining national access targets, about 1.5 billion people in the world still lack ready access to potable water, and 2.4 billion people have no water-sanitation services. Those denied basic water supplies are forced to lead qualitatively diminished lives, with little prospect of pulling themselves out of poverty.

The deteriorating quality of water—polluted by industrial wastes, agricultural runoff, and sewage discharges—has aggravated the water crisis, making it harder in many nations to provide the basic right denied to many. In the developing world, nearly two-thirds of the wastewater is discharged, with little or no treatment, into rivers and other watercourses. Such discharge seriously compounds resource pollution and public-health problems. With more than a million tons of wastewater being dumped daily into the world's waterways, diseases caused by the use of contaminated water are on the rise.

More than half of the world's major rivers are polluted, imperiling the health and livelihood of those for whom these waterways are the lifeblood.

Roughly 3.5 million people die every year from waterborne diseases, such as cholera—an acute intestinal infection caused by ingestion of contaminated water or food—and schistosomiasis (caused by contact with freshwater containing flatworm eggs). The United Nations estimates that a child dies every eight seconds from one of the diseases associated with lack of clean water. And according to the World Health Organization, thirty-five new waterborne disease agents were discovered between 1972 and 1999 alone; some long-dormant agents have also resurfaced with a vengeance.[47]

All this shows that water-resource degradation and depletion constitutes first and foremost a humanitarian crisis. Water indeed is the common denominator in the health, development, and environmental challenges facing the world. The water-pollution scourge and growing hydrologic variability arising from the disruption of natural water flows due to river fragmentation and other anthropogenic factors have seriously affected fluvial ecosystems. That, in turn, has impinged on traditional agriculture and grazing, devastated fisheries, and marginalized rural communities dependent on those waterways.

Through various initiatives, the United Nations has sought to increase international awareness of water-related challenges and the attendant need to find innovative and sustainable solutions. Yet the initiatives have yielded modest results. For example, the UN General Assembly since 1993 has dedicated March 22 of each year as World Water Day. Over the years, a different theme has been chosen for every World Water Day.

The 2007 theme was "Coping with Water Scarcity" to help shine a spotlight on water shortages plaguing the world, while 2008 was designated the "International Year of Sanitation" in order to highlight the plight of those who do not have sanitation facilities and thus are vulnerable to disease and early death. The 2009 World Water Day focused on transboundary watercourses—with the theme "Sharing Water, Sharing Opportunities"—to help spotlight serious issues that serve as the springboard for conflict over shared resources. With the pollution of rivers, lakes, and other sources becoming a pressing problem, the 2010 theme was "Communicating Water Quality Challenges and Opportunities." The focus in 2011 turned to urban water systems in response to the rapid growth of cities and the resulting urban water shortages. While the 2012 theme was designed to draw attention to the nexus between water and food security, 2013 was declared the "International Year of Water Cooperation."

The United Nations also designated 2005–2015 as the International Decade for Action, with the theme "Water for Life," to help encourage the international community to meet the targets on freshwater and sanitation already agreed upon.[48] Little traction, however, has been secured to realize the water-specific Millennium Development Goals by the agreed 2015 deadline, including the reduction by half of the proportion of global population "without sustained access to safe drinking water and basic sanitation."[49] At the 2002 World Summit on Sustainable Development in Johannesburg, it was even agreed to promote integrated water resources management (IWRM) and water efficiency—commitments that are being met by the international community with little more than lip service.[50]

Indeed, no specific new commitments were made at the 2012 Rio+20 Summit—so named because it came twenty years after the first Rio Earth Summit in 1992. However, by redefining "sustained access to safe drinking water" as "the use of improved drinking water sources," the United Nations claims that one of the Millennium Development Goals has been achieved before the deadline, even though it acknowledges that improved sources may not necessarily yield safe drinking water.[51]

Decisions on land use affect water, but decisions on water affect both land and the broader environment. That is why the IWRM concept—which is concerned with ways to sustainably manage water demand as well as supply—must be embedded in national policies. The Global Water Partnership has emphasized three principles as being integral to IWRM: economic efficiency to help optimize scarce water resources, social equity in distribution, and environmental sustainability to safeguard water sources and aquatic ecosystems.[52] Although the IWRM concept has received growing international support since it was unveiled at the 1992 Rio Summit, efforts toward a holistic, integrated approach at the basin and subbasin levels leave much to be desired. Some experts, in fact, have suggested expansion of the concept to include land so as to put the accent on integrated land and water resource management (ILWRM), given that any land-use decision is effectively a water decision too.[53]

THE FUNDAMENTALS OF RESOURCE PRESSURES

Water, food, mineral ores, and fossil fuels like coal, oil, and gas are resources of the greatest strategic import. They hold the key to human development and, in the case of water and food, even human survival. Food production is closely intertwined with water and energy. Energy-resource extraction and refining and electric power production demand ample water, while the supply of water requires energy. Water and energy are strongly connected to climate change. While the way humanity produces and uses energy contributes about two-thirds of all anthropogenic-related greenhouse gases, the availability of water resources will be directly affected by global warming.

Add to this picture the growing strain on water resources due to population and developmental pressures, illustrated by the fact that the world population has more than doubled and the gross domestic product (GDP) has grown tenfold since 1960 alone.[54] The ready availability of abundant water was a key growth driver in the industrial era, but now the renewable supply of freshwater has fallen behind global demand. And unlike the earlier localized and moderate impacts on aquatic and other ecosystems, the human effects now are extensive and severe. Dwindling water resources and increasing water pollution have created a potent nexus between water quality and quantity.

The future of our world will be shaped by several key issues, including natural resources—both biotic (obtained from the biosphere, especially forests and the wildlife on land and in water) and abiotic (free of living things, including water, land, mineral ores, and hydrocarbons)—food production, demographics, sustainable economic growth, and climate change, besides political factors such as quality of governance, great-power rivalries, religious zealotry, conventional and unconventional conflicts, and accelerated weaponization of science. Greater technological advances could open the path to alleviating resource constraints but could also create potential opportunities for developing information dominance on the battlefield, building a new generation of weapons of mass destruction, and extending the arms race to outer space.

A combination of these factors will create winners and losers in the world. Over the next half century, many states will progress rapidly, and some will even accumulate considerable economic and military heft. But some states will also be left behind and risk marginalization. Weak states saddled with serious resource constraints in the face of burgeoning populations and internal strife could actually fail.

The geopolitics of natural resources promises to become murkier. Resource constraints are already hobbling development in poor countries and are raising questions about future growth trajectories in the advanced and emerging economies. Abiotic resources are central to economic progress and modern lifestyles. These resources are nonrenewable other than freshwater, a replenishable natural capital whose rate of utilization now exceeds nature's renewable capacity. The various resources that constitute natural capital help to generate natural income through goods and services, including marketable commodities (like bottled water, timber, fossil fuels, and fish) and ecological support services, such as those provided by forests in preventing riverbank erosion and in serving as carbon sinks.

Use of nonrenewable (or "stock") resources is tantamount to a slow, incremental liquidation of natural stock, because any current rate of utilization proportionately diminishes some future rate of use. As for renewal resources, their use must not exceed rates of natural-capital renewal or else ecosystems may suffer irreparable damage. In this light, the depletion of freshwater and other abiotic resources carries long-term implications for the world, with the overexploitation of both renewable and nonrenewable resources accelerating environmental degradation and contributing to climate change.

Our Changing World

As the growth of human population, consumption, and development has risen and the power of technology has expanded, human alteration of earth has reached levels threatening the future well-being of societies. Humanity is not only changing earth more rapidly than its own understanding of the implications of change, but human activities are also threatening international security by degrading and depleting natural resources and "causing rapid, novel, and substantial changes to Earth's ecosystems."[55]

Over the twentieth century, the world witnessed a surge in total population by a factor of 3.8, in urban population by 12.8, in water use by 9, in irrigated area by 6.8, in oil production by 300, in energy use by 12.5, in industrial output by 35, in fertilizer use by 342, in fish catch by 65, and in atmospheric carbon dioxide by 30 percent.[56] The twentieth-century expansion of farmland was at the expense of wildlife habitat and biodiversity; unsustainable soil- and water-use practices continue to cause land degradation at a time when rapid urbanization and industrial development have made further increases in cropland size difficult.

In addition to altering or degrading between 39 and 50 percent of the earth's land as well as driving one-quarter of all bird species and many large mammal species to extinction, human actions have resulted in the natural flow of about two-thirds of all rivers being controlled, the creation of countless artificial lakes, and the modification of stream, lake, estuarine, and coral-reef ecosystems.[57] About half the world's wetlands have been destroyed in the past century alone. And the fast-paced degradation of water resources has resulted in aquatic ecosystems losing half of their biodiversity since just the mid-1970s.[58]

The geopolitical change is profound too, with tectonic power shifts unfolding, not because of any military dynamic—such as battlefield victories or new military alliances—but due to a peaceful factor: rapid economic growth. The speed of economic progress in some countries has been exceptional, contributing to the reemergence of the East on the global stage. China—the world's largest producer and exporter of manufactured goods—has risen rapidly as the world's largest consumer of copper, tin, zinc, iron ore, aluminum, lead, nickel, and coal, and also as the biggest emitter of greenhouse gases. The metals demand in India doubled in just the first decade of this century. Rapid economic growth, along with the global population explosion, rising per capita consumption, and soaring food and energy demand, have put unprecedented pressure on water and other natural resources.

International power shifts have now become part of an evolutionary, rather than a revolutionary, process. Typically in history, decisive international change came in radical style—after a major bloody war, as in 1945, 1919, and 1815, with those triumphant on the battlefield fashioning the new order. Such revolutionary power shifts resulting from a war among great powers now seem unlikely. With the nature of power changing, even if subtly, rapid economic growth has helped spur qualitative power shifts, even as the importance of military power remains intact. Given that non-Western civilizations have become important actors through a gradual accretion of skills, technology, and wealth, the new order of the twenty-first century will be determined both by old and new powers.

More fundamentally, the greater role for economic power in shaping global geopolitics helps drive the mounting demand for natural resources and serves as a reminder that the world is at a defining moment in history. Today's manifold global challenges and power shifts actually epitomize the birth pangs of a new world order. Many of the challenges the world confronts are new—ranging from degradation of water resources and accelerated global warming to uncontained international terrorism and pandemics—and their reach is truly global. The challenges unmistakably signal the rise of nontraditional security issues at the core of international concerns.

Although the world clearly is in transition, with the age of Atlantic dominance in retreat, the contours of the new order are still not visible. This has only helped to promote greater international divisiveness. The divisiveness, in turn, has hindered effective action on international challenges, including containing resource degradation, discouraging mercantilist policies to lock up long-term supplies of raw materials, stemming anthropogenic climate change, and addressing the financial, water, food, and energy crises.

On the positive side, the spread of prosperity in the world is creating more stakeholders in international peace and stability. But it is also making wide-ranging institutional reforms inevitable in the coming years. Despite a systemic shift in the global distribution of power, the international institutional structure has remained largely static since the mid-twentieth century. A twenty-first-century world cannot indefinitely stay encumbered with twentieth-century institutions and rules. It is no longer possible for any single power or a single international institution, even if it is the United Nations Security Council, to dictate terms to the international community.

Changes in the international institutional structure are beginning to occur, but rather modestly. Carrying out major reforms that mesh with the new power realities and create more effective institutions will not be easy. One reason is that while the present ailing international order emerged from the ruins of a world war, its replacement has to be built in an era of international peace and thus be designed to reinforce that peace. This means that it will need to be more reflective of the consensual needs of today and have a democratic decision-making structure. The world, however, has little experience in establishing or remaking institutions in peacetime.

The implications of the ongoing power shifts for international peace and security also remain unclear, even though the changes may symbolize a process whereby the world is to gradually return to the more normal conditions of its history—uneasy coexistence among several great powers, with Asia at the international center of gravity, as it had been for two millennia before the advent of the industrial revolution. It is uncertain whether the new order will be rules based or pivoted on a classical balance-of-power politics of the great powers.

Uncertainty arises from the advent or return of authoritarian great powers, which seemingly implies that the new order, far from being liberal or rules based, may be centered on balance of power. Even to the great powers that are democracies, rules tend to matter on many issues only when they can serve the great powers' interests; otherwise the rules appear bendable or expendable. This may explain why the present order is centered on both rules and balance-of-power diplomacy. This blend could also characterize the new order but with greater reliance on balance of power to help underpin equations among several great powers. The recrudescence of territorial and maritime disputes, largely tied to geopolitical competition over resources, as in the East and South China seas, also has a bearing on international peace and security.

Indeed, at a time when the qualitative reordering of power is challenging strategic stability and reshaping major equations, a new Great Game has unfolded, centered on securing a larger share of strategic resources, building novel alliances, balancing competition between other players, and gaining greater market access. From war exercises on the high seas to the use of bilateral aid and arms sales for securing resources, this Great Game is likely to influence the new power dynamics.

Competition over metals and minerals (including hydrocarbons) constitutes an important dimension of this Great Game. The degradation of natural ecosystems is directly linked to the profligate use of, and competition over, resources. Not only is the integrity of fluvial, aquatic, and other ecosystems under increasing threat, but human-engineered land transformation represents the main driver of biodiversity loss worldwide.

It is in this context that water, of all the natural resources, looms as a source of growing competition and conflict. Just as energy-related geopolitics has sharpened, with some countries employing their energy-import needs to expand their navies and cozy up with scofflaw states, water is becoming a source of greater interstate competition in situations where streams, rivers, lakes, and aquifers traverse international boundaries. Transboundary water resources, instead of linking countries in a system of hydrological interdependence, are often fostering competition for relative national gain.

Resource Acquisition as a Geopolitical Driver

Resources historically have played a key role in the ascension and decline of civilizations. Access to resources has also been a critical factor in war and peace. From the rise of Portugal in the fifteenth century to the rise of the United States in the twentieth century, resources have served as a key determinant of foreign, defense, and trade policies. As brought out by Paul Kennedy, Pax Britannica was made possible by a nimble navy that secured vital commodities from resource-rich lands overseas.[59] Paradoxically, it was Britain's failure to gain preeminence in Europe, where it faced other major powers like Russia, Germany, and France and where no state was strong enough to impose its will, that motivated it to concentrate on distant lands. That is how Pax Britannica was established.

More broadly, gaining access to resources has been a major driver of armed interventions and wars in history, including the European colonial conquests in Asia, Africa, and the Americas and many of the wars of the nineteenth and twentieth centuries. Mercantilism was an early modern European economic theory and system that actively supported the establishment of colonies to supply materials and provide markets so as to relieve home countries of dependence on other nations. There are many other examples in history of how attempts to gain control over the resources of others have led to interventions and fierce wars. Many believe the U.S. invasion and occupation of Iraq in 2003 was more about controlling large oil reserves there than about noble principles like freedom.

When a country's resource supplies are blocked by a hostile state, it can go to war. Take the 1941 Pearl Harbor attack: Although it took the United States by complete surprise, the attack was triggered at least in part by the U.S.-British-Dutch oil embargo against Japan, which was then relying on imports largely from the United States to meet its oil needs. The oil embargo indeed marked an escalation of the economic squeeze of Japan through denial of essential resources in a U.S.-led campaign that began in 1939 with partial bans on supplies of scrap metal and gasoline for aircraft.[60] As economic historian Niall Ferguson has written, "Western powers had no desire to relinquish their mastery over Asia's peoples and resources. Even when they were comprehensively beaten by Japanese forces in 1942, the Europeans and Americans alike fought back with the aim of restoring the old Western dominance" in Asia.[61] The United States, in fact, has signaled since 1941 that its security begins not off the coast of California but at the western rim of the Pacific Ocean and beyond.

The reemergence of economic giants in the East has sharply accelerated the global resource hunt, changing the pattern that prevailed for long after the rise of Western powers. The new economic powerhouses are competing for resources not only against the old economic giants but also against each other. For Japan, whose ruinous decision to go to war with America in 1941 appeared designed to secure peace on its terms, the more recent use of rare-earth exports by China as a trade weapon against Japan has served as a bruising reminder of its dearth of natural resources and thus of its continuing vulnerability. A rare-earth embargo against Tokyo became China's weapon of first resort in response to the brief Japanese detention of a Chinese fishing-trawler captain in September 2010.

China, which dominates the global production and supply of these minerals that are critical to the manufacture of a vast array of high-tech products, persisted with its unannounced embargo against Japan for about seven weeks while continuing to blithely claim the opposite in public—that no export restriction had been imposed. Yet its own subsequently released trade data showed that its rare-earth metal shipments to Japan fell to zero in October 2010, and to nearly zero for rare-earth oxides, which are more processed chemical compounds.

That embargo was followed by China's reduction of rare-earth export quotas to all countries in early 2011, prompting the United States, the European Union, and Japan to file a World Trade Organization complaint alleging that Beijing was using its rare-earth monopoly as a political and economic weapon. China's move, however, provided other major economies an advance notice to find ways to reduce their dependence on imports of Chinese rare-earth minerals so as to offset Beijing's leverage. Indeed, the Chinese actions set in motion concerted efforts by others to develop new international sources of supply and break China's chokehold on the market. Production of rare earths is now expanding at plants outside China, undercutting the Chinese monopoly.

In keeping with the historical lessons about the centrality of adequate resource availability in powering a country's rise, China has calculatingly cultivated cozy ties with countries that can supply raw materials for its rapidly growing economy, regardless of their human-rights record. Sudan, Zimbabwe, Iran, North Korea, and Burma, among others, have all enjoyed, to some degree or other, Chinese political protection at the United Nations Security Council, where Beijing wields veto power. China's rapidly growing trade and investment in Africa, for example, has been spurred by its ravenous need for hydrocarbons and mineral ores. State-run Chinese companies, flush with cash from their country's economic boom that has generated unparalleled and still-growing foreign exchange reserves, have been encouraged by the government to enter the oil, gas, and mining sectors in Africa, where Chinese investments in infrastructure, real estate, and manufacturing are often designed—in the old colonial style—to facilitate resource extraction, processing, and shipment. China indeed is now widely seen by Africans as the new colonial power on their continent, with the influx of Chinese investors, traders, and laborers stoking tensions over the external control of Africa's natural resources.

Acquisition and control of resources is a key goal of Chinese policies. In securing overseas supplies of fuel and minerals to meet the soaring demand of its cities and factories, China is emulating what dominant powers have done for more than two centuries. In fact, the world's most assertive policies today to gain control of strategic resources are arguably being pursued by China, which employs aid and other diplomatic tools to secure commodity deals while placing its state-owned corporate behemoths at the vanguard of such an outreach. In the case of iron ore and some other important minerals, China is relying on greater imports to help conserve its own reserves.

While going into overdrive overseas to corner energy resources, metals, and other raw materials, China at home is aiming to control transnational river outflows by accelerating what already has been for three decades the world's largest dam-building program.[62] Such a focus has resulted in water becoming a new divide in its relations with several of its neighbors, including Russia, Kazakhstan, Vietnam, Burma, and India.

Its broader resource strategy seems aimed at gaining a long-term strategic advantage that its competitors would find hard to neutralize. By buying hydrocarbon and mineral-ore reserves in distant lands at a time when the planet is running short of natural resources and by building a capacity at home to manipulate cross-border river flows through dams, reservoirs, and other diversions, China has embraced an overtly mercantilist approach to lock up long-term supplies as a vital strategic interest. Its far-flung assets and growing resource imports actually serve to rationalize its focus on building a more powerful navy and playing a maritime role far from its shores.

China is just one key example of a wider resource-gaining drive which suggests that, in contrast to Samuel Huntington's theory that civilizations will increasingly clash along cultural and religious fault lines, wars in this millennium are likely to be fought more over resources than over ideologies, as states battle to control, or secure access to, dwindling supplies of commodities.[63] This is not to suggest that differences in basic values, attitudes, beliefs, and cultures will cease to be a source of conflict, or that "resource curse"— the institutional inability to employ an abundance of natural endowments to lift citizens out of poverty that afflicts a number of major resource-exporting nations—is not as serious a source of conflict as resource scarcity.

Take resource curse: A similar profile linked the oil-producing countries battered by violent upheavals that rapidly spread like a contagion across borders in 2011—burgeoning young populations; high unemployment; growing income disparities; water stress; aging, autocratic, and kleptomaniac leaders ensconced in power for more than a generation; and rising public expectations. The explosive way that popular discontent led to civil disorder in the Arab world, with external powers playing a role in regime change or regime reinforcement, made the international geopolitics of oil only murkier. While it is now apparent that much of the Arab world is in transition, the end point is still not clear, making external factors important in influencing internal and regional developments. Long after the political tumult has subsided, the underlying challenges centered on spreading water scarcity, growing food insecurity, and the ticking population time bomb will continue to sharpen.

Resource wars are in no way inevitable. Yet it is important to recognize assertive mercantilism or water appropriation as a source of regional tensions and potential armed conflict. The competition for scarce resources likely carries greater conflict potential than the rivalries between disparate cultural blocks. After all, underneath the cultural and religious divides are universal aspirations for a better life.

The danger of resource wars was acknowledged by the international community way back in 1982 while adopting the World Charter for Nature, which declared that "competition for scarce resources creates conflicts, whereas the conservation of nature and natural resources contributes to justice and the maintenance of peace."[64] Not all resources, of course, are in danger of becoming scarce. For instance, there are enough recoverable reserves of coal—the main culprit in global warming—to meet the present world demand for at least the next two centuries. Technological innovations could lead to substitutes for some resources now under pressure, although perhaps not in time to ward off scarcity or greater competition.

Geostrategist Michael Klare has postulated that the escalating demand for finite natural resources, the geopolitical contests over ownership, and resource shortages combine to make key resources—from water and oil to gems and timber—the engines of power struggles and conflicts, while increasing the likelihood that more and more supplies of some strategic resources will come from places that are remote, unstable, and politically unreliable.[65] Indeed, there is a well-documented link between resource scarcity and violent conflict.[66] Conflicts, even when they are rooted in resource scarcity, are often camouflaged as civil wars or political or sectarian hostilities. The effects of water scarcity are particularly insidious on internal and regional security.

The fact that water cannot be imported like coal, oil, food, and mineral ores creates incentives for states to commandeer internationally shared waters before they leave their national borders. This allure, and the resultant focus on large hydroengineering projects, has contributed to the resurrection of some key territorial disputes and long-festering border demarcation issues. Like the phenomenon of resource nationalism in oil-rich countries, upper-riparian states are seeking to profit from the geopolitical advantage they hold vis-à-vis downstream nations. Major hydroengineering projects can enable an upstream state to regulate the volume of transboundary river flows, thereby affecting the livelihoods of those resident downstream. By manipulating cross-border flows, an upper riparian can fashion water into a major instrument of political leverage. Tensions over unilateral overexploitation can also extend to aquifers straddling international borders.

The broader international competition over natural resources is only set to sharpen, increasing the risks of severe market disruptions and interstate and local conflicts in hot spots.[67] As the North Atlantic Treaty Organization has warned, "Environmental degradation as a result of depletion of natural resources, [and] transboundary issues arising from shared water sources, pollution, etc., can lead ultimately to regional tensions and violence."[68] According to a study on the future of the world by the European Union's Institute for Security Studies, "Longer-term extrapolations to 2030 and 2040 show that many countries will be faced, in different ways, with severe environmental distress and growing shortages of water, food, and energy."[69]

As more people step up to the middle class and seek everyday comforts of modern life, per capita energy and natural-resource consumption patterns in the developing world are bound to grow substantially. Slaking the tremendous craving of the emerging economies for water, mineral ores, hydrocarbons, and other resources, while meeting the huge demands of the old economic giants, is at the core of the great resource dilemma the world faces in the twenty-first century—one that carries crucial long-term political and socioeconomic implications. Although geopolitical transformations do not move along linear, projectable trajectories, the present trends are a pointer to the conflict potential inherent in resource competition, including overtapped water resources. This underscores the national and regional imperative to make the right policy choices now to help secure the future.

Key Factors behind Spiraling Resource Consumption

Societies historically have fought more often over nonrenewable resources than renewable ones. This is largely because nonrenewables like oil, iron ore, gold, and other mineral wealth can be more directly converted into state power than renewables such as freshwater, forests, fish, and croplands. The nonrenewables also can be cost-effectively transported across long distances, unlike most renewables. Another key reason is that the availability of renewables was not under the same pressure in olden times as in contemporary times. Freshwater availability, for example, has emerged as a central concern for an increasing number of countries only in recent times.

One important factor that has brought freshwater and most other renewables, as well as some nonrenewables, under growing strain has been the population explosion. Such has been the rapid demographic growth that the world population climbed from 1 billion around 1804 to 7 billion by 2011, in just over two centuries. The more the human population has swelled, the more the number of animal, bird, fish, and tree species has dwindled. Rapid population growth—besides serving as a key driver in bringing resource availability under increasing pressure—has traditionally been a central factor in the conflict spiral.[70]

The first billion people accumulated over thousands of years, from the origins of humans to the early 1800s. After 1804, the world began adding a billion new people to its population in a succession of record periods continuing until the end of the twentieth century, as table 1.1 shows. The unparalleled rate of population increase coincided with economic-boom times: global economic output per person jumped elevenfold between the dawn of the industrial age in 1820 and the end of the twentieth century.

The second billion population milestone was reached in just 123 years in 1927 and the third billion in 33 years by 1960, after which the world actually went into a procreation overdrive. The global population then doubled between 1960 and 2000, as new antibiotics, better immunization, clean water, and improved food availability produced major improvements in infant, child, and maternal mortality and a sharp rise in average life expectancy.[71] During the twentieth century's phenomenal population explosion, the world

Table 1.1.   World Population Milestones

World population reached:

1 billion in 1804
2 billion in 1927 (123 years later)
3 billion in 1960 (33 years later)
4 billion in 1974 (14 years later)
5 billion in 1987 (13 years later)
6 billion in 1999 (12 years later)
7 billion in 2011 (12 years later)
8 billion in 2025 (projected; 14 years later)
9 billion in 2043 (projected; 18 years later)
10 billion in 2083 (projected; 40 years later)

Sources: United Nations Population Division, 2011; and U.S. Census Bureau, 2011.

population skyrocketed from 1.6 billion to more than 6 billion, yet per capita income grew faster than ever before, rising nearly fivefold.

Maternal mortality—the rate at which women die in childbirth or soon after delivery—is a key gauge of any population's health and wealth. It has continued to decline in recent decades, falling by about 40 percent just between 1980 and 2008.[72] The highest rate in 2008 was in Afghanistan (1,575 maternal deaths for every 100,000 live births), and the lowest was in Italy (4 maternal deaths per 100,000). As for average life expectancy, it has continued to rise, and is projected to increase globally from sixty-eight years during 2005–2010 to eighty-one by 2095–2100.[73] Because low-fertility countries tend to have higher average life expectancy, they also boast the fastest population aging.

So, given this explosive population growth and increasing longevity, demand for resources has grown rapidly, bringing both renewables and nonrenewables under intense pressure, even as the already large global population is projected—despite a slowing growth rate—to rise to 8.32 billion in 2030, 9.30 billion in 2050, and 10.12 billion by 2100, according to the "medium variant."[74] The average number of children per woman, however, has fallen worldwide from 5 in 1950 to about 2.5 now.

Can the earth sustainably support such a projected huge population, which will mean enormous increases in the size of cities, the volume of material consumption and waste, and pressures on the environment? Just between 2011 and 2025, an extra billion people will join the world population.

Population and economic growth are fostering rapid urbanization, as underlined by the sharp increase in the number of cities in the world with more than a million residents—from 86 in 1950 to 387 in 2000, with the biggest one hundred of them serving as home to more than 6 million people each.[75] Between 2011 and 2050, the global population is expected to increase by 2.3 billion, but the number of urban dwellers is projected to grow faster, rising 2.6 billion to reach 6.3 billion—or two-thirds of the total world population, up from slightly more than 50 percent at present.[76] Such exponential growth is making cities a major driver, for example, of increased water consumption. Yet, to meet the increased demand of fast-growing urban populations for various products, water use by industry and agriculture will continue to grow faster than direct use by urban areas in the coming decades.

The earth's atmospheric and renewable-freshwater capacity and land size have remained the same since the time humans evolved, yet anthropogenic demands and impacts have grown phenomenally. At least 850 million people still live in slums, and nearly twice that many lack ready access to potable water. Most of the upcoming additions to the global population will be concentrated in the developing world, large parts of which are already crowded and water stressed; the population there is likely to increase from 5.7 billion in 2011 to 7.9 billion in 2050.[77]

The geopolitical imbalance of numbers will increasingly challenge the relatively resource-poor developing world. For example, sub-Saharan Africa, which did not have even a third of Europe's population in 1950, is projected to have nearly five times more people than Europe by 2100. The United Nations has identified thirty-nine countries in Africa, nine in Asia, six in Oceania, and four in South America as high-fertility states that will contribute the bulk of the additions to the global population; this list includes the hotbeds of terrorism, extremism, and internal conflicts, such as the AfPak belt, Yemen, Iraq, Sudan, Ethiopia, and Nigeria. The aggregate population of the high-fertility countries is projected to more than triple by 2100.

Whereas the populations of the low-fertility and intermediate-fertility countries are likely to peak much before the end of this century, those of the high-fertility states would still be increasing at the turn of the century. National censuses show that fertility rates are on the decline in the vast majority of countries, with the cultural preference for boys and the resultant gender imbalance in many Asian and other societies also carrying important implications for reproduction and child care. This, in turn, suggests that the world population could peak earlier than anticipated by the United Nations Population Division.

Consumption growth is another principal culprit in the sharply rising global resource demand. Economic growth and rising prosperity are fueling consumption growth, which in turn is aggravating the environmental impacts of human activities. Even as a growing number of countries have reduced their birth rates, they have become more consumptive.

Declining fertility rates indeed are correlated with rising prosperity and greater consumption levels. An early economic boom can accompany a still fast-growing population. Yet it doesn't take long for an economic boom to help depress fertility rates while raising per capita income and consumption levels. State policies can accelerate such trends. China's use of state muscle to enforce a one-child policy in urban areas—along with promarket national reforms—coincided with its dramatic economic rise. As per capita income in China has risen more than eightfold since 1980, escalating domestic consumption levels, coupled with its export drive, have wrought major environmental impacts, exemplified by the serious degradation of watercourses in the Han heartland and the country's emergence as the largest emitter of greenhouse gases.

Globally, even as population growth has slowed, consumption is soaring, making it a leading factor behind increasing resource degradation and depletion. In this light, the discussion on overpopulation must not be limited to head counts or rate of population growth but must also take into account other key factors behind rising consumption levels—growing prosperity and the increasing body mass of the average global citizen.

Water, food, energy, and other resource-consumption levels depend not just on the number of people but also on their actual body weight and income levels. For example, even small population increases in wealthy nations have a disproportionately high impact on global resources and the environment because these countries' relatively heavier citizens make greater demand on food, water, energy, and other resources. Fat people tend to be more consumptive than those who are slender. The global obesity epidemic, however, is not just about being rich; in adults, corpulence can also be related to low economic status or poor eating habits.

Put simply, the competition for natural resources isn't just about how many mouths there are to feed but also about how much excess body fat there is on the planet. Unfortunately, many in the world are getting fatter, with the worldwide prevalence of obesity almost doubling between 1980 and 2008.[78] Once considered a problem only in advanced economies, obesity now is on the rise in middle- and low-income countries. Persons are considered to be overweight if their body mass index (BMI)—an estimate of body fat based on height and weight—is between 25 and 29.9, and obese when their BMI is 30 or more.[79]

A net population increase usually translates into greater human capital to create potential innovations, power economic growth, and support the elderly. But a net increase in added weight only contributes to state liability and added resource consumption. The higher any individual's BMI, the higher the risk of disease. According to an estimate cited by an Institute of Medicine report, the annual expense of treating obesity-related illnesses and conditions in the United States has reached $190 billion.[80]

The increasing global prevalence of obesity, however, carries implications extending beyond individual health to the environment, including water, with the weight problem a serious drain on global resources. Given that food production is the main drawer of water, overweight or obese people—with their higher food-energy needs—significantly add to the global water demands, besides causing much greater greenhouse-gas emissions through their bigger food and transport needs. For example, greater car use is common among the overweight.

A study published in the *International Journal of Epidemiology* found that compared with a population made up of people generally fit, a community with 40 percent obese citizens requires 19 percent more food energy for its total energy requirements.[81] In its model, the study compared a population of a billion lean people, with weight distributions characteristic of a country like Vietnam, with a billion people from richer countries, such as the United States, where about 40 percent of the adult population is classified as obese and another 25 percent as overweight. One-third of American children are also overweight or obese.[82] The prevalence of obesity, according to the International Association for the Study of Obesity, is particularly high in some other countries too, such as the Persian Gulf sheikhdoms, where citizens have discarded the physically demanding life of the desert for air-conditioned comfort and fast food.

But at a time when Americans are gaining weight as a nation—weighing 18.7 kilograms heavier on average than the rest of the world—it is the United States that stands out for the dubious distinction of having the largest global population of overweight or obese people.[83] A 2012 study published in the British journal *BMC Public Health*, using global data on BMI and height distribution to estimate average adult body mass in each country, determined the global adult human biomass to be 287 million tons on the basis of population size and average body weight. In other words, this is how much the entire adult human population would weigh if it stepped on a scale. Of that total weight, 18.5 million tons is attributable to people being overweight or obese.

The study found that North America has 6 percent of the global population but 34 percent of the world's biomass due to obesity, while Asia, in sharp contrast, has 61 percent of the world's population but accounts for only 13 percent of global excess weight due to obesity. Its conclusion: If every nation had the same fatness level as the United States, it would be equivalent to adding an extra 935 million people of average weight to the global population, with major implications for the world's water, food, and energy situations.[84]

Environmental stress also tends to rise with income level because the rich are far more consumptive. Given that the wealthiest fifth of the world population boasts more than three-quarters of the total global income and the poorest fifth has just 1.5 percent, according to World Bank estimates, resource-consumption levels contrast starkly between the rich and the impoverished. The United States, the world's third-most populous nation and the fastest-growing industrialized nation despite a slightly slowing population growth rate, remains the most heavily consuming society in per capita terms.

The aging of many developed countries, paradoxically, is also aiding higher per capita consumption, because people over sixty, on average, consume more than those under fifteen.[85] In the coming years, population aging will accelerate in the low-fertility nations, which are mostly advanced economies, with the number of their citizens aged sixty-five and over increasing from 11 percent of aggregate population in 2010 to 26 percent by 2050.[86] In the developing world, as growing numbers of people advance to the middle class, high-protein diets, gasoline-fueled transportation, and water-guzzling, energy-hogging home appliances (including dishwashers and laundry washers) will become more common.

The plain fact is that the average person in the world today is consuming more resources, including water, food, oil, and energy. Even the per capita utilization of metals and minerals has shot up. Yesterday's luxuries are becoming today's necessities, putting greater demand on natural resources. While nonrenewable resources are being extracted and consumed without considering their availability for future generations, renewable resources like freshwater, land, and biomass are being used in excess of their natural capacity for replenishment or regeneration. Consumption growth is laying ever-increasing pressure on the environment.

Take freshwater: Global water consumption has grown at more than double the rate of the demographic explosion in the past century. World water use, after soaring from approximately 770 billion cubic meters per year in 1900 to 3,840 billion cubic meters in 2000, is projected to climb further to over 5,000 billion cubic meters by 2025.[87] In other words, water consumption is set to multiply 6.5 times in just 125 years. Is it thus any surprise that a resource whose availability until a few decades ago was considered adequate to meet global needs is now increasingly under pressure and attracting sharpening competition?

Significantly, because the spiraling per capita consumption levels have been accompanied by human ingenuity in discovering and extracting new nonrenewable reserves, the distinction between nonrenewable and renewable resources is being turned on its head: several resources that have a continuing natural process of renewal and supply are becoming more finite than many nonrenewables, whose reserves have swelled because of major new discoveries. For example, since 1970, known reserves of copper have quadrupled, including the quantity already mined, while total gold reserves have increased 12 times in size.[88] Major additional supplies of aluminum, lead, oil, natural gas, silver, tin, tungsten, uranium, and zinc have been opened by tapping new sources.

As a result, it is the renewables more than the nonrenewables that paradoxically are in danger of getting critically depleted. In addition to the demand-induced scarcity of renewables, such resources, including freshwater, are highly vulnerable to degradation, a factor that can exacerbate the supply-linked paucity. Degradation of one renewable resource tends to adversely impact another renewable. For example, land degradation can have wide-ranging impacts on water resources, including altering watercourse regimes and causing silting of reservoirs and estuaries, as well as water pollution.

Reserves of some nonrenewables, such as coal and several types of metals and stones, seem almost inexhaustible at present from an economic standpoint, whereas a number of animal, bird, fish, and tree species are clearly exhaustible, with some already becoming extinct. Iron and aluminum, the two most widely used metals, are second only to silicon in their abundance on earth. In fact, unlike energy resources such as oil, gas, and coal, metal ores like iron, lead, and copper and even rare-earth elements can actually be released for further use through industrial recycling. By contrast, freshwater consumption in excess of natural replenishment or dumping of untreated wastewater can cause irreparable damage to ecosystems. Water resources are most susceptible to pollution, and the process of reversing contamination is expensive and energy intensive.

To lessen shortages, freshwater can be stored in dam reservoirs in the wet season for release in the dry season—a common practice in many countries, although the per capita storage capacity varies widely, with the storage in water-stressed developing economies, as a whole, below the global average of nine hundred cubic meters per head yearly. Increasing water storage by channeling excess surface water to artificially recharge aquifers holds promise for coping with droughts. Adequate water storage can help moderate seasonable and spatial variability of water availability within national borders. But when aggregate freshwater demand exceeds the natural replenishment potential, storage alone cannot obviate scarcity.

Given the pervasive yearning for a better life and the wide income disparities between and within nations, the consumption factor is difficult to manage. That difficulty is compounded by the fact that the measure of economic growth in our world is ever-increasing production and consumption. Global commodity prices, as figure 1.3 reveals, have risen steeply since the early 1970s, a period that has coincided with dearer oil and with economic growth accelerating in the non-Western world. Per capita global resource consumption, despite the soaring prices of many commodities, could actually double within the next half century, seriously increasing the environmental costs.

THE INCREASED RISK OF WATER CONFLICTS

When water resources are not owned by a single nation or province, they must be shared—often a thorny issue in an era of scarcity and at the root of the increasing risks of violent water conflicts. Peace and environmental sustainability demand that the resources of any basin be managed cooperatively and holistically. Yet, more often than not, it is difficult to work

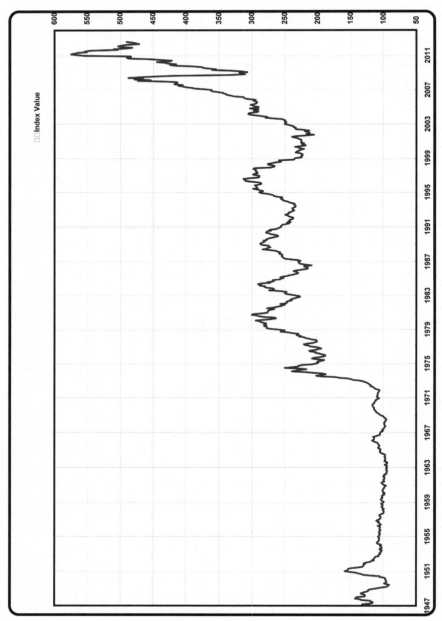

Figure 1.3. The Sharp Rise in Global Commodity Prices since the 1970s

Note: This is the CRB BLS Spot Market Price Index, a measure of the price movements of twenty-two sensitive commodities, mainly raw materials or products close to the initial production stage, ranging from rubber and zinc to wheat and cottonseed oil. The index uses 1967 as the base year.

Source: Commodity Research Bureau, 2012.

out equitable and reasonable sharing arrangements that satisfy all parties. In fact, even when a water institution exists, many divided basins tend to preclude joint water management.

Watersheds and watercourses often traverse political boundaries. Such is the extent of shared surface-water resources in the world that there are 276 transnational river and lake basins, extending to the territories of 148 countries and accounting for three-fifths of all river flows in the world.[89] In addition, at least 274 underground freshwater basins straddle international borders, with the exploitable flow paths of some aquifer systems stretching across several countries.

Nature helps to store much more of the annual precipitation in aquifer systems than in rivers and lakes, and many ecosystem services critically rely on groundwater. Although groundwater has now become the world's most extracted natural resource, the arrangements for managing or governing groundwater remain internationally far weaker than for surface waters. There are indeed few interstate agreements that address issues relating to transboundary aquifers.

Given that the true volume of groundwater is often unclear, drawing heavily on these resources can create unforeseen situations, such as serious aquifer depletion. The hidden nature of subterranean water resources not only hinders demarcating user domains and collecting reliable, widely accepted data, but also makes governance issues complex and conflictive when the competing users are sovereign states or even provinces or communities. Examples extend from the "war of the well" in drought-battered parts of northeastern Africa to the intracountry groundwater disputes in the United States between Nevada and Utah as well as between the state of Mississippi and the city of Memphis in Tennessee.

As many transboundary river basins exemplify, sharing and sustainably managing even surface waters remains a challenging task. Nearly half the earth's land surface—excluding Antarctica (dubbed "the Ice")—is covered by river basins that extend beyond one country. Multinational basins include the world's great rivers, such as the Amazon, Brahmaputra, Congo, Danube, Euphrates, Ganges, Indus, Mekong, Nile, Rhine, Salween, and Tigris. Bangladesh, Belarus, Hungary, and Zambia are among at least thirty-three countries that have more than 95 percent of their territory within one or more international basins. At least six countries—Botswana, Bulgaria, Egypt, Hungary, Mauritania, and Turkmenistan—rely almost entirely on inflows from neighboring countries. In sub-Saharan Africa, the principal surface-water resources of practically all the countries are transnational in nature.

Several river basins actually encompass five or more nations. The Danube is the world's most international river because its basin includes the territories of a record nineteen countries. Five other river basins—the Congo, the Niger, the Nile, the Rhine, and the Zambezi—are shared by between nine and twelve countries. The eight-nation Amazon, which drains the biggest basin on earth, including the largest tropical rainforest, is the greatest of great rivers, with its average flow larger than the combined flows of the world's next ten largest rivers.

The twenty-first century will be a defining epoch for how humanity manages and addresses its grave water challenges. Ensuring adequate freshwater availability to underpin continued progress has become critical to the future well-being of human civilization. Yet rival claims over shared water resources have given rise to numerous festering conflicts between and within states. The sharpening competition over shared resources carries significant security and business risks. The growing dangers have prompted even the United Nations Educational, Scientific and Cultural Organization (UNESCO) to become involved in water-conflict management issues, after the thirty-six member states of the International Hydrological Program Intergovernmental Council (IC-IHP) in 1996 approved the inclusion of "water conflicts" as a major theme in the work plan of this UNESCO-assisted program.

When freshwater issues get linked with national security, the water-conflict potential is known to increase.[90] And when intracountry water disputes are already rife in a region, the risks of shared resources turning into a flash point in intercountry relations are often higher. That water is getting increasingly tied to security in many parts of the world holds major long-term strategic ramifications, especially given the prospect of world climate turning measurably drier and warmer. One report has identified China, Central and South Asia, the Middle East, Southern Africa, and Mexico as the likely places "to be affected by violent conflict over water rights."[91] Peacefully managing intrastate and interstate disputes has become a paramount concern.

The growth of water stress and insecurity is an unambiguous reminder of the rise of nontraditional security challenges. In a world characterized by extraordinary technological, economic, and geopolitical transformation since the 1980s, international security challenges have also fundamentally changed. Princeton University professor Richard H. Ullman presciently foresaw in the early 1980s that "nonmilitary tasks are likely to grow ever more difficult to accomplish and dangerous to neglect."[92] Alleviating water scarcity and insecurity is now one such critical task.

Water indeed is where the old security issues of freedom of fluvial navigation, security of sea lanes of communication, and prevention of ocean piracy intersect with new security challenges, such as framing international rules on shared watercourses, regulating the building of large storage dams on transnational rivers, and containing the international effects of deteriorating freshwater quality and the degradation of fluvial, coastal, and marine ecosystems. The U.S. State Department has now identified freshwater as a central foreign-policy concern because of the potentially far-reaching effects of increased interstate competition.

Weak Water Accords and Institutions

At a time when hydropolitical issues in several major international basins bedevil interriparian relations, especially in the developing world, it seems unlikely that humanity's past can influence or predict its future. Conflict over land, at the center of some prominent interstate disputes, also often includes conflict over the water that flows through the disputed land.[93] The struggle for freshwater resources threatens to overshadow today's more conspicuous struggle for energy resources, prompting a former U.S. Central Command commander, Anthony Zinni, to warn, "We have seen fuel wars; we are about to see water wars."[94] Because water cannot be physically secured like gems and minerals, a water war must aim not to seize a flowing resource but to control a water-rich land or to divert water flows.

With interstate river and lake basins home to two-fifths of the global population and covering 47 percent of the world's land surface, the leitmotif of riparian relations ought to be interdependence, not competitive hydroengineering projects and strident assertions of national sovereignty. Whenever any upriver state has sought to unilaterally exploit its riparian advantage through a new water-diversion project, it has created water tensions in the region. Examples of such contemporary schemes, intended to meet new water needs upstream, include China's building of giant dams on the international rivers originating in its ethnic-minority homelands and Turkey's Southeastern Anatolia Project, which covers nine of its provinces in the Tigris-Euphrates Basin (a program referred to as GAP, the Turkish acronym for Güneydoğu Anadolu Projesi). As illustrated by India's Farakka Barrage—a water impoundment completed four decades ago to flush silt and protect the port of Calcutta—even projects to serve narrow purposes can be controversial and trigger an interriparian dispute that may take many years to resolve or continue to fester.

Despite the promotion of cooperation on the environment and natural-resource management taking center stage in global diplomacy, international water cooperation still faces major challenges, including managing disputes over the sharing of transnational water resources, building institutionalized cooperation and collaboration, and dealing with limited compliance with international norms and limited funding support for basin-level initiatives. If anything, there is increased mistrust and divisiveness at the regional and international levels. The international community's ability to avert water

wars in the coming decades will depend on its "collective capacity to antici-
pate tensions and to find the technical and institutional solutions to manage
emerging conflicts."[95]

Yet, even as some states exploit their riparian advantage to capture resources through new projects and thereby present a fait accompli, it has
proven difficult to stop such moves or to establish genuinely cooperative institutions at a time when the majority of countries are chasing limited water
resources to meet their growing needs. In an international system pivoted
on national security, not collective security, the assertive pursuit of relative
national gain is common, even at the expense of the planetary interest.[96]
The *doctrine of prior appropriation*, under which the first appropriator
(user) of river waters gains a priority right in customary international law,
actually serves as an invitation to resource capture, especially by the more
powerful. Resource capture, in turn, helps build greater political leverage
over co-riparian states.

Water has long been used as a tool of political bargaining between rival
states. The history of averting conflicts over freshwater resources actually
dates as far back as 2500 BCE, when the two Sumerian city-states of Lagash
and Umma in the region now called southern Iraq signed a treaty to end their
war over the resources of the Tigris River.[97] More than 3,600 treaties related
to water resources have been concluded since 805 CE, according to the Food
and Agricultural Organization of the United Nations.[98]

However, a review of water treaties—as opposed to nonbinding accords,
agreements-in-principle, or memorandums of understanding—shows that
most of them relate to navigational uses of transnational rivers, territorial-
demarcation matters, fishing rights, building of a specific project, or other issues unrelated to a clear division of shared waters. It is only since the last century that treaty-making efforts have sought—with mixed results—to focus on
the use, development, and management of water resources. Since the 1950s,
some enemy countries in Africa, the Middle East, and South and Southeast
Asia have entered into water agreements, however weak or incomplete.

Yet how many internationally shared basins are now governed by broad-
based, legally binding rules and institutions that can forestall escalation of
disputes to violent conflict? The startling answer is only a tiny number. There
are, as appendix B shows, just eighteen genuine intercountry water-sharing
agreements covering one or more basins currently in effect, plus an Israeli-
Palestinian interim groundwater accord. About half of these agreements actually lack conflict-resolution and institutionalized-cooperation mechanisms.

Notwithstanding glaring examples of conflictive issues between basin
states, a vocal "water peace" school of thought, contending that water is a
vector of cooperation, has pitted itself academically against the dominant

school that emphasizes the conflict potential of water scarcity.[99] In one sense, this is a clash between idealistic and realistic views, which is not uncommon in a wide range of disciplines in the social sciences, humanities, and natural sciences. In another sense, such an intellectual divide can act as a brake on attempts to turn water into a pressing strategic concern demanding concerted, integrated action at the subbasin, basin, and international levels.

The water-peace school tends to look back at the past, rather than at how the new security challenges in an era of growing water scarcity could shape the future. Science journalist Wendy Barnaby has asserted that in the five decades between 1948 and 1999, water-related cooperation far outweighed conflict over water. "Of 1,831 instances of interactions over international freshwater resources tallied over that time period (including everything from unofficial verbal exchanges to economic agreements or military action), 67 percent were cooperative, only 28 percent were conflictive, and the remaining 5 percent were neutral or insignificant," she wrote.[100] Geographer Aaron Wolf has also argued that water cooperation has generally prevailed over water conflict, pointing out that of the 412 interstate crises between 1918 and 1994, there were only seven cases where water was a cause.[101] Some other scholars, however, paint a different picture of the past. While one analyst has contended that "history is replete with examples of violent conflict over water,"[102] several other scholars have cited the 1967 Israeli-initiated Six-Day War—in which the water-rich Golan Heights and the aquifer-controlling West Bank were captured—as an example of how water can be an underlying driver of armed conflict.[103]

Fundamentally, the peace-school arguments are pivoted on the premise that common threats and challenges inexorably drive states toward cooperation and collaboration. Because harmful impacts on any shared resource damage regional communities, there can be, it has been argued, no zero-sum security at play.[104] If anything, water "can serve as a cornerstone for confidence building and a potential entry point for peace."[105]

Even the long-standing theory of rational choice, which postulates that people weigh the costs and benefits of their actions and make decisions in their best interests, has been applied to state actions on water resources to rule out increased risks of conflict. Yet resource grabs through land capture or through the tools of hydroengineering have been perfectly rational actions for the executing states, while weaker nations have rationally resorted to asymmetric warfare.[106]

Few will dispute that water scarcities offer opportunities for collaborative international efforts to develop political and technological solutions. Water scarcity can compel state parties to cooperate, but it can just as well exacerbate conflict. What if water shortages, instead of promoting interriparian cooperation, spur greater resource grabs and a zero-sum game behind the banner of win-win national hydroengineering programs, as exemplified by China and Turkey? As one assessment has cautioned, "In the postmodern world, conflicts over water may contribute to more instability than has heretofore been the case. Inadequate or unsafe water is seen as a security issue, much as one might view a heavily armed military force mobilizing across the border. Both threaten survival."[107]

It is true that subnational groups conflict with each other over access to resources more than nations do. It is also correct that water disputes in the past did not trigger many interstate military conflicts.

The world, however, is entering a new era of increasing water scarcity, which threatens to become a significant constraint on human activities. Few had foreseen water scarcity as a potential constriction on human development at the time the present international institutions were established in the post–World War II period. But now lack of access to stable water supplies, as the U.S. National Intelligence Council has warned, is "reaching unprecedented proportions in many areas of the world and is likely to grow worse owing to rapid urbanization and population growth."[108] The number of countries that are freshwater or cropland scarce is projected to more than double between 2008 and 2025, to include, among others, Burundi, Colombia, Ethiopia, Eritrea, Malawi, Pakistan, and Syria.[109]

Against this background, what factors are likely to make states pursue a more confrontational or unilateralist path with regard to shared water resources than what has been the historical record? Let's be clear: Past history can hardly serve as a guide to the future. That water has not served as a major instigator of war in history does not preclude water-related armed conflicts within and between states in the future. After all, just a few decades ago, water scarcity was relatively unknown, other than in desert lands. The water situation has changed dramatically since the 1950s. In the twenty-first century, as environmental scholar Thomas Homer-Dixon has emphasized, "the renewable resource most likely to stimulate interstate resource war is river water."[110]

Some current realities already raise concerns. For one, the vast majority of transnational rivers, lakes, and aquifers lack *any* legal water-sharing or co-operative-management framework. Where treaties do exist, they are in most cases structurally anemic or skip over a key basin state that has obstinately chosen to stay out of the institutionalized arrangements. For another, highly unilateralist hydroengineering policies are already being pursued in a number of basins. The United Nations, however, proudly asserts that more than two hundred international water agreements or memorandums of understanding have been signed in the period after World War II, with "only 37 cases of reported violence between states over water."[111]

Given that global water resources have started coming under intense pressure only in recent times, the reported occurrence of thirty-seven cases of interstate water violence since 1946 is scarcely a comforting statistic. The cases of violence involved festering water disputes, while the vast majority of accords or agreements-in-principle signed in the post–World War II period seek to address one or more of a narrow range of transboundary issues, including flood control, hydropower development, hydrological data sharing, joint research, irrigation, watercourse protection, storage, use of river islands, and interim arrangements. Moreover, can we ignore the fact that the number of treaties that actually share out basin waters and set up meaningful water institutions or joint water management remains disappointingly very small? In most of the UN-listed water agreements, water allocation—the most contentious and conflictive issue—has been dispensed with or not spelled out in a manner that would help obviate disputes or conflicts.

In fact, a clearly spelled-out sharing formula figures in very few treaties that cover transboundary basin resources. Even in the accords that do specify sharing quantities, water allocations have been rigidly set, with little room to adjust to hydrological variations and changing basin dynamics, thus raising concerns about their durability under rapid-growth or climate-change conditions. Water-quality obligations, moreover, have been left out of most agreements, although degradation and contamination of shared resources are increasingly becoming important transnational concerns. Multinational basins, at best, have bilateral accords, because only in the exception do water-related agreements rope in all riparian neighbors, thus ruling out the pursuit of integrated basin management.[112]

To compound matters, most existing water agreements are toothless. The bulk of them lack enforcement and conflict-resolution mechanisms, or even elementary monitoring provisions. A growing number of new agreements, however, tend to incorporate at least some elements of water-quality control, information sharing, monitoring, and conflict resolution in their provisions. But few of them are genuine water-sharing treaties.

The plain truth is that most existing accords hardly stand out as examples worthy of emulation in transnational basins lacking any institutional arrangements. Lumping all pacts, provisional accords, and memorandums of understanding together under a single heading of water agreements simply distorts reality by presenting an inflated picture of cooperation. In fact, some agreements stand out as examples of asymmetrical cooperation dictated and driven by the strongest, as in the Nile Basin.

Add to this picture the weak, underdeveloped international legal framework for transnational river basins; there is a dearth of comprehensive and well-accepted international water laws. The norms in relation to internationally shared aquifers are even weaker, even though groundwater has emerged as a critical transboundary resource subject to competitive overexploitation. More fundamentally, the 1997 United Nations Convention on the Law of the Non-Navigational Uses of International Watercourses—which took more than a quarter of a century to develop but whose entry-into-force is still not within sight—has become a symbol of both the international community's desire for rules to govern common waters and its failure thus far to put its money where its mouth is.

The majority of existing sharing treaties, especially those centered on far-reaching arrangements, were signed when serious water shortages were uncommon. For example, the 1960 Indus Waters Treaty, under which India benignly agreed to reserve four-fifths of the total waters of the six-river Indus system for downriver Pakistan, was concluded when water shortages were unknown in India's Indus Basin and policy makers had not anticipated the rapid advent of water scarcity. This pact—hailed by a 2011 majority staff report prepared for the U.S. Senate Foreign Relations Committee as "the world's most successful water treaty" for having withstood four wars[113]—leaves for Pakistan's use more than ninety times greater volume of water than the 1.85 billion cubic meters guaranteed to Mexico yearly from the Colorado River under the 1944 U.S.-Mexico Water Treaty. The Indus pact stands out as the world's most generous water treaty in terms of both the sharing ratio in favor of a downstream state (80.52 percent) and the total volume of basin waters reserved for it.

Among more recent pacts, there are no water-sharing allocations in the 1995 Mekong Agreement, while the water arrangements under the 1994 Israeli-Jordanian peace treaty, although significant in the bilateral context, are relatively modest in scale. Compared with the Indus treaty's 167.2 billion cubic meters of water per year for Pakistan—the aggregate average yearly flows of the three western rivers reserved for that country—the water allocations under the Jordan-Israel treaty are really small: Israel secured the right to pump 45 million cubic meters from the Yarmouk River yearly, and in return it agreed to transfer to Jordan 20 million cubic meters from the Jordan River every summer and also to accept that country's entitlement to store for its use in the winter at least 20 million cubic meters of the Jordan River floodwaters. Jordan, in addition, got access to an annual quantity of 10 million cubic meters of desalinated Israeli spring water.[114]

In fact, the bonus waters Pakistan receives every year from the India-earmarked eastern rivers are far greater than the total volumes spelled out in the

Israeli-Jordanian arrangements, or the quantity of water—28.6 million cubic meters yearly—Israel has granted the West Bank Palestinians access to until the contentious issue of water rights can be settled in the future. Waters of the eastern rivers not utilized by India—estimated to aggregate to 10.37 billion cubic meters yearly according to Pakistan, and 11.1 billion cubic meters according to the Food and Agriculture Organization—flow naturally to the downstream basin in Pakistan.[115] These extra outflows amount to six times Mexico's total water share under its treaty with the United States.

The 1996 Ganges Treaty—signed on the twenty-fifth anniversary of the India-Pakistan war that resulted in an independent Bangladesh—stands out for guaranteeing specific quantities of transboundary water deliveries, especially in the dry season, a new principle in international water law. The Ganges, a river sacred to Hindus, holds up to twenty-five times more dissolved oxygen than any other river system—a retention that scientists say prevents the rotting of organic matter and forms unique, germ-killing bacteriophagic conditions in its waters.[116] During the driest period between March and May, India has guaranteed Bangladesh an equal share of the Ganges waters, with the assured cross-border outflow averaging 991 cubic meters per second (or 31.27 billion cubic meters yearly). In other seasons when the downstream Ganges flow exceeds 2,265 cubic meters per second (or 71.48 billion cubic meters per year), Bangladesh's share is larger than India's.

The Ganges allocations, while not comparable to the cross-border flows under the Indus treaty, are much larger than the combined allocations set out in other accords signed since the 1990s, including the Jordan-Israel arrangements, the Komati River sharing between South Africa and Swaziland, the Lebanese-Syrian accord over the Al-Asi/Orontes River, and the twenty-first century's first water-sharing treaty, the 2002 Syria-Lebanon division of the waters of the small El-Kabir River, which has sources in both countries and forms part of their border. The Indus treaty is a colossus among existing international pacts, yet its division of rivers and lopsided 80/20 water formula favoring the subjacent state shows it represents more of a partition of whole rivers than an evenhanded or true sharing arrangement.

Most of the existing water-sharing treaties in the world relate to surface waters, despite the critical importance of transnational groundwater resources in many regions. Furthermore, experience in several basins shows that the mere existence of a water-sharing treaty does not preclude conflict. Cooperation and conflict indeed coexist in the vast majority of the river basins that boast a water-sharing treaty.

In this light, it will be useful to put forth, as this book does, descriptive hypotheses as to what the future may bring and how the risks of water conflict can be contained. It is apparent that to help underpin regional peace and

stability, interstate cooperation must be anchored in institutional arrangements that promote cooperation for the sustainable sharing and management
of basin resources. Yet instituting such collaboration remains a daunting
challenge in many basins. Even where vaunted water-sharing treaties exist,
water disputes surface every so often, irrespective of whether the parties are
friends (e.g., the United States and Canada, the United States and Mexico)
or old adversaries (Jordan and Israel, Pakistan and India). Rapid economic
growth, growing population and consumption pressures, increasing resource
constraints, changing climate, and sharpening geopolitical competition promise to make regional water management more challenging than ever.

Waging Water War by Another Name

Water, while transcending traditional security disputes between states, poses
an increasing security risk in politically divided basins. It is already a cause
of distress or a source of perceived injustice in several prevailing interstate
situations. As an issue, water usually lurks at the interstices of larger regional
matters ranging from territorial disputes to intercountry relations. So it often
tends to manifest itself in a broader context rather than as an issue by itself.

Given the extent of water feuds and conflicts in our present-day world, one
question that arises is, what circumstances are necessary for such conflicts to
spill over into violence? A false sense of grievance over water vis-à-vis a rival state can easily be instilled among citizens by a country employing its media and other tools, especially when the concerned nation is nondemocratic.
A host of factors, however, determine whether a dispute escalates to armed
conflict or low-level violence, including the relative economic and military
weight of the actors involved and the existence of any alternative means for
overcoming water scarcity.

Generally, a dominant riparian state—whether located upstream or downstream—can successfully assert its perceived rights, even if its control of resources deprives others of access to freshwater in sufficient quantity or quality. For example, when a downriver country is much stronger militarily than
an upstream state, but the latter attempts to materially diminish cross-border
flows through new projects, the danger is bound to arise of water becoming
a casus belli between the two.[117] But when the uppermost basin country is
ascendant and assertive—as exemplified by China's frenetic dam building on
international rivers and its ambitious interbasin and interriver water transfer
plans—subjacent states can do little more than use the tools of diplomacy to
protest moves that aim to unreasonably appropriate transboundary resources.

A dominant downriver state eyeing a small, upstream neighbor's water resources can resort to covert use of its power. In 1986, unable to secure greater

access to Lesotho's rich water resources, white-minority-ruled South Africa aided a military coup that helped oust that impoverished country's tribal government. Having engineered the political change while ostensibly engaged in operations against transboundary antiapartheid guerrillas, South Africa was able to conclude an agreement within months with the new military rulers in Lesotho to jointly develop a mammoth, multiphase Lesotho Highlands Water Project. The venture was designed to increasingly channel Orange River waters to South Africa from the upper catchment region, with the diversion scheduled to reach a level of 2.2 billion cubic meters per year when the project is fully completed by 2020 or later.[118] Earlier in 1975, South African forces moved into southern Angola and occupied the Ruacana hydropower complex, including a major dam on the Cunene River, whose protection against guerrilla attacks served as justification for the first deployment of regular South African military units inside Angola.

Under the Lesotho Highlands Treaty, Lesotho secured hydropower from the South African–financed project in return for delivering cross-border freshwater supplies, including drinking water for Johannesburg. The multidam Lesotho Project, despite its environmental costs, has indeed yielded important electricity benefits for that tiny landlocked kingdom, which is located entirely within the Orange River Basin and comprises mostly highlands and some spectacular canyons.

For South Africa—which geographically surrounds the kingdom from all sides and has developed a major strategic stake in the Lesotho Project—water is a central issue in the bilateral relationship. In fact, it has repeatedly intervened in Lesotho's politics, including in 1998 when—seeking to ensure uninterrupted cross-border freshwater inflows, among other things—it helped crush unrest by military intervention.

It is true that people cause wars, not any issue, even if it is resource scarcity. Although military conflicts are fostered by deep-seated grievances, those at the helm—as history attests—have in numerous cases triggered war either by acting recklessly or by presenting their state as weak and vulnerable and thus a tempting target for a preemptive attack. Water wars, however, can be waged by various means—military or nonmilitary—employed either overtly or covertly. One form of water war, for example, is the use of hydroengineering tools to change natural flows to the detriment of downstream neighbors while at the same time gaining leverage over their behavior. China's megadam projects in its borderlands, for example, will likely give it control over the transboundary flows of rivers that are the lifeblood for countries such as Kazakhstan, Russia, Vietnam, Burma, India, and Nepal.

As water resources become scarcer in the world and the interstate competition to control them intensifies, water wars will become more likely without an open resort to force. They could take the form of reengineering cross-border river flows, a pumping competition over shared aquifer resources, and the use of surrogates, including armed militants or irregular forces, to undermine a rival state's control of water resources.

Water wars waged by diplomatic, economic, or political means can actually exact costs higher than a military conflict, as is apparent from cases in Africa, the Middle East, and South Asia. Perceived water interests can even be used as justification to wage proxy wars by supporting terrorist or separatist groups, as Syria did against Turkey and as Pakistan continues to do against India in spite of having secured a water-sharing treaty on very favorable terms. More people have been killed in the Pakistan-India proxy war than in all the military conflicts on the subcontinent.

But when there is an actual resort to force, will it qualify as a water war in the military sense only if one or both sides admit it is a water conflict? Or will it still be a water war even if fought by another name? Yet another question is whether scarce water resources can prompt a nation to launch a cross-border military attack with the aim of gaining control over the sources of transboundary waters.

These are more than academic questions because they go to the heart of some of the military conflicts fought in the past half century. For example, Israel's 1967 Six-Day War against Jordan, Syria, and Egypt stands out for the successful Israeli grab of water resources, while Pakistan's five-week-long military attack on the Indian part of Jammu and Kashmir in 1965 failed to alter the political control of water resources. Water was as key a driver as land in the initiation of both wars. Yet few called them water wars.

Take the 1965 Pakistan-India war: It was waged barely five years after the bilateral water treaty generously reserved the largest three of the six transboundary rivers for Pakistani use, leaving upstream India with barely one-fifth of the aggregate waters of the Indus Basin. India agreed to give away the lion's share as part of an effort to trade water for peace with Pakistan at a time when the China-India relationship was on the rocks following the Dalai Lama's 1959 flight across the Himalayas. But the 1960 treaty only whetted Pakistan's desire to gain control of the land through which the three largest rivers reserved for Pakistani use flowed. Water security became synonymous with territorial control, with Pakistan's then military ruler contending that gaining Pakistani control over the Indian-administered region of Jammu and Kashmir—the center of the largest runoffs into those three rivers—had become crucial to his country's national security.[119]

Earlier, Pakistan had seized more than one-third of the original princely state of Jammu and Kashmir in the 1947–1948 war. China, for its part, gained control of one-fifth of Jammu and Kashmir in stages, with its land grab in the

1962 invasion of India being supplemented by Pakistan's gifting of a strategic slice of its own Kashmir territory the following year.

However, only one of the three rivers set aside for Pakistani use originates in the Indian part of Jammu and Kashmir; the other two start elsewhere—one in Chinese-held Tibet and another in India's Himachal Pradesh state. Yet Pakistan yearned to control the Indian portion of Jammu and Kashmir because that was where the three rivers make their largest water collections. Water thus was a key instigator in the abortive 1965 attack—codenamed "Operation Gibraltar"—to capture India's remaining Jammu and Kashmir territory.[120]

In this light, is it any surprise that the much-trumpeted Indus treaty has failed to resolve the underlying water tensions between Pakistan and India? Not only did the attempted trade-off of water munificence for peace backfire on India, with the country getting stuck with an iniquitous water-allocations pact of indefinite duration, but water has also remained a core divide between the two neighbors. In fact, after having failed to achieve its water designs by military means, Pakistan has continued to wage a water war against India by other means, including politically, diplomatically, and by proxy, especially by rearing transnational extremists on its soil to strike terror in India's heart and bleed it in Jammu and Kashmir. In other words, its covetous, water-driven claim to the Indian-administered portion remains intact.

By contrast, the Israeli-Arab war in 1967 lasted less than a week but changed the political control of regional water resources in a decisive and spectacular way. Apart from China's 1950–1951 annexation of the Tibetan Plateau—Asia's "water tower" that supplies freshwater to multiple countries—no other post–World War II war has so dramatically changed a regional water map as the brief hostilities in 1967. Just as the gobbling up of the then-independent Tibet enabled China to emerge as the principal source of cross-border flows in its region, Israel went from transboundary water dependency to transboundary water leverage through its key territorial gains in 1967—an outcome in sync with an archetypal water war.

The events that led to the Six-Day War were actually rooted in the rival water-diversion plans of Israel and its Arab neighbors, which set off water wars between 1964 and 1967—a period that witnessed military clashes and the founding of the Palestine Liberation Organization and its increasing, Syrian-backed guerrilla attacks against Israel. After Israel commissioned its National Water Carrier by diverting Jordan River waters, a summit meeting of Arab states in early 1964 finalized an opposing plan to redirect flows from the river's headwaters to Syria and Jordan. The subsequent start of diversion works in Syria as part of this plan prompted the Israeli military to repeatedly target that project during 1965–1966, forcing Syria to suspend work in July 1966, before Israel completely destroyed the project in the full-fledged 1967 war.

Water clearly was the trigger of the Israeli air and artillery attacks on the so-called Arab Headwater Diversion Project and a key, even if unpublicized, instigator of the Six-Day War.[121] By seizing control of the Golan Heights from Syria, the West Bank and East Jerusalem from Jordan, and the Gaza Strip and the Sinai Peninsula from Egypt, Israel achieved a swift and dramatic victory. More important, it reaped tremendous water spoils: the war left it in control of sizable groundwater resources and all of the Jordan River's headwaters.

Israeli leader Ariel Sharon, who as commander of one of the southern divisions played a lead role in the rout of Egyptian tank forces in 1967, candidly acknowledged in his memoirs—published soon after he became prime minister in 2001—that the Six-Day War was as much about water as anything.[122] "People generally regard June 5, 1967 as the day the Six-day War began," Sharon said. "That is the official date. But, in reality, it started two-and-a-half years earlier, on the day Israel decided to act against the diversion of the Jordan."

In fact, the Israeli prime minister during the Six-Day War, Levi Eshkol, was fixated on gaining control of water sources, harking back to the Zionist dream of "making the desert bloom." As David Ben-Gurion, Israel's founding father, put it on why he moved to the Negev Desert after stepping down as the first prime minister, "After all, there is room for only one prime minister, but for those who make the desert bloom there is room for hundreds, thousands, and even millions."[123]

Ben-Gurion returned as prime minister in 1955 and was succeeded in 1963 by Eshkol, who had served as the chief executive of Israel's main water company, Mekorot, which constructed the country's lifeline, the National Water Carrier.[124] Eshkol was determined that Israel seize the sources of water supply, or else "the Zionist dream could not be realized."[125]

Through its capture of the Golan Heights, Israel secured not only effective control of the Jordan River, but also the leverage to employ water as a bargaining weapon. Crisscrossed by multiple streams, Golan, the headwaters of the Jordan River, is a principal water source for the Israeli lake—Tiberias, also known as Lake Kinneret, or the Sea of Galilee—which is Israel's largest freshwater reservoir and the point of origin of its National Water Carrier (map 1.1). The Jordan River's key tributary, the Banyas, originates on the Golan Heights, a 1,800-square-kilometer sloping plateau whose drainage system sends most water westward to the basins of the Jordan River and Tiberias. It was the Arab-backed Syrian project to siphon off the flows of the Banyas and its sister tributary, the Hasbani, that prompted the Israeli airstrikes and, according to an Israeli scholar, "precipitated the outbreak of the 1967 war."[126]

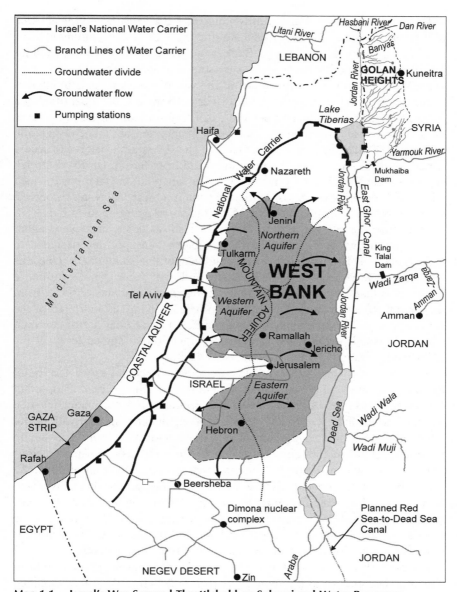

Map 1.1.   Israel's War-Secured Throttlehold on Subregional Water Resources

During the war, Israel also gained control of the West Bank's water-rich multiaquifer system and the Coastal Aquifer basin portion extending into the Gaza Strip. But whereas the West Bank is the provider of substantial ground-water and some modest surface water to Israel, it is Israel that is the source of water outflow to the Gaza Strip, one of the world's poorest areas in internal water resources. However, the groundwater outflow to Gaza, quantified at a mere 25 million cubic meters per year, is a tiny fraction of the West Bank groundwater resources entering Israel or in Israeli control.[127] The occupation of the West Bank was motivated, at least in part, by Israel's resolve to secure water supply from the underlying aquifers and from the Jordan River stretch separating the West Bank from Jordan.[128] The key point is that the war ended with Israel acquiring de facto control of the main subregional water resources, with the Syrian-Jordanian plans for water diversion going up in smoke.

Two other Israeli armed interventions in subsequent years were either overtly tied to water or had water as a subtext, and appeared designed to consolidate Israel's newfound riparian and geopolitical advantages. One was Israel's dual 1969 raids on Jordan's newly built East Ghor Canal in response to concerns that the Jordanians were diverting excess amounts of water from the Jordan River's tributary, Yarmouk, to irrigate the East Ghor area of the Jordan Valley. The attack put most of the East Ghor Canal out of commission, with Israel permitting its repair only after Jordan agreed to expel the Palestine Liberation Organization from its territory.

The other was the 1978 invasion of southern Lebanon up to the Litani River. That intervention put Israel temporarily in charge of an important Jordan River Basin source, the Wazzani Springs. In 1982, Israel invaded Lebanon a second time, with water again an underlying motivation—an intervention that became the longest and most controversial military action in Israeli history. In 2002, it was Lebanon's turn to aggravate water tensions in the region by constructing a pumping station at the Wazzani Springs, the source of the transboundary Hasbani River that, along with the Banyas and Dan rivers, helps create the upper Jordan River, which in turn feeds Tiberias, Israel's main freshwater reservoir. Israel receives about 138 million cubic meters of water annually from the Hasbani, and the Lebanese pumping station was built to divert water a few hundred meters from the Israeli border.[129]

Although that crisis was defused after the United States and the European Union rushed envoys to the region, the pumping-station issue has remained a bone of contention, with threatened Israeli military reprisal hanging like a sword over the Lebanese plans to route a portion of Wazzani's annual flows for local consumption. In 2006, Israel—using its war against Hezbollah—actually carried out airstrikes that damaged the Wazzani pumping station and destroyed the Lebanese irrigation canals supplying water from the Litani

example, the territorial issues in the Middle East, or between India and Pakistan over Kashmir, or between China and India over Tibet and Arunachal Pradesh, may center on land, but they are also about water. When territorial and water issues intersect at both the intercountry and intracountry levels, an explosive situation can arise, as illustrated by Central Asia's divided Fergana Valley, where the bloody anti-Uzbek violence in the Kyrgyzstan-controlled portion in mid-2010 was triggered by local ethnic Kyrgyz' fears that Uzbekistan was seeking to grab the entire water-rich region.

Identifying the extent of water-related interstate violence admittedly hinges to a large extent on how such conflicts are defined: whether water has to be the central element or can be an auxiliary yet significant factor in armed actions. Given that academics have debated endlessly the nature and causes of war without being able to reach agreement even on what actions amount to waging a war, the debate on water wars is likely to be similarly never ending. But understanding the connections between water scarcity and interstate conflict is essential if the international community is to avert water wars by reining in the rasping hydroconflicts.

Water as an Object, Target, or Tool of Warfare

Nations cooperate over shared water resources when it serves their national interest. But when a nation pushes a sense of grievance against a riparian neighbor, or when the transboundary water competition over economic development becomes fierce—with water stress inflaming public passions—the potential for conflict grows. The relationship between freshwater and conflict indeed is age-old; river basins were the cradle of the world's great civilizations, and water conflicts thus were unpreventable. The Pacific Institute's water-conflict chronology lists hundreds of examples of water-related conflicts, with recent trends indicating an increase in subnational conflicts relative to international conflicts.[139]

The greater frequency and intensity of subnational water disputes only increases the risks of such conflicts spilling over across national borders, especially when such boundaries crisscross natural water flows. After all, widespread or serious subnational discord over water often translates into tensions or conflict with a co-riparian country.

In the intercountry context, water can turn into a political or military instrument of use, a target of control or attack, or a victim of purported collateral damage in armed conflict. Water resources or water systems can be used for achieving political or military objectives against a rival state, including by turning them into a weapon, or they can become targets of military or covert actions.

For example, dams were routinely hit as strategic targets in World War II, with the May 1943 British aerial bombing of the Mohne Dam in Germany unleashing flash floods that destroyed smaller dams downstream and left about 1,200 residents dead.[140] Similarly, dams and irrigation systems were bombed during the Korean and Vietnam wars, and Baghdad's water-supply network was targeted by U.S.-led coalition forces in the war that followed Iraq's 1990 invasion of Kuwait.

Upstream dams, barrages, and other diversions can help shape water as a political instrument—a weapon that can be wielded overtly in a war, or subtly in peacetime to signal dissatisfaction with a co-riparian state. Even the denial of hydrological data in a critically important season can amount to the use of water as a political tool against a downriver state. Nonstate actors, with or without state support, are known to have used water as a political tool. Nonstate actors, through acts of terror, have also employed water resources or systems as targets or tools of violence or coercion. A 2012 grenade attack by pro-Pakistan militants in the Indian-administered part of the disputed region of Jammu and Kashmir halted work on a lake project Pakistan had objected to. Dynamite was employed in a 1990 terrorist strike to abortively blow up another disputed project on the same freshwater lake, Wullar.

Although aggression against, or by means of, water resources can be broadly construed to be incompatible with the principles of international humanitarian law and with customary international law, water has repeatedly become an object, target, or tool of warfare. To be sure, there are no clear-cut rules regarding water in international humanitarian law. Water resources and facilities are covered by the provisions of the 1949 Geneva Conventions only under a liberal interpretation.

The Geneva Conventions—which emerged in response to the bloodbaths and mass destruction of the two world wars—incorporate prohibitions against, among other things, the "extensive destruction and appropriation of property not justified by military necessity and carried out unlawfully and wantonly." Customary rules enshrined in the much-violated 1907 Hague Regulations, which entered into force in 1910, and the 1977 Protocols Additional to the Geneva Regulations do contain definite prohibitions against attacks on objects indispensable to the survival of civilian populations, including installations containing "dangerous forces," such as dams, dikes, and nuclear power plants, and against the use of poison as a means of warfare.[141] A more recent principle, which is part of the nonbinding 1982 World Charter for Nature, provides that "nature shall be secured against degradation caused by warfare" and that "military activities damaging to nature shall be avoided."[142] Yet water has "often been the object, target, or weapon of military or terrorist action."[143]

Despite the international norms on discrimination, proportionality, and avoidance of damage to the environment or innocent civilians in armed conflict, attacks on water infrastructure in combat operations have at times caused extensive suffering to local populations. The impunity with which such attacks have been carried out reflects in part the continuing gaps and weaknesses in international principles and norms. For instance, the customary principles of proportionality are so ill defined that the use of water as a weapon or the targeting of water-resource systems can be militarily justified.

The concept of collateral damage, moreover, offers a convenient cover to fob off deliberate attacks on waterworks as an unintended consequence of armed conflict. When water is overtly and repeatedly employed as a military tool or target, however, such an explanation may hardly sound credible. The rise of transnational terrorism is another complicating factor in the protection of water infrastructure, as international humanitarian law was not designed to cover acts of terror by nonstate actors, even if such actors have been reared by a scofflaw state. In fact, the most serious terrorism-related threats to regional or international security now often emanate from state-sponsored nonstate actors.

At the root of the growing threat of water wars is a classic security predicament: one country's efforts to build water security through the tool of hydroengineering can cause water insecurity in a co-riparian state, with the insecurity extending to the food, energy, and economic fronts and fueling regional tensions. Transnational resources, by their very nature, demand institutionalized cooperation, including reasonable and sustainable sharing, so as to stanch their conflict-inducing characteristics.

When water institutions are robust and able to withstand shocks from political or hydrological change, including government toppling and natural disasters, their resilience helps to minimize the conflict potential. But when institutions are nonexistent or weak, the conflict risks can scarcely be dismissed. Conflict dynamics, moreover, seldom follow set patterns, and just because water as an overt and exclusive cause of intercountry war has been a rarity in modern history does not mean that the same will continue to hold true in the future. The world has little experience in managing what the future holds—widespread water shortages.

Chapter Two

The Power of Water

Water—literally the bloodstream of the biosphere—is a source of human livelihoods, yet also a source of risk and vulnerability. Water is a life preserver, but also a potential life destroyer when it becomes a carrier of deadly bacteria and microorganisms. With its ubiquity, water is even a potent force behind extreme weather events and natural disasters, including flash floods, hurricanes, and tsunamis. Fukushima has become a symbol of how water can turn into a force of brutal destruction.

The world's great river systems serve as lifelines, but they can also become major hazards when they overrun their banks and leave a trail of death and destruction. In the humid tropics and monsoon regions, flooding of the rivers is often an annual feature, causing enormous hardship and damage, yet delivering nature's yearly gift of silt. The fresh nutrient-rich silt helps to replenish the fertile but overworked soils.

Water generates cooperation but also conflict, especially when the interests of stakeholders clash. With the new era of resource depletion and degradation eclipsing the availability of cheap, plentiful, and clean water, one alarming trend is that water gets laced with organic and inorganic chemicals discharged from industries, applied in agriculture, or used by consumers in their homes.[1] Deaths from waterborne infectious diseases remain rife worldwide, with the majority of the victims being infants and children.

Water, like religion and political ideology, has the power to unleash grassroots passions and cast a major influence on the millions of people caught up in such tides. Since the dawn of civilization, humans have sought to settle close to water. "People move when there is too little of it. People move when there is too much of it. People journey down it. People write and sing and dance and dream about it. People fight over it. And all people—everywhere and every day—need it," former Soviet leader Mikhail Gorbachev has written, bemoaning the advent of a global water crisis. "Modern technologies have allowed us to harness much of the world's water for energy, industry, and irrigation—but often at a terrible social and environmental price—and many traditional water conservation practices have been discarded along the way."[2]

When seen from space, Earth is literally a blue planet, covered with water. Water covers more than seven-tenths of the planet's surface. All the continents are encircled by water and, other than Antarctica, drained by major rivers. Symbolizing its connection to human survival, water also makes up about 75 percent of the human body. Water may be the most common substance on earth, yet the world already confronts a growing water crisis. The abundance of water in the world, after all, is just an optical illusion.

Only a tiny fraction of the planet's water—less than 1 percent—is drinkable. Most of it (97.5 percent) is seawater, which is expensive to desalinate—an energy-intensive, greenhouse-gas-emitting process that also leaves a lot of toxic residues. The oceans, however, are the principal source of precipitation; the seawater lost through evaporation returns as rain and snow. The oceans thus play a pivotal role, along with the sun, in making freshwater.

Barely 2.5 percent of the earth's water is potentially potable, but more than two-thirds of that is locked up in polar icecaps and glaciers. All this leaves a fraction of 1 percent of the total water on earth for consumption by human beings and other species, which range from insects to vertebrates, such as birds, fish, mammals, reptiles, and amphibians. According to United Nations estimates, just 42,370 cubic kilometers of the 1.4 billion cubic kilometers of the water on the planet are notionally available for drinking, hygiene, cooking, agriculture, and industry. Glaciers, which store about 69 percent of the world's freshwater, are important elements in the hydrologic cycle because they control the volume, variability, and quality of runoff into streams, rivers, lakes, and other bodies of water.

Water is indispensable to continued socioeconomic advances. Water grows food and sustains manufacturing, generates electricity directly or serves as turbine-driving steam and coolant in power plants, supports ecosystems and biodiversity, and holds aesthetic or recreational values. Humanity needs water to produce virtually all the goods required for its daily existence.

Of the three resources directly critical to human survival—air, water, and food—only air is more critical than water. Without air, a person will asphyxiate within minutes. Without water, a person will die within days. And without food, a person will shrivel and perish within weeks. In fact, like air, water is an ambient resource that neither knows nor respects human-set boundaries, with the hydrologic cycle freely traversing political frontiers.

The power of water has been acknowledged in different cultures and religions since ancient times. In Hindu and ancient Greek cultures, gods and goddesses related to water are powerful characters tied to fertility or to new beginnings or destructions. "By means of water," says the Koran, "we give life to everything." Because the Bible was written amid water scarcity, water is a recurring theme in the Old and New Testament, with drought associated with the wrath of God and water connected both with the gift of life and with baptismal cleansing from sins.

Historically, water has determined the centers of civilization. It was not a coincidence that the early civilizations bloomed around rivers. And the assured access to such water resources, by helping to initiate agriculture, fostered an evolution from nomadic to sedentary and well-developed civilizations. In fact, early trade wholly relied on water for transportation of goods, with water even serving as the catalyst for the first cross-cultural interactions.

As technology developed, communities gained easier access to water. Instead of people following the water by settling near rivers, lakes, and springs, communities moved the water to their settlement centers by constructing aqueducts, reservoirs, and other works.

The twentieth century marked the high point in history in moving water in increasingly large quantities over long distances, thereby allowing communities to settle and thrive in remote, once-waterless regions. Human efforts to control water resources have extended from the damming and redirecting of rivers, thus channeling water to arid or semiarid regions, to creating artificial rain and snow by seeding clouds with chemicals—a practice that has ignited debate about the wisdom of tinkering with nature, given that efforts to bring about precipitation in one region could have unpredictable consequences elsewhere. Most of the world's dams have been built in the post–World War II period. Despite such engineering and scientific activities, more than half of all countries today are classified by the United Nations as lacking sufficient water resources to rapidly expand agricultural and industrial development in a sustainable fashion.

The world's renewable freshwater reserves—concentrated in mountain snows, lakes, aquifers, and rivers—have roughly remained the same since time immemorial, whereas population, consumption, and economic activities have expanded exponentially. As the ultimate renewable resource, water naturally recycles, rising to become rain or snow, except when pollutants hold it down. Water is thus never lost but merely gets recycled. The loss due to evaporation from surface-water bodies, evapotranspiration (plant-related water subject to the processes of evaporation and transpiration), and sublimation (ice changing into water vapor without first becoming liquid) returns in the form of precipitation.

The water crisis springs from the growing human demand and pollution of a finite resource. Global per capita freshwater availability has unstoppably declined for more than a century, plummeting over 60 percent since 1950 alone, according to United Nations data. While equatorial regions and some northern latitudes have a surfeit of water, many other parts of the world have too little water in comparison to their population and consumption levels. In some regions, including one-third of Africa, much of the Middle East, parts of central and southern Asia, and sections of northern and northwestern China, low water availability, coupled with high water withdrawals, has actually fostered acute scarcity. With global demand outstripping sustainable supply, the struggle over freshwater portends a new turning point in history.

THE GREEN/BLUE WATER EQUATION

One key challenge is to maintain the natural balance between "blue water"—the resources in rivers, lakes, and aquifers channeled for irrigation, urban and industrial use, and environmental flows—and "green water," or the water from precipitation that naturally infiltrates the soil and supports rainfed agriculture and other plants and trees. Terrestrial-ecosystem diversity is dependent on green water, whereas aquatic ecosystems dwell in blue-water habitats. If natural ecosystems are to be protected, maintaining the balance between the blue and green waters is indispensable.

Agriculture, whether irrigated or rainfed, often employs both blue and green waters. Irrigated agriculture relies to some extent on rain, while rainfed agriculture in several regions seeks to beat dry spells through supplemental irrigation. According to one study, 20 to 35 percent of all water consumption by irrigated cropland from 1971 to 2000 consisted of green water, and almost half of the world's irrigation water apparently originated from sources that were nonrenewable (fossil groundwater) and nonlocal (i.e., blue water brought in from distant regions).[3]

Yet at least three-fifths of the world's food is produced with green water—that is, on rainfed land.[4] The production of meat from grazing and wood from forestry also relies on green water. In sub-Saharan Africa, where irrigation remains underdeveloped, nearly all food production depends on green water.

The larger the proportion of green water consumed by crops, plants, and trees, the less such water there will be to generate surface runoff into rivers and lakes or to recharge groundwater. By contrast, degradation or depletion of blue water imperils aquatic ecosystems. Keeping the green/blue water balance for nature and society is becoming increasingly difficult, largely because of the growing hydrological impact of human-induced land cover changes and intensive irrigation. Although irrigation dates back to Mesopotamia and other ancient civilizations, it is in modern times that the dramatic spread of irrigated agriculture, including to arid areas, has created serious problems along the green-to-blue water continuum—that is, from purely rainfed to purely irrigated.

The conversion of large tracts of natural-vegetation land to cropland has resulted in significant green-to-blue water alterations. The reverse type of alteration—from blue to green water—is represented by the growing channeling of waters from human-made storage structures like reservoirs, or direct withdrawals from watercourses, for production of biomass (food, timber, and biofuels), setting in motion the process of large-scale river depletion and falling groundwater levels.[5]

There is now increasing international awareness that the focus on water quantity and distribution must be balanced with an equal emphasis on protecting water quality so as to maintain the integrity of the water-supported ecosystem services and to safeguard fluvial, aquatic, and marine ecosystems. For example, many fish species, such as salmon, depend early in their life on steady river flows to flush them to the estuary, from where they gradually travel into the open ocean before returning to spawn in freshwater. Diminished river flows directly affect such species. Higher salinity in bay areas caused by significantly reduced river discharges into oceans also affects oysters and other species native to coastal waters, as is apparent in Matagorda Bay, where the Colorado River empties into the Gulf of Mexico.

There is also the imperative to balance the policy focus on blue-water availability and demand with a better international appreciation of the critical importance of green water by including such flows in assessments of water resources and scarcity.[6] Green water has been largely ignored in policy assessments since it cannot be piped or priced. A combined green-and-blue water accounting framework can actually help broaden the scope of options for decision makers seeking to boost farm production.[7]

It is possible to optimize local green-water utilization through terracing, contour plowing, mulching, conservation tillage, small-scale rainwater harvesting, and other cost-effective techniques. Effective green-water management, by allowing more rainwater to infiltrate the soil and thus preventing fertilizer and pesticide toxins from entering surface waterways, can mitigate the risks of flooding and water contamination.

Instead of focusing solely on blue-water investments in costly irrigation networks, governments must encourage farmers to profit from green-water management techniques. This need is underscored by the greater role green water is expected to play in meeting global food demand because irrigated cropland is running into limits of further expansion.

"Give a man a fish and you feed him for a day," according to an ancient Asian proverb. "But teach him how to fish and you feed him for a lifetime."

In the contemporary world, however, blue-water depletion and contamination, along with overfishing, is making that maxim difficult to apply. Such is the extent of overexploitation of blue-water resources and unsustainable practices in many fisheries, including large-scale bottom trawling, that in some places even traditional fishermen are being driven out of business. Upstream retention of nutrient-rich river sediment loads by dams and other water-diversion projects, for its part, is seriously affecting various species of fish, which are a key source of protein for more than a billion impoverished people worldwide. Such impact actually extends beyond downstream basins and deltas to the oceans, because marine life also depends on the nutrients and minerals disgorged by rivers when they empty into the seas.

Actions to control or divert blue waters are being spurred by increasing national demands for water. With several major arid regions—from China's north to Israel's south—having been turned into breadbaskets, the diversion of waters from rivers, lakes, and aquifers for irrigation has increased phenomenally. Thanks to such long-distance rerouting of blue waters, the amount of water used for irrigation in excess of the locally available, renewable resources already appears to make up nearly half of the aggregate irrigation volumes worldwide. Growing cereals, oilseeds, cotton, and other water-rich crops in arid or semiarid regions entails the use of almost three times more water than in fertile lands. Poor irrigation practices, for their part, have resulted in the accumulation of soluble salts in more than a fifth of the world's irrigated lands, thereby affecting crop yields.

WATER STRESS AND RISING FOOD DEMAND

Water is such a key element in plant growth that the production of biomass, especially food for direct human consumption, easily ranks as the largest freshwater-utilizing activity. Food production accounts for 70 percent of all global water consumption, compared with 19 percent by industry and about 11 percent by cities and towns (figure 2.1). Still, the world confronts a food crisis, underscored by spiraling prices and widespread hunger and malnutrition.

Indeed, the continuing population growth and sharply increasing daily calorie and protein intake, largely due to the rising affluence of once-poor countries, mean that the world must boost food production significantly. However, the task of growing more food—and at affordable prices—is beginning to be constrained by water stress, which demands that agriculture use less, not more, water.

Even as the global food system struggles to meet present requirements, with millions of people not getting enough food to lead healthy, active, and

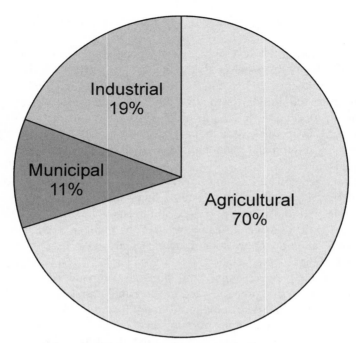

Figure 2.1. Global Water Use by Different Sectors
Source: United Nations, 2013.

productive lives, food demand is projected to rise sharply in the coming years (figure 2.2). The rapid growth in global farm output since the 1960s has now considerably slowed, falling behind the rate of increase in food demand. Emblematic of this trend are the four key staples that supply most human calories—wheat, rice, corn, and soybeans; their buffer stocks have dipped to

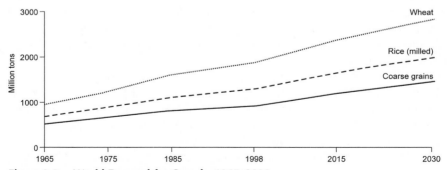

Figure 2.2. World Demand for Cereals, 1965–2030
Source: Data and projections by the Food and Agriculture Organization.

alarming levels because production has lagged demand since the beginning of the century.

In 1970, a third of the developing world's population was undernourished, but by the mid-1990s—thanks to the fruits of the green revolution—that share had fallen below 20 percent, and the absolute number of hungry people dropped under 800 million for the first time in modern history.[8] Those gains are being reversed by the international food crisis, with the number of people lacking access to adequate food to meet basic nutritional needs jumping to an estimated 925 million and malnourishment contributing to at least one-third of all child deaths.[9]

Yet about a quarter of all food grown or raised globally ends up in the trash, with Americans the most wasteful, dumping up to 40 percent of their food, or $165 billion worth of produce and meats.[10] The first of the Millennium Development Goals—to halve by 2015 the number of undernourished people—has fallen by the wayside. In fact, as many as thirty-six nations still need international food assistance.

The imbalance between supply and demand, compounded by drought in some food-exporting regions, has led to three major spikes in international food prices since 2007, with some grains more than doubling in cost and the United Nations food-price index reaching an all-time high.[11] The price jumps have triggered not only panic buying, hoarding of grains, and national export restrictions, but also food riots in more than thirty countries, with soaring prices contributing to the Arab Spring uprisings since 2011.

The food-price crisis is linked with the freshwater crisis. For example, the 2009 ouster of Madagascar's president was rooted in water problems in South Korea. To help ease the problem of increasing food production in water-stressed South Korea, the giant South Korean business group Daewoo entered into a deal to lease as much as half of Madagascar's arable land to grow cereals for the home market, triggering a local backlash and a military intervention that eased out a democratically elected president. In 2008, food riots helped topple the Haitian government. The combination of political and economic mismanagement, internal strife or repressive governments, and recurrent natural disasters has even contributed to famine, as in North Korea and some African states.

Significantly reducing the global hunger gap, in essence, is a water-resource challenge, although new drought- and flood-resistant crop varieties and new farming techniques are also required. Rising food prices, while increasing world hunger, have indeed created strong incentives to develop new farm varieties and to utilize water more wisely in order to grow additional food. But the incentives come with several constraints, including degradation of water and land resources and a decline since the late 1980s in global investments in agricultural research and in farm-related international developmental aid.

Moreover, food production, which thus far had to vie with industry and municipalities for water, must now also compete with biofuels for both water and land. The web of subsidies, mandates, and tariffs to promote the manufacture of biofuels from crops in the West has come atop the long-standing general farm subsidies in the developed world that have placed farmers in poor countries at a severe disadvantage.

Today, "fuel farming" is claiming greater volumes of water, even as more cropland gets diverted for biofuel production.[12] Natural habitats in some regions have been converted into farmland to take advantage of the generous biofuel subsidies, leading to the destruction of vast swaths of the world's tropical rainforests.

Misguided energy policies, by redirecting vast amounts of grains and other farm produce away from the food chain, are eroding food security and contributing to widespread inflation, more hunger, and political and economic instability, especially in developing countries. The fastest growth in corn demand in this century has come from corn-derived ethanol production. This has led to a major surge in corn prices, just as biodiesel manufacture from soybean and other vegetable oils has made oilseeds dearer. Rising biofuel production, expectedly, has accentuated water shortages in already water-stressed subregions.

Yet food continues to be increasingly turned into fuel in the main grain-exporting zones—North America, the European Union, Australia, Russia, and Argentina. State-subsidized biofuels produced from crops like corn, sugar, cassava, rapeseed, and palm oil more than tripled in the world between 2000 and 2010. A quarter of all U.S. cereals grown in 2012 went into fuel farming, including more than two-fifths of the corn crop.

Such is the biofuels' water intensity that as much as 2,500 liters of water are needed to grow enough corn to refine just one liter of ethanol.[13] Adding ethanol only increases the price of blended fuel because it is more expensive to transport and handle than gasoline. Worse still, mileage drops off, as ethanol's energy yield is lower than gasoline's.

The competition between cars and people for the world's oilseed and grain output has triggered an "oil shock" of a different kind—shortages and soaring prices of palm oil, soybean oil, and other types of vegetable oils used in cooking, despite the oilseeds sector having grown in recent decades at double the pace of global agriculture as a whole. About half the increase in the worldwide demand for vegetable oils between 2007 and 2012 was tied to biofuel production. The United States, Brazil, and Argentina rank as the world's biggest exporters of oilseeds, some of which like soybean are also used as animal feed or turned into retail food products.

The United States and the European Union each provide billions of dollars in annual subsidies for biofuels in the name of green investment, although energy-intensive biofuel production is hardly a net saver on carbon emissions. The EU has set a target for plant-based biofuels to constitute 10 percent of transportation fuel by 2020, while the U.S. Congress has mandated that biofuel use must reach 136.3 million cubic meters annually by 2022. Emerging economies like China, India, Indonesia, and Thailand have also adopted biofuel targets, though more modest ones. But it is the rigidly ambitious biofuel targets in the West, coupled with the slowing output growth of some major crops, that have contributed to record international food prices. Such targets have encouraged farmers to switch from growing crops for human consumption to planting the different crops used for fuel production.

To compound matters, cotton—the biggest nonfood crop whose cultivation accounts for more than 3 percent of global agricultural water use and 6 percent of all pesticide use—is increasingly competing with grain for arable land and water. The pressures on agriculture owing to the rising global demand for food and cotton have resulted in overworked and thus degraded soils, conversion of forest ecosystems to cropland, and overexploited river and groundwater resources.

Water-supply limitations have reached critical proportions for agricultural growth in a number of counties, as rising affluence and a growing middle class promote the adoption of Western-style diets and lifestyles. The water crisis stands out as a major threat to greater food production and to continuing advances in poverty alleviation.

Although three-quarters of the global population lives in developing countries, the developed world still consumes almost half of the world's agricultural products. The largest growth in food demand is thus taking place in the developing world. As more people take the first steps out of poverty and shifts in middle-class dietary preferences become more pronounced, the per capita consumption of water-intensive food products is set to soar in developing countries, a trend likely to exacerbate the water crisis. Without efficiency gains, reduced food waste, and other water-saving measures, global agriculture's water requirements will likely jump 45 percent in two decades to 4,500 billion cubic meters annually by 2030.[14] This will promote gross overexploitation of water resources by all sectors because industrial and municipal water demand is projected to increase even faster.

In this light, dramatic improvements have become necessary in crop yields per unit of water and land utilized. Agriculture is now straining the sustainable supply of not just water but also land resources, with increases in food production per hectare of cropland not keeping pace with population and consumption growth.

Expanding urban structures, transportation systems, and industrial parks have eaten into existing farmlands, including prime agricultural land, thus allowing only small net increases in cultivated land between 1960 and 2010. With the rapid growth of the global population, this trend has translated into a sharp downward slide in per capita cropland, which has shrunk by more than half since 1960—from about 0.4 hectares to 0.2 hectares (table 2.1). Unless cities grow in density, not extent, this decrease in per capita farmland availability will only accelerate because the urban population is projected to climb by 2.6 billion, growing from 3.6 billion in 2011 (or just over half the total global population) to 6.3 billion in 2050.[15]

If there has been any good news, it is that increases in yields and, to a much smaller extent, cropping intensities have more than matched the fall in per capita cropland, thus allowing global food production to increase significantly in the past half century. Such impressive productivity growth has translated into considerable improvement in overall nutrition levels, even as many people remain seriously undernourished. Yet, in the face of growing land and water degradation, the big question is whether further productivity growth can cope with the fast-rising food demand that has already outstripped supply.

Although yields of grains—the staple human food and the key input to produce meat, eggs, dairy products, and farmed fish—are still increasing, the rate of yield increase has slowed since the late 1980s. Furthermore, per capita grain production has been on the decline since 1984, even as the world remains decades away from stabilizing its population. Rising affluence has also boosted food and feed demand. To meet such increased demand from

Table 2.1. Global per Capita Cropland and Irrigated Land Area

Year	Population (billions)	Cropland Area		Irrigated Land Area	
		Total (mha)	Per Capita (ha)	Total (mha)	Per Capita (ha)
1900	1.5	805	0.536	40	0.027
1950	2.5	1,170	0.468	94	0.038
1970	3.9	1,300	0.333	169	0.043
1980	4.5	1,333	0.296	211	0.047
1990	5.2	1,380	0.265	239	0.046
2000	6.3	1,360	0.216	277	0.044
2030	8.1	1,648	0.203	300	0.037
2050	9.2	1,673	0.182	318	0.034

Source: R. Lal, *Crop Science* 50 (March–April 2010): S-121.

Note: mha = million hectares; ha = hectares.

the projected arrival of 2 billion more people by 2045 and the changing dietary preferences of a burgeoning middle class, food production must nearly double over the next half century.[16]

The centrality of the terrestrial environment as the source of human food is underscored by the fact that barely 0.3 percent of human calorie intake originates from the oceans and other aquatic ecosystems.[17] Yet without further human encroachment on ecologically sensitive areas like rainforests, peatlands, and savannahs or the clearing of more forests, there is little room to significantly increase the current size of croplands, which account for 11 percent of the earth's total land.

In South Asia, the Middle East, and North Africa, almost all land suitable for crop growing is already being cultivated. Much of the global expansion in cultivated land is projected to take place in sub-Saharan Africa and South America. However, with water supplies tightening, temperatures rising, and the weather becoming erratic, global food markets are already stretched.

Water is relatively scarcer than land. The availability of water and land resources often bears little relationship to a region's population size or food-production needs. China and India together are home to 37 percent of the global population, yet they have to make do with 10.8 percent of the world's water and 19.3 percent of its arable land. Some traditional Indian farmers have found it more profitable to sell their land to developers of industrial parks and residential buildings than to plant crops because a long, inefficient supply chain ensures that the average grower receives less than a fifth of the price the consumer pays. China, the world's largest grain producer, faces important challenges to achieving food security. Although it has declared its intent not to let cultivated land fall below 120 million hectares, China's reliance on heavy inputs of fossil-fuel-based fertilizers to make up for poor soils, cropland shortages, and bad-quality water has only opened the path to accelerated resource degradation.

The three countries with the world's largest aggregate freshwater resources—Brazil, the Russian Federation, and the United States—also have the largest stocks of land potentially suitable for further farmland expansion. Excluding poor-quality soils and environmentally protected or vulnerable acreage, the land stocks in Brazil, Russia, and the United States alone make up 28.7 percent of the world's potentially cultivable areas; but in terms of land size actually under cultivation, the United States ranks number one, followed by India, which, however, is much poorer than America or China in water resources.[18]

National paucity of water resources and arable land is driving some countries to produce food for their home markets on farmland acquired overseas, especially in sub-Saharan Africa, in what has been described as a twenty-first-century land grab by outsiders. Some twenty states in sub-Saharan Africa have sold or leased fertile land measuring more than double the size of Britain to outside governments and agribusiness firms to grow food for export, although many of their own citizens remain hungry or undernourished.

Because of agricultural underdevelopment, sub-Saharan Africa actually has among the world's hungriest societies. Sub-Saharan land is attractive to outsiders because much of it is fertile yet largely fallow and inexpensive to lease and cultivate. Labor is cheap and water supply in many areas is subsidized or free.

Petrodollar-rich Arab states and international agribusinesses have been in the lead in this land rush, which began after food prices began spiraling in 2007–2008. To be sure, the land acquisitions mean employment benefits for local communities and reduced pressure on international food markets by providing capital and technology and by consolidating small plots of land into more productive large farms. But such acquisitions by outsiders are also fostering greater corruption among African governments and shrinking Africa's own food supply. In addition, they exact significant environmental costs, including contributing to water-resource depletion and degradation.

Soaring international food prices and domestic water stress, however, have only increased the incentive for food-importing nations flush with cash to secure sources of food supply, thereby accelerating farmland acquisition in impoverished states, with those evicted from their traditional lands often becoming workers in the new foreign-owned farms. European and North American hedge funds, private equity funds, and institutional investors—just like Middle Eastern and Asian sovereign wealth funds and commercial firms—are making long-term bets that the fast-rising demand for food will yield major profits for those owning farmland and other parts of the agricultural business. So, Western investors too are buying sub-Saharan and other farmland, as well as grain elevators and fertilizer-supply depots. A British investment fund, Emergent Asset Management, for example, has focused on sub-Saharan Africa because "land values are very, very inexpensive, compared to other agriculture-based economies," and its microclimates, labor availability, and logistics are attractive.[19]

East Asian companies have also signed major African farmland contracts, best illustrated by the highly controversial deal between South Korea's Daewoo Logistics and Madagascar that triggered widespread local protests, leading to the 2009 collapse of the tropical island nation's government and the deal itself. The megadeal, involving a ninety-nine-year lease of 1.3 million hectares, or nearly half of Madagascar's arable land, would have effectively turned the poor Indian Ocean country that relies on UN food aid into a South Korean breadbasket and allowed South Korea, the world's third-largest corn buyer, to replace more than half the corn it imports from the Americas. Many other such deals, however, have been implemented.

The land grabs are effectively water grabs because with the acquisition of land comes the right to appropriate local water resources for cultivation. Given that investments to acquire farmland overseas are often driven not so much by arable-land shortage as by water paucity at home, the real story is about water. For example, aquifer depletion has squeezed irrigation in several of the petrodollar-laden countries making investments in foreign farmland. Water appropriations through overseas land acquisitions, however, threaten to alter regional flow regimes by significantly diminishing cross-border river outflows to downstream states.

In the Nile River Basin, for example, Arab and other foreign investors have acquired considerable arable land in upstream Sudan and Ethiopia. Among others, South Korea has leased 690,000 hectares and the United Arab Emirates 400,000 hectares in Sudan to produce grains for consumption back home. Such intensive cultivation of huge tracts of upstream land—equivalent to the total land area of Jamaica or Lebanon—means that utilization of Nile waters will likely surpass Sudan's relatively modest share under its 1959 Nile Waters Agreement with densely populated Egypt, which critically depends on the river as its lifeline. More than dam building, such water appropriations through land acquisitions threaten to engender greater intrastate and interstate water conflicts in African basins.

Many scientists believe that genetically engineered (GE) food could be useful in meeting the world's rapidly expanding food needs. Some of the most common GE crops today include alfalfa, canola, corn, cotton, papaya, soybeans, sugar beets, and zucchini, with 88 percent of all corn and 94 percent of all soybean output in the United States in 2011 grown from GE seeds. Many processed foods include ingredients, such as high-fructose corn syrup, made from GE crops.

GE food offers important environmental and economic benefits compared to conventional food. For example, the adoption of GE plants that are more resistant to pests or better able to withstand the use of herbicides potentially helps to control pesticide-related runoff into rivers and thereby reduces water pollution. There is no scientific evidence, however, that GE crops require substantially less water than conventional crops.

In the years ahead, genetic-engineering technology could help develop crops that are resilient amid changing climate conditions, slow the depletion of local water resources by demanding less water, and prevent the contamination of waterways through improved use of nitrogen and phosphorous fertilizers. The benefits of GE crops at present, however, are not universal and indeed may decline over time as biotechnology is applied to modify more crops with genes from other species, as opposed to the selective breeding process used in nearly all crops.[20]

More biotech-related research is also needed to understand potential long-term health and ecological risks, including those posed by possible "gene flow" via wind or insect pollination from GE crops to related nontarget species nearby. Cross-pollination threatens the rapidly growing organic food industry because it could contaminate organic crops, including those fed to cows.

The Role of Irrigation in Resource Depletion

Water stress is mostly concentrated in subregions where food production relies on large-scale irrigation. In modern world history, however, major irrigation projects have played a critical role in poverty alleviation and economic development. Such projects indeed have helped turn irrigated agriculture into an engine of economic growth and food security. For example, Asia, once a continent of serious food shortages and recurrent famines, opened the path to its dramatic economic rise by emerging as a net food exporter on the back of unparalleled irrigation expansion, which has more than doubled its irrigated acreage since 1970 alone and made it the global irrigation hub. Without large-scale irrigation, there would have been no green revolution in Asia, which receives much of its rain in the relatively short monsoon season.

The very technological advances that have contributed to spectacular improvements in global food production—including chemical fertilizers (manufactured from natural gas), pesticides (made from petroleum), farm machinery, long-distance channeling of water, modern food processing and packaging, and rapid transportation—have contributed to making agriculture more water and carbon intensive. Agriculture, in addition to being the world's leading water-guzzling sector, emits more greenhouse gases than all cars, planes, trains, and boats combined worldwide. Even dams, which are often central to large-scale irrigation, produce significant amounts of methane and carbon dioxide.[21]

The greatest expansion in irrigation has occurred since the early 1960s, with the total global irrigated area almost doubling since then, increasing from 160 to 318 million hectares. Today the area equipped for irrigation makes up more than 20 percent of the world's total cultivated land, according to the Food and Agriculture Organization. But because irrigation makes higher yields possible and also increases cropping intensity by permitting multiple cropping, irrigated farmlands account for more than three-fifths of the world's grain production. However, at least 60 percent of the world's total food is produced on rainfed land, which may in some areas require supplemental irrigation. Irrigated crop yields are approximately 2.7 times higher than from rainfed cultivation.[22]

For agriculture in any region to be blue-water dominated, the water-diversion infrastructure must be well developed to facilitate extensive irrigation. For example, Israel's distinctive National Water Carrier links its three main freshwater sources—Lake Tiberias, the Mountain Aquifer, and the Coastal Aquifer—to take irrigation water deep into arid areas, besides supplying more than half of the nation's drinking water. Climate, of course, is a key determinant of the importance of blue water in agriculture. Developed countries, by and large, have temperate climates and longer rainy periods, thus facilitating greater reliance on green water in agriculture.

Developed countries generally also enjoy more favorable land-to-population ratios and per capita water availability than developing nations. In fact, such is the widespread prevalence of rainfed agriculture among rich nations that industry, not agriculture, is their leading water consumer, except in Australia and New Zealand. Green-water agriculture is also widely practiced in South America, which boasts the highest per capita water availability among all continents and holds 28.9 percent of the world's freshwater reserves—virtually equal to the share of the much larger and more densely populated Asia.[23]

It is thus scarcely a surprise that most of the world's large irrigation systems are concentrated in the water-deficient regions, stretching from Asia to the Texas–Southwest–California belt in the United States. Almost 72 percent of the world's total irrigated area is located in Asia alone, followed by 14.3 percent in the Americas. The countries with the largest irrigated areas are India (66 million hectares), China (63 million hectares), and the United States (25 million hectares).[24]

More than two-fifths of the cultivated land in Asia is equipped for irrigation. By contrast, irrigation remains relatively underdeveloped in Africa, especially sub-Saharan Africa, where only 3.2 percent of the cultivated land is irrigated. Yet Africa ranks number one—just ahead of Asia—in agricultural water withdrawals as a percentage of total renewable water resources.[25]

Farmers who traditionally practiced rainfed systems in difficult conditions, with their crops dependent on the vagaries of rain, have greatly benefited from the introduction of irrigation systems. This in turn has led to the rise of new grain-producing giants like China and India, which have morphed from net food importers to net food exporters. The developing world now accounts for about 55 percent of the world's annual grain harvest. But with the era of cheap, abundant water resources having been replaced by intensifying water stress, those gains are now under threat, raising the prospect of such food exporters again becoming important importers—a development that would roil the already tight world markets.

In several parts of the world, the virtual lack of surplus water and arable land has become a constraint on sustainable increases in irrigated acreage. Indeed, in per capita terms, global irrigated land, after peaking at 0.046 hectares in 1990, is projected to decline to 0.034 hectares by 2050.[26] North Africa, the Middle East, Central Asia, and sizable parts of South Asia are already using more than 40 percent of their total renewable water resources for agriculture—a danger threshold that signals serious water scarcity.

Given that more than two-thirds of the water removed from all sources worldwide is channeled for agriculture, irrigation is a principal culprit in water-resource depletion. Indeed, between 15 and 35 percent of the present water withdrawals for agriculture, according to one assessment, are unsustainable, as they exceed the hydrological cycle's renewable capacity.[27] Aquifers recharge at slow rates of 0.1 to 0.3 percent per year on average, and the rapidly falling groundwater table in several parts of the world attests to the reckless resource extraction for irrigation.

Groundwater is the source of irrigation for 37.8 percent of the total global irrigated acreage. Significantly, in terms of reliance on groundwater for irrigation, the U.S.-Canada belt ranks ahead of even the Middle East. Whereas 62.9 percent of North America's irrigated acreage (concentrated mainly in the United States) is serviced from subterranean water reserves, the figure is 46.2 percent in the Middle East, where the total irrigated land size is about the same.[28]

Such is the extent of inefficiency that characterizes global irrigation systems that, on average, barely 45 percent of the water withdrawn reaches crop roots; the rest is lost in transmission, distribution, and application, according to the Food and Agriculture Organization (figure 2.3). In arid regions, the water loss is even greater. The Soviet-era irrigation systems in Central Asia, for example, are notorious for their inefficiency. Evaporation and seepage cause significant water losses, but some of the water lost in conveyance can be available for reuse if it percolates to aquifers.

Overirrigation represents a serious problem because, coupled with inadequate drainage, it creates waterlogging and soil salinity. An accumulation of salt reduces yields in irrigated farms; sodic soils can even become infertile and uncultivable. About 10 million hectares of cropland on average are abandoned every year due to soil erosion.[29] Actually, the environmental costs of overirrigation, and overintensive agriculture involving the heavy use of technological inputs, include not only degraded soils but also the contamination of drinking-water sources with pesticides and nitrogen fertilizers.

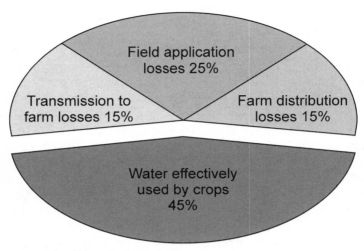

Figure 2.3. Water Lost in Irrigation
Source: Food and Agriculture Organization, 2013.

Growing amounts of fixed nitrogen—compounds such as ammonia and nitrogen oxides—end up in surface waterways, harming fisheries and causing, among other things, coastal algal blooms, including the infamous red and brown tides.[30] Atmospheric deposition of nitrogen, for its part, is a cause of acid rain, smog, and other ill effects on terrestrial ecosystems. The degradation of land and water resources due to the human alteration of entire ecosystems, if unchecked, threatens crop-yield growth and increases the risk of the world running acutely short of its total food requirements.

Microirrigation systems—such as drip-feed irrigation, a tangle of plastic veins that direct water to each plant's root zone—can efficiently apply water (and even fertilizers and pesticides). These systems not only reduce significantly the use of water, fertilizers, and pesticides but also boost output considerably. But only a small number of countries employ drip irrigation on a significant scale, with the global area under microirrigation estimated at between 10 to 11 million hectares, or slightly over 3 percent of the total irrigated acreage. Indeed, a sizable number of countries do not line their irrigation canals to prevent seepage.

Surface-water paucity, groundwater overdrafting, and salinization of soils pose significant challenges to present irrigation practices at a time when the worldwide growth of irrigated acreage has appreciably slowed. Even without factoring in the effects of climate change, water withdrawals for irrigation will need to rise by 11 percent in the next three and a half decades to meet food production demands.[31] However, as expanding cities and industries claim a larger share of the available water resources, supplies for irrigation will inevitably lessen.

In this light, if major improvements in crop yields per unit of water and land used are to be achieved to help advance the goal of food security, overirrigation and overintensive agriculture must give way to sustainable practices. Agricultural water productivity can be enhanced by adopting more efficient irrigation systems and ingeniously exploiting the green-to-blue water continuum, including through better water management, rainwater harvesting, terracing, and other techniques to channel rainwater to increase soil moisture and cut reliance on blue water. The aim should be to maximize the green-water component and reduce losses by runoff and evaporation.

Changing Diets Accentuate Water Scarcity

Spreading prosperity is boosting global demand for water just when the pressing imperative is for water conservation and efficiency. A rapidly expanding middle class is eating more meat (which requires prodigious amounts of water to produce) and also seeking high water-consuming comforts like dishwashers and washing machines. Changing diets are particularly adding to the water woes, as more and more people seek a diverse, healthy diet of plant and animal products. Consequently, even as nearly a billion people experience chronic food insecurity or undernourishment, the bulk of the world's corn and soybean output and a growing share of wheat are channeled to feed cattle, pigs, and chickens for meat production.

As much as seventy times more water is required on average to grow a person's food than to meet his or her household needs. Even greater quantities of water are required to maintain ecosystem services, without which the human lifestyle is not sustainable. In addition to the spiraling global demand for water-intensive meat and cereals, lifestyle changes are leading to greater per capita water use. In China, for example, per capita daily household water consumption increased from less than 100 liters in 1980 to 244 liters in 2000, a period during which the Chinese also began eating more meat, poultry, fish, eggs, and dairy products.[32]

To meet the fast-growing demand for meat, the diet of industrially raised cattle has been changed to make them speedily gain weight through grain rather than grass feed. This development, coming with the routine administration of antibiotics to artificially confined cattle, carries environmental, animal-health, and human-health impacts.

In the United States, food animals consume about 80 percent of all antibiotics sold, with the antibiotics being administered not to sick animals but to healthy animals for growth-enhancing and prophylactic purposes.[33] The European Union, however, banned the use of antibiotics on healthy food animals in 2006 over concerns that this practice contributes to creating drug-resistant bacteria (or "superbugs") that threaten human health.

The plain fact is that meat-based diets are a much greater driver of water withdrawals for farming than vegetarian food, with meat production multiple times more water intensive than plant-based calories and proteins. Beef has the largest water footprint of all meats. For example, it takes up to 2,400 liters (2.4 cubic meters) to produce just one standard hamburger patty because of all the water required to grow feed for the cows. By contrast, 1,500 liters of water on average can grow one whole kilogram of cereal—or ten times less than to produce an equivalent quantity of beef.[34]

Beef production is also very energy and carbon intensive. A kilogram of beef production, according to a study by Japanese scientists, emits as much carbon dioxide as an average car does every 250 kilometers and consumes energy sufficient to light a one-hundred-watt bulb for nearly twenty days.[35]

The reason meat production is much more water intensive than plant-based calories and proteins is that animals transform only 5 to 15 percent of the plant calories they consume into meat. So it takes several times more cereal to produce the same amount of calories through livestock as through direct grain consumption by humans. Put simply, growing biomass to feed animals takes far more water, energy, and land than growing biomass for direct human consumption.

The actual water footprint of different food products varies significantly with climate, soil conditions, irrigation methods, and crop and livestock genetics. Table 2.2 provides merely the most basic comparison of the water required to produce different food products. Chickens and pigs convert grain to meat more efficiently than cattle and thus have relatively smaller water footprints. Among humans, the wide difference in individual water footprints is driven by the type of diet consumed.

Table 2.2.   Average Water Requirements of Main Food Products

	Water Required (m^3) per Unit (kg)
Beef	15
Lamb	10
Poultry	6
Palm oil	2
Cereals	1.5
Citrus fruits	1
Pulses, roots, and tubers	1

Source: Food and Agriculture Organization, 2013.

The striking point that emerges from this comparative picture is that the water withdrawal differential between a meaty diet, especially one rich in beef, and a wholly vegetarian diet is large enough to seriously affect national water supply/ demand ratios. If there were no wastage of water in food production, it would take about 712 liters of water on average to feed one adult daily with a purely vegetarian diet. (Water inefficiency in food production is common, so the actual quantity of water needed would likely be more than double that.) The greater the meat content in the diet, the higher the associated water consumption.

If meat were to account for just 20 percent of a diet, the water use for producing the requisite food for one person would be almost twice that of a wholly vegetarian diet.[36] Using estimates from several studies, the average water requirement for a meaty diet works out as follows: if the share of meat in diet is 27 percent, say in Burma, the water withdrawal would be 2,964 liters per person per day; and if the share is 64 percent, say in the U.S. Midwest, the water withdrawal would total 5,908 liters per head per day.[37]

If just a third of the cereals fed to livestock were instead put directly onto human plates, the average calories of food available globally for human consumption per person daily would shoot up significantly. In 2009–2010, for instance, the world produced 2.3 billion tons of grains—enough calories to sustain a much-larger-than-present global population of 9 to 11 billion people. Yet barely 46 percent of that output went into human mouths, with livestock being fed 34 percent of the production and another 19 percent going into biofuels, starches, and plastics.[38]

If every citizen in the world became vegetarian, water security and environmental sustainability would be ensured. By significantly contributing to water conservation, a vegetarian diet is eco-friendly. A question that arises in this context is whether governments should play a role in encouraging their citizens to eat healthier diets by reducing their meat intake and consuming more plant-based calories and proteins. What matters for human health is not how much food an individual consumes but a balanced and varied diet containing essential nutrients. For example, a diet that includes whole grains, vegetables, fruits, low-fat dairy products, lean meat or protein-rich legumes, seeds, and soy-based meat substitutes tends to be high in fiber and protective phytonutrients and low in saturated fat.

Meat consumption, at the current rate of growth, threatens to outpace, and destabilize, the earth's replenishable water capacity.[39] The water-withdrawal difference between a 2,500-kilocalorie (kcal) intake per day with meat and a similar caloric intake without meat is huge, as figure 2.4 indicates. Actually, meat-based diets in a number of Western nations now exceed the 2,500-kcal figure. In the United States, where hamburger arguably is the cheapest convenience food, average daily intake has risen to almost 2,800 kcal.

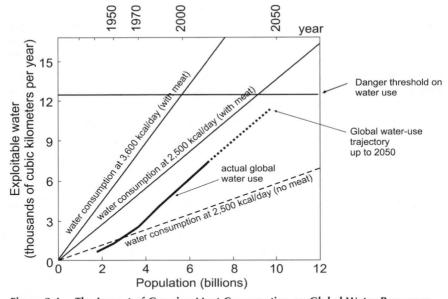

Figure 2.4.   The Impact of Growing Meat Consumption on Global Water Resources
Source: Adapted from P. Brabeck-Letmathe, Swiss-American Chamber of Commerce, Zürich (2008); and A.
J. B. Zehnder, Swiss Federal Institute of Aquatic Science and Technology ETHZ (1999).

World meat production more than quadrupled in the past half century—
from 70 million tons to almost 300 million tons—as changing diets and
a 225-percent growth in global population significantly boosted meat
demand. Global per capita meat intake jumped more than 50 percent just
between 1963 and the beginning of the twenty-first century—from twenty-
four to forty kilograms per person per year on average; the annual per capita
consumption in the developed world, however, averages much higher at
approximately eighty kilos.[40]

Significantly, with the rise of the East, the fastest global growth in meat
intake has been occurring since the 1990s, with world meat production
projected to reach 465 million tons by 2050.[41] Meat demand in developing
countries is increasing at twice their population growth rate. But in the West,
per capita meat consumption, after steadily climbing between 1945 and 2000,
has stabilized or even begun to decline in response to nutritional concern
about fat and cholesterol. Meat consumption is projected to double over the
next three decades in the developing world, where many countries, extending from Mexico and Brazil to Vietnam and Iran, have emerged as important
meat consumers.

As communities grow wealthier and urbanize, the new middle class mani-
fests a shift toward Western dietary preferences. This is best illustrated by

the trends in Asia, where the portion of meat consumed per capita has grown dramatically in just one generation, ever since the advent of economic reforms opened the path to rapid growth. China, Vietnam, and Thailand almost doubled their production of pigs and poultry during the 1990s alone.[42]

The shift from traditional rice and noodles to a meatier diet resulted in the quadrupling of China's meat consumption between 1980 and 2010, with the number of calories in the Chinese diet from meat and other animal products more than doubling just during 1990–2003. There was not much of a beef industry in China in 1980, but in a little over one generation, its beef industry became the world's third largest by 2010. Boasting the world's fastest increase in meat demand, China also accounts for almost half of the world's consumption of pork, once a rare luxury but now within the reach of its relatively poor.[43]

The only thing going in India's favor in an otherwise dismal water situation is that a vast portion of its population is vegetarian. Even those Indians who are nonvegetarian eat much less meat than Europeans, North Americans, or even East Asians. India's per capita meat consumption, at 3.2 kilograms per year, indeed ranks the lowest in the world. The diet of an average American, by contrast, is so rich in calories derived from grain-fed livestock that it includes "enough grain to support the diets of four typical Indians."[44]

Yet, despite such a remarkable national dietary profile, India is an increasingly water-stressed economy because of an imbalance between its population size and available water resources. India's population, set to surpass China's by 2030 and to peak at 1.7 billion around 2060, lives on one-third the land size of China or the United States.[45]

The United States maintains the highest per capita meat consumption in the world, with Americans consuming three times the global average of beef, poultry, and pork.[46] Meat consumption by Africans, in contrast, is less than half the global average of 40 kilograms per year, a figure that also includes lamb, mutton, and other meats.[47] With barely 4.5 percent of the world population, the United States, according to its own statistics, accounts for 21.8 percent of the global beef and veal consumption, 18.2 percent of broiler meat, and 8.9 percent of pork.[48] American consumption of fish, eggs, and dairy is also significant. Coincidentally, Americans also lead the world in per capita water and energy consumption and carbon emissions.

An important reason for their high meat consumption is that meat is subsidized in the United States (including through grain subsidies), with almost 10 billion animals slaughtered every year. Health-related changes in the American diet, however, are promoting not only a shift from red meats to poultry but also an overall decline in meat eating per person.[49] Still, if citizens in emerging and developing economies were to eat as much meat as Americans, this planet would likely confront an environmental catastrophe.

The prime driver of global water stress is more the growing consumption of meat than population growth. Rising meat consumption means more demand for animal feed, which in turn brings water supplies under mounting strain. More meaty diets not only accelerate the cumulative national water demand but also implicitly complicate intrastate and interstate water management and sharing.

About 20 percent of the protein in human diets is currently animal based, and, according to one study, unless that figure drops to 5 percent by 2050, the world will face a severe food crisis.[50] The world could stay on the sustainable path, even with a projected 10 billion population by 2083, if all citizens ate vegetarian food or returned to the nineteenth-century European diet of eating meat, at best, twice a week. Yet no national government thus far has tried to encourage citizens to switch to a diet dominated by less thirsty forms of protein, even though consumption of more protein from plant than animal sources is healthier.

More fundamentally, the expanding livestock sector is bringing virtually all natural resources under stress, with underpriced inputs and feed encouraging environmentally damaging practices. The proliferation of assembly-line meat factories has fostered the consumption of huge and growing amounts of water, energy, and grains; the pollution of waterways from animal wastes and the fertilizers and pesticides used for producing feed; the large-scale injection of antibiotics and hormones into livestock; and the problem of manure "lagoons," tannery chemicals, and sediments from land erosion.

Livestock production also contributes to eutrophication (or the process by which a body of water acquires a high concentration of inorganic nutrients, especially phosphates and nitrates, resulting in excessive algae growth); to deforestation and the ensuing increased runoff and global warming; to degradation of coastal ecosystems, including salt marshes, mangroves, and coral reefs; and to human-health problems, such as the emergence of antibiotic resistance.

Besides consuming a rising share of the world's water resources, livestock production generates 18 percent of the world's greenhouse gases—more than the transportation sector—and directly or indirectly utilizes 30 percent of the earth's land surface that was once wildlife habitat.[51] Belching cattle are responsible for emitting 37 percent of the anthropogenic methane, a greenhouse gas whose global-warming potential (GWP) is twenty-three times higher than carbon dioxide.[52] Researchers have found that adding omega-3 fatty acids—naturally found in pasture grasses—to the high grain-based diet of cattle on factory farms helps to reduce methane emissions due to fermentation in their digestive systems. Animal dung, for its part, generates nitrous oxide, another greenhouse gas that has 296 times the GWP of carbon dioxide.

From the local to the global level, the livestock sector is one of the most significant contributors to environmental problems, including water degradation. Indeed, it is responsible for probably the greatest biodiversity loss.[53] If further major damage is to be stemmed, this sector's wide-ranging environmental impacts per unit of production must be significantly cut, including by phasing out perverse subsidies that promote environmentally injurious practices and by making the livestock industry pay market prices for using water, land, and other resources.

Tackling the growing problem of human obesity, whose global prevalence more than doubled just between 1980 and 2008 because of labor-saving technologies and societal and dietary factors, is also critical to world food and water security and to environmental sustainability. After all, according to the World Health Organization, two-thirds of the world's population lives in countries where overweight or obesity kills more people than underweight.[54] Excess body fat, as discussed in the previous chapter, is an important factor contributing to water and environmental stresses. Staying slim is good for individual health and for the health of the planet—a fact highlighted by the more than doubling of the world's diabetes-stricken adult population since 1980.

In the first detailed global analysis of the effects of diet and lifestyle changes, a study published in the journal *Lancet* found that adult diabetes prevalence increased from 8.3 percent in men and 7.5 percent in women in 1980 to 9.8 percent and 9.2 percent respectively in 2008, with the result that the total diabetic population aged twenty-five years and older jumped from 153 million to 347 million in this period.[55] Diabetes prevalence is particularly high in the South Pacific Islands, the Middle East, North Africa, South and Central Asia, Latin America, and the Caribbean. Five of the ten countries that top the list are Persian Gulf sheikhdoms, where oil wealth has created an explosion of diabetes through major diet and lifestyle changes.[56] The United States has the dubious distinction of registering the steepest rise in mean glucose levels in the high-income world.[57]

Nearly a tenth of the world's adults have diabetes, raising the danger of national health systems coming under serious strain if the prevalence of this disease continues to grow rapidly. The diabetes epidemic is a reminder that people may be eating better but they are not necessarily eating healthy food.

As a consequence of changing diets, excess body fat, and continued population and consumption growth, the lack of access to stable water supplies is already "reaching critical proportions, particularly for agricultural purposes."[58] Water stress indeed is forcing some countries to decrease their domestic food production and rely more on imports. Consequently, North Africa and the Middle East have emerged as the fastest-growing food import markets. The worst effects of rising international food prices will be borne by the world's poor because of the greater purchasing power of wealthy and middle-income states.[59] A number of low-income countries, mostly in sub-Saharan Africa, do not grow enough food to feed their populations, yet they lack the financial resources to make up the deficit with imports.

More broadly, growing more food with less water, land, and energy is at the heart of the global challenge that tests humanity's ingenuity. This challenge is underscored by the end of the era of ample water and land resources and cheap fossil energy—factors that helped to dramatically boost food production in the second half of the last century. The world has now entered a new era where resources critical to sustaining increased food production, especially freshwater and land, are coming under growing strain. Without stemming water and soil degradation, the yield potential of improved crop varieties and elite germplasm cannot be fully realized.

THE INTERSECTION OF WATER AND ENERGY

Water and energy are inextricably interdependent: moving water, or providing a supply of clean water to consumers, demands energy, while, without water, energy cannot be produced to keep our plugged-in world humming. Such is the intimate nexus between water and energy that the water industry is energy intensive and the energy industry, in turn, is water intensive, needing copious quantities of water for energy-resource extraction, conversion (refining and processing), energy transportation, and power generation. The expanding output of biofuels from irrigated crops is another important source of energy-related water consumption.

Adequate water availability is a must for electric power generation because of the thirsty nature of power plants, most of which are located within easy reach of a surface-water source—a river, lake, or ocean. Plants lacking such access or situated in areas where surface-water resources are scarce turn to groundwater, with some of them causing aquifer depletion. The amount of water required to produce electricity to light a home and power its various appliances and electronic devices is generally higher than average household water use.

Water shortages throw up a range of challenges and risks in the energy sector, including threatening the viability of power generation projects, crimping the development of new sources of energy, affecting the reliability of existing operations, and imposing additional costs. The processes of energy mining and production (including resource extraction and refining and electric power production) actually pose a potential threat to the quality and quantity of water resources. Local water sources can be contaminated, for example, by untreated wastewater from oil-and-gas refining and processing, or by the chemical-laced wastewater from the extraction of shale gas and oil, or by the runoff from uranium and coal mine operations and ore tailings. Fossil-fueled plants turn more than a quarter of their water intake into steam; the remaining water is channeled for cooling and discharged with excess heat into waterways and oceans, potentially killing aquatic life, affecting ecosystems, and creating harmful algal blooms.[60] Diminished water supplies, on the other hand, threaten energy production: the impact of a drought on water availability can decrease generation by power plants just when electricity demand has peaked, leading to brownouts or even blackouts during the summer and beyond.

Energy powers the national water infrastructure, which channels surface water and groundwater for agricultural, industrial, and household uses, besides controlling floods and storing water for the dry season. Treating water to make it potable and making wastewater safe for discharge into watercourses or for reuse also demands energy. Energy intensity is indeed the principal downside of currently available clean-water technologies. The United States uses 4 percent of its national electricity for water treatment and supply.[61]

The greater the distance for conveying water, the greater is the energy use and cost, with China's ambitious interbasin and interriver transfer programs seeking to transport water even over mountains. An increasingly water-stressed world is likely to use more of its energy to treat water and move surface water, and even groundwater, around.

The advantage of groundwater is that it is often cleaner than surface water and, if not brackish, needs less treatment and thus less energy for purification. Groundwater extraction, however, is energy intensive, and the sinking water table in many countries has significantly increased the energy needed to bring the same quantity of water to the surface. One study in California found a fourfold rise in energy requirements when the depth of extraction increased from 37 to 122 meters.[62]

The relationship between national incomes and national consumption of water and energy—the two key resources for development and prosperity—is revealing. Wealthier countries, by and large, have greater per capita energy- and water-consumption levels than poorer states. This is an unpalatable reality that ties higher per capita income levels to higher per capita resource consumption within nations—a link that needs to be snapped to help build global resource stability and to combat the growing buildup of greenhouse gases in the atmosphere. That the United States has the world's largest per capita water and energy footprints sets the wrong model; it is an example that, in any event, cannot be emulated by other states in a resource-stressed world.

In an era in which water demand is growing exceptionally fast to produce energy and to power economic modernization, the major strategic implications of increasing water-supply limitations are obvious. Reduced water supplies threaten both energy production and economic growth. In poor countries, shortages of water and energy actually constitute a double recipe for continued underdevelopment. Poverty and slow economic progress, in turn, foster a cycle of environmental decline that helps degrade water and land resources, including through overcultivation or overgrazing of land and human encroachment on natural ecosystems.

In policy terms, water and energy are so closely intertwined that, without their integrated management and planning, moves aimed at addressing the challenges related to one resource can make the problems concerning the other resource worse. The nexus of water/energy challenges demands a holistic, long-term approach so that national policies on water, energy, and even food are harmonized to help achieve greater water efficiency. After all, falling water tables, surface-water degradation and depletion, and climate change mean that water resources must be used prudently and to their full value.

Water Appropriations for Energy Production

Resources such as coal, oil, natural gas, and uranium cannot be harnessed for energy or other productive purposes without a heavy dependence on local water resources. Water is essential to generate electricity from nearly all energy sources, and changes in water resources can impact the reliability of power generation. The energy sector is so water hungry that it is the largest user of water in Europe and the United States.

Water consumption by petroleum refineries, for example, is usually larger than the quantity of gasoline or diesel fuel actually manufactured. Extraction of hydrocarbons from shale and other "tight rock" fields involves the loss of much of the initial water used to stimulate the release of gas and oil because the chemical-laced wastewater turns so toxic that it must be buried in special deep wells. In the United States—the world's second-largest producer of coal after China—natural-gas processing and pipeline operations consume more water daily than coal mining. It takes about eighty-seven liters of water on average to produce one kilowatt-hour of electricity to run an energy-efficient refrigerator for a day.

Whereas fast-flowing water can be directly tapped to generate hydroelectricity, great amounts of water are required for cooling and steam-cycle processes at thermoelectric power plants that receive heat from sources such as coal, nuclear, natural gas, oil, biomass, concentrated solar energy (also called concentrating solar power, or CSP), and geothermal energy. In using steam to drive turbine generators, thermoelectric technologies require cooling to condense the steam back to water at the turbine exhaust. Steam-based power systems are widely prevalent across the world.

The actual water withdrawal and consumption rates of these water-hogging electricity stations depend on the cooling technology they employ (figure 2.5). In the United States, the thermoelectric power industry accounts for 41 percent of all freshwater withdrawals—far higher than irrigation—but the entire energy sector represents about half of all freshwater withdrawals and 27 percent of all freshwater consumption (the volume withdrawn but not returned to its source).[63] In Europe, where the levels in rivers, lakes, and reservoirs fell to historic lows during the 2003 and 2006 heat waves, thermoelectric generation accounts for 44 percent of total water abstractions and about one-third of all river-water withdrawals.[64] Globally, according to the International Energy Agency, water withdrawals for energy production totaled 583 billion cubic meters in 2010, of which 66 billion cubic meters were consumed.[65]

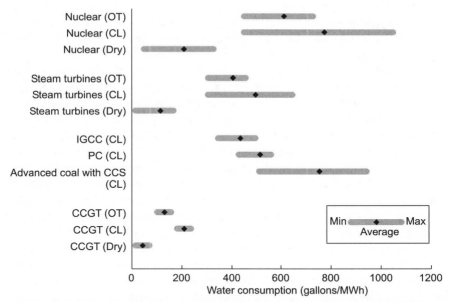

Figure 2.5. Water Consumption in Thermoelectric-Power Generation Using Different Cooling Technologies (OT = once through; CL = closed loop; IGCC = integrated gasification combined cycle; PC = pulverized coal; CCS = carbon capture and sequestration; and CCGT = combined cycle gas turbine. Includes water consumed in fuel extraction and processing)

Source: Adapted from Erik Mielke et al., "Water Consumption of Energy Resource Extraction, Processing, and Conversion," Harvard Kennedy School, October 2010; and U.S. Department of Energy, *Energy Demands on Water Resources*, 2006.

The location of thermoelectric power plants, such as nuclear, coal-fired, CSP, and open cycle gas turbine (OCGT) facilities, is determined by adequate availability of local water resources. These plants return much of the water they draw back to the source but at a higher temperature and with other quality changes (such as a trace of residual chlorine) that affect freshwater ecosystems. Fish and other aquatic organisms can get trapped and killed by the inflows, as well as by the warm outflows because they are cold-blooded. Furthermore, air emissions of mercury, sulfur, and nitrogen oxides from fuel combustion can alter downwind water quality and affect aquatic ecosystems.[66]

In determining the extent of water abstractions by a thermoelectric plant and its consumptive water use relative to total withdrawals, the cooling system type is often more important than the particular technology employed to generate electricity. In the older type of system known as "once-through" (or "open-loop") cooling, water is withdrawn from a source, circulated through the heat exchangers, and then returned to a surface-water body. This cooling type appropriates large quantities of local water resources, but the consumptive rate is low relative to the withdrawals. There is a water loss of just about 1 percent due to evaporation.[67] Once-through cooling is still well liked because it offers thermal efficiency and reduced fuel and other costs, including eliminating the need for cooling towers.

In the newer type built since the 1970s and known as "closed-loop" (or "recirculation") cooling, water is withdrawn from a source, circulated through heat exchangers, cooled using towers or ponds, and then recycled. Plants employing recirculated cooled water demand on average barely one-twentieth of the water needed by the once-through-cooled plants. Whereas once-through cooling systems continuously withdraw water from a water source and return almost the same quantity of water to the source, closed-loop cooling systems (after the initial filling) withdraw water only to replace the loss from evaporation and blowdown. Yet this loss is often substantial because the evaporative cooling in closed-loop systems consumes much more water than once-through cooling systems.[68]

Some renewable-energy technologies, even if they demand water withdrawals for their operation, cause no evaporative losses. The non-freshwater-consuming technologies include wind power, run-of-river hydropower, solar photovoltaics, solar dish engines, air-cooled geothermal binary systems, and ocean energy systems. Wind and solar photovoltaic plants actually need no water even for their normal operations, although photovoltaic arrays need to be intermittently sprayed with water and kept clean to maintain optimal output. But solar thermal power and geothermal steam plants are so thirsty that it is essential to locate them where water resources are ample.

Much of global electricity is now produced through closed-loop, steam-based (Rankine) power cycles, which lose considerable quantities of water to evaporation while condensing low-pressure steam back to water for return to the plant's heat source for reboiling. The employment of a cooling tower or pond to remove heat from the steam during the condensation process leads to such large evaporative losses that slightly more than four-fifths of all the closed-loop generating units in the United States reported consumptive-use rates of 50 percent or greater in 2005, according to government data.[69]

Such plants, by consuming up to 70 percent of their water intake, can exacerbate local water stress. Yet water-scarce regions such as the American Southwest prefer to rely on closed-loop systems, despite their greater evaporative losses, than on the highly water intensive once-through systems, whose share in the U.S. generating capacity has declined to 43 percent.[70]

New technologies can slash water use in electrical-energy production through "dry cooling" (which employs more expensive air-cooled condensers for the steam cycle) and "hybrid cooling" (which fuses wet and dry cooling). However, the plant-efficiency penalties and higher investment costs of the alternative cooling technologies, among other factors, have limited the extensive use of these technologies.[71]

Dry cooling, for example, reduces plant efficiency, especially in hot weather. Hybrid cooling is better at shielding efficiency in hot-weather conditions, but, like dry cooling, it is more expensive. The annual performance penalty for switching from wet cooling to dry cooling for fossil plants like coal and natural-gas steam stations has been estimated at 6.9 percent and for nuclear plants at 6.8 percent.[72]

Still, as companies try to overcome water-supply constraints, a growing number of plants are being built with alternative cooling. Dry-cooled, as opposed to wet-cooled, CSP plants, for example, cut water usage by 90 percent.[73] A hybrid cooling system in, say, a nuclear plant has the flexibility to switch from wet cooling in the wet season (the energy-conservation mode) to dry cooling in the dry season when river and lake levels fall (the water-conservation mode). Utilizing alternative cooling technologies indeed precludes some of the risks associated with drought and climate-change scenarios. For example, drought-triggered reduced watercourse levels, or sharp increases in watercourse temperature, can force thermoelectric power plants with wet cooling to shut down or operate at lower capacity.

The rise of combined cycle gas turbine (CCGT) plants is also good news on the water front. Such plants, which boast relatively high thermal efficiency, source two-thirds of their power from the gas turbine (Brayton) cycle that generates energy from hot, pressurized gases, not steam; the remaining one-third of their power output comes from the conventional steam (Rankine) cycle. Some of the new CCGT plants actually need little cooling water because they employ air-cooled condensers for their steam cycle.

Water use, however, is central to emissions scrubbing in coal-fired plants, as well as to carbon capture and sequestration (CCS) technologies. The thermoelectric power sector's daily water consumption would jump significantly if carbon-mitigation policies forced all new and existing pulverized-coal plants to have scrubbers and plants based on the integrated gasification combined cycle (IGCC) to utilize CCS technologies.[74]

Going green thus will impose significant water costs. In fact, pulverized coal with carbon capture technologies as well as CSP systems utilizing cooling towers have among the highest water-consumption factors.[75] Other new energy technologies also pose water-related challenges.

National programs to tap nonconventional energy—extending from shale gas and oil to tar sands and "synfuels" made from coal-to-liquids (CTL) technology—are, in fact, bringing local water resources under greater pressure. Water consumption for producing alternative fuels, such as synfuels or hydrogen from methane, is up to three times greater than for petroleum refining.[76]

However, no energy source matches the water prolificacy of ethanol and biodiesel manufactured from corn or soybeans that have been grown with the help of irrigation. The production of biofuels is estimated to be at least three to five times more water intensive than traditional fuels.[77] Yet corn-derived ethanol, which tops in water intensity, is an increasingly important part of the U.S. liquid fuel mix for transportation.

The shale hydrocarbon boom—a game changer—has come with water depletion and pollution problems because the extraction method uses a high-pressure mix of water, sand, and hazardous chemicals to blast through rocks to release the oil and gas inside. Known as hydraulic fracturing, or "fracking," this technique was first developed to extract natural gas from shale and has now been extended to oil shales. Thanks in large part to the use of this method to expand production of previously inaccessible shale gas and oil, the United States has virtually surpassed Russia as the world's largest gas producer, with cheap and plentiful supplies of natural gas fueling an American industrial revival and the U.S. industry now seeking to initiate gas exports. The United States is projected to overtake Saudi Arabia as the world's top oil producer by about 2020, which will put it on track to become a net oil exporter by 2030.[78]

To initially stimulate a well, millions of gallons of water must be shot into it to crack the shale rock and get crude oil, natural gas, or natural-gas liquids flowing. About 1,590 cubic meters of water are used per million cubic feet of gas produced from shale.[79] Shale oil development is typically several times more water intensive than shale gas. To compound the water challenges, oil-shale deposits are largely located in areas where water resources are already scarce or under pressure, such as the Colorado River Basin, which boasts the world's largest oil-shale reserves.[80]

Fracking is not just water intensive but also a water-contamination hazard. One concern is the safe disposal of the toxic drilling mix, which must be buried in sealed wells. The controversial combined use of fracking and improved horizontal drilling to unlock vast quantities of oil, gas, or gas liquids also carries the risk of the slurry or gas escaping through rock fractures and contaminating aquifers, especially when the drilling is improper.[81]

Fracking waste has been held responsible for drinking-water contamination in several places, including Pennsylvania.[82] Because only a limited number of geologically suitable sites are available for deep-well burial, the wastewater often has to be trucked a considerable distance for disposal, entailing both significant costs (up to three dollars per liter) and environmental risks.[83]

The deep injection of such chemical-laced wastewater into the earth has actually been blamed by authorities for lubricating a previously unmapped geological fault in northeastern Ohio and triggering a series of local earthquakes.[84] This has prompted Ohio to unveil new rules for drillers, including electronic monitoring of wastewater. In Estonia and elsewhere, oil-shale mining has caused groundwater and air pollution.[85] Well blowouts and other surface impacts tend to compromise water quality. Upon completion of the drilling and fracking processes, 20 to 80 percent of the water used returns through the bore hole as flowback, thus needing to be treated or safely disposed of.[86]

After the intensive, front-loaded water consumption to drill and hydraulically fracture the shale and stimulate the well, production-related water consumption tends to be lower than for conventional oil and coal, although it is possible that as the wells age, fracking may have to be carried out again.[87] Yet the high water usage and loss at the drilling-and-fracking stage, coupled with the contamination hazards, means that shale hydrocarbon development can adversely affect local water supply, particularly in areas already water stressed.

Energy-resource mining and processing generally demands a lot of water while posing a threat to water quality. In a number of cases, runoff from coal and uranium mine operations and tailings piles, as well as untreated wastewater from oil and gas refining and processing, have contaminated local water sources. For example, more than 1.1 million cubic meters of coal sludge spilled in a 2000 Kentucky incident. Large-scale Chinese energy-resource extraction and processing in Xinjiang, Tibet, and Inner Mongolia has been identified as a source of local water pollution.

Hydropower: Strengths, Impacts, and Vulnerabilities

Whereas price volatility governs electricity generation from extracted resources like coal, gas, and uranium, the fuel expense is zero for producing hydroelectricity, a renewable energy that is one of the oldest methods of power generation. This fact, coupled with a hydroelectric station's relatively low operation and maintenance expenses, means that the cost of hydropower generation progressively declines over a plant's lifespan.

Hydropower plants also supply continuous baseload electricity and can meet peaking needs, as in hot summers, when thermoelectric-generation efficiency declines due to high humidity and temperatures. These factors make hydropower attractive for governments and private investors, despite the high up-front capital costs involved in the erection of any such project.

Hydropower, however, often carries environmental costs. These range from inundation of land and wildlife habitat and displacement of local residents to dam-induced changes in stream water quality and quantity and modification of fish habitat. Large hydropower projects tend to affect fluvial and aquatic ecosystems and alter dissolved oxygen and nitrogen levels in downriver waters both by discharging waters at a higher temperature from their once-through cooling system and by changing the natural river-flow characteristics through water impoundment.

There are essentially two types of hydropower projects: the run-of-river class, which stores little water because it employs a river's natural flow energy and elevation drop to produce power, and the larger, storage-centered kind that usually impounds an enormous amount of water in a reservoir behind the dam and then diverts it through a penstock and turbine to produce electricity. A variant of the storage type is based on pumped-storage technology, which demands two reservoirs (one more highly elevated than the other so as to maneuver a waterfall) and uses a reversible turbine to pump water back to the upper reservoir in off-peak hours.

Much of the world's hydroelectricity is produced by storage-type plants from water sequestered in large dam reservoirs. Because such plants function independently of natural water flows, they are ideally suited for balancing fluctuations in national and provincial electricity generation and consumption; they can be used both to cover the electrical base load and for peak load operations. These large plants, however, reengineer river flows, cause river fragmentation, and affect downstream communities. Storage-type hydroelectric plants, in fact, account for more than 90 percent of the world's total renewable electricity.[88]

The smaller run-of-river projects, due to their reliance on the river's flow velocity and drop, face a constraint on their location and size. Their energy output usually cannot be increased to meet any seasonal upsurge in power demand, as in summer. Their electricity production indeed declines in dry, hot months if river flows ebb. Because they have very limited water storage, such plants experience fluctuations in power output due to changes in weather and local hydrology, with less river flow in the lean season translating into less electricity production. This makes them less cost effective. Run-of-river projects, however, can be built in an environmentally considerate way to have minimal impact on ecosystems, despite their infrastructure's effects on wildlife and plant populations.

Conventional storage-type projects, unlike the energy-centric run-of-river plants, are often built for multiple purposes, ranging from power generation and irrigation to drinking-water supply and river navigation. They can also stockpile snowmelt and other runoff for release in the nonwet seasons. Storage dams enable the production of more reliable and economical electricity than run-of-river projects. Advances in science and engineering, as well as moves to ramp up production of hydroelectricity, have actually resulted in the building of larger and larger dams and reservoirs.

As a result, the scale of major dam projects has progressively increased over the past century. But with about 50,000 *large* dams already built, most of them since the mid-twentieth century, the good damming sites across much of the world have largely been taken.[89] Such was the worldwide rush to build new dams between 1950 and the mid-1980s that about 885 large dams were completed on average every year.[90] Since then, dam building has slowed in the West but picked up momentum elsewhere in the world, with China in the lead and boasting no less than half of the world's total number of large dams.[91]

Not surprisingly, China is the world's largest producer of hydroelectricity, followed by Brazil, Canada, and the United States.[92] The thirty-two turbine generators at the world's largest hydropower station, China's Three Gorges Dam on the Yangtze River, have a combined generating capacity of 22.5 gigawatts—nearly three and a half times greater than America's leading hydroelectric project, Grand Coulee. The world's biggest coal-fired plant, by contrast, is the 4.1-gigawatt Kendal station in Mpumalanga, South Africa. Rivers like the Jinsha (the name of the upper section of the Yangtze), the Mekong, the Salween, and the Brahmaputra are at the center of the current Chinese dam-building program to carry out the world's largest expansion of hydropower-generating capacity in the coming years.

The construction of large, storage-type projects on shared rivers can carry political costs. Despite the continuing attraction of supply-side approaches centered on large-scale storage to meet rising energy and water demands and to address spatial and seasonal variability in intracountry water availability, such projects are often at the root of interstate or intrastate water disputes because they can significantly alter downstream flows. Such is the sharpening competition over water resources that even run-of-river projects have become a source of interriparian tensions—such as between India and Pakistan—although, unlike the reservoir-linked storage projects, they generally do not materially change cross-border flows.

The hydropolitics of dams is compounded by the fact that most of the good dam sites in the world—with high hydrostatic heads, ideal perennial flow velocity, and close proximity to load demand—have been taken, leaving only politically and environmentally sensitive sites for new projects, except in some water-rich underdeveloped countries, such as in sub-Saharan and Himalayan regions. Large hydropower projects, by submerging land, displace local residents. The grievances of relocatees or those attempted to be moved out have in many cases engendered unrest and litigation, slowing or stalling projects, especially in democracies. In Japan, India, and Nepal, for example, organized protests by nongovernmental organizations have driven up project costs and acted as a damper to hydropower expansion.

Large hydropower projects also tend to affect river hydrology, sediment load, riparian vegetation, patterns of stream bank erosion, migration of fish, and quality of water. A chain of dams and reservoirs on a river can cause potentially wide-ranging effects on the natural and human environments through river fragmentation and depletion. Degraded watersheds and altered natural-flow characteristics of rivers have emerged among the most serious problems for sustainable development. A growing recognition of the significant environmental impacts of large dams, coupled with the grassroots resistance they often set off, has increased the attraction of run-of-river plants in many countries.

Evaporation from storage reservoirs results in considerable consumptive loss of water. Although large-scale hydropower production appropriates vast quantities of water, quantifying water withdrawal and consumption rates is not easy because reservoirs often serve multiple purposes, including irrigation, dry-season water supply, flood management, tourism, and recreation. When a project serves several different purposes, hydropower generation can be held responsible for only part of the evaporative losses.

Still, such losses attributable to hydropower production range from 17 to 34 cubic meters per megawatt-hour on average in the United States, according to two separate estimates.[93] And although hydropower generation is nonpolluting and helps to stem greenhouse-gas emissions, hydroelectric reservoirs produce significant amounts of methane and carbon dioxide.[94]

One major vulnerability of hydropower springs from its reliance on hydrology. Hydropower demands predictable river flows or reservoir levels, and prolonged drought conditions created by meager rainfall in the upper catchment can affect output even at storage-type hydroelectric plants.

Hydropower is vulnerable not just to low flows in the hot season or low water availability in drought but also to the long-term impact of climate variability on water supplies. Limiting the contribution of fossil-fueled electricity generation to the buildup of greenhouse gases in the atmosphere demands greater use of renewable energy sources like hydropower. Yet, paradoxically, climate change itself threatens to adversely affect existing and potential hydropower projects. As the Intergovernmental Panel on Climate Change has warned, "Widespread mass losses from glaciers and reductions in snow cover over recent decades are projected to accelerate throughout the twenty-first century, reducing water availability, hydropower potential, and changing seasonality of flows in regions supplied by meltwater from major mountain ranges."[95]

Nuclear Power's Water Risks

The central dilemma of nuclear power in an increasingly water-stressed world is that this source of energy is a water guzzler, yet vulnerable to water volatility. Nuclear power generation, with the highest water usage among all thermoelectric technologies, is enormously water intensive and strikingly vulnerable to drought, extreme weather events like tsunamis and hurricanes, and rising ocean levels. As an energy source that makes large demands on water resources and is susceptible to water-related natural disasters, nuclear power poses special challenges and risks. The reactor meltdowns at the Fukushima Daiichi plant, ranking alongside Three Mile Island and Chernobyl as history's worst nuclear accidents, are a testament to such vulnerability.

The water intensity of nuclear power begins with uranium mining and milling and extends beyond the steam and cooling requirements of power generation to discharged fuel storage. Uranium mining and processing consumes far more water than the mining and supply of fossil fuels. Enriching uranium to fabricate fuel also consumes significant quantities of water, with enrichment by gaseous diffusion consuming more water than enrichment with centrifuges. The entire nuclear fuel cycle indeed is water intense and requires the protection of freshwater quality. For example, the millions of gallons of water used daily in uranium mines must be carefully handled and disposed of to avoid wider water contamination.

Light water reactors (LWRs)—which use water as a coolant and low enriched uranium as fuel—produce most of the world's nuclear power. These reactors, like those at the stricken Fukushima plant, are also the leading users of water in the energy sector. The twin-reactor Indian Point nuclear power plant north of New York City withdraws up to 9.5 million cubic meters of water a day from the Hudson River, or more than double New York City's entire daily water use. New York State authorities want the aging plant closed because of its role in the depletion and degradation of water resources. The huge quantities of local water that LWRs appropriate for their operations become hot-water outflows, which are pumped back into rivers, lakes, and oceans. This can damage plant life and fisheries.

LWRs include the two most common models: the pressurized water reactor (PWR) and the boiling water reactor (BWR). Both of these types of reactors, employing the once-through or the closed-loop cooling process, have the highest water withdrawal and consumption rates among all thermoelectric technologies. The reason is that, thermodynamically, LWRs have lower steam conditions than fossil-fueled plants, which results in reduced turbine efficiency and higher cooling-water intake, reaching up to 227 cubic meters per megawatt-hour of electrical production.[96]

Because they generate less electricity per kilogram of circulating steam than fossil plants, they need greater steam circulation rates. This translates into larger cooling-water requirements for the steam-condensation process as well as higher evaporation rates.

According to an assessment of U.S. power plants using freshwater for cooling in 2008, a typical nuclear power reactor withdrew nearly eight times as much water—and consumed three times as much water—as an average natural-gas plant per unit of electricity produced.[97] Nuclear-generated power thus tends to put local freshwater resources under serious pressure.

Consequently, many nuclear plants are located along coastlines so that they can draw more on seawater. Yet the projected greater frequency of natural disasters like storms, hurricanes, and tsunamis due to climate change, along with the rise of ocean levels, makes seaside reactors particularly vulnerable. Before an earthquake and tsunami triggered the nuclear calamity at Fukushima, the occurrence of more than one natural disaster simultaneously had not been considered by plant builders and operators.

The risks that seaside reactors face from water-triggered disasters actually became evident more than six years before Fukushima, when the Indian Ocean tsunami in December 2004 inundated India's second-largest nuclear complex, shutting down the Madras Atomic Power Station. But the reactor core could be kept in a safe shutdown mode because the electrical systems had been ingeniously installed on higher ground than the plant level. Unlike Fukushima, which bore a direct impact, the Indian plant happily was far away from the epicenter of the earthquake that unleashed the tsunami.

In 1992, Hurricane Andrew caused significant damage at the Turkey Point nuclear power plant in Biscayne Bay, Florida, but fortunately not to any critical system. And in a 2012 incident, an alert was declared at the New Jersey Oyster Creek nuclear power plant—the oldest operating commercial reactor in the United States—after water rose in its water intake structure during Hurricane Sandy, potentially affecting the pumps that circulate cooling water through the plant.

Because reactors located inland put a major strain on freshwater resources, water-stressed countries that are not landlocked try to find suitable seashore sites for their reactors. But whether located inland or on the coast, nuclear power is vulnerable to the likely effects of climate change. For example, all of Britain's nuclear power plants are located along the coast, and a government assessment has identified as many as twelve of the country's nineteen civil nuclear sites as being at risk due to rising sea levels.[98] Several nuclear plants in Britain, as in a number of other countries, are just a few meters above sea level.

As global warming brings about a rise in average temperatures and undermines the reliability of river flows, inland reactors will increasingly contribute to, and be affected by, water shortages. Such shortages will impinge on LWR operations, with such plants becoming unable to generate electricity at their rated capacity. During the record-breaking 2003 heat wave in France, operations at seventeen commercial nuclear reactors had to be scaled back or stopped because of rapidly rising temperatures in rivers and lakes. Spain's reactor at Santa María de Garoña was shut for a week in July 2006 after high temperatures were recorded in the Ebro River.

In fact, during the 2003 heat wave, Électricité de France (EDF), which operates fifty-eight reactors—the majority on ecologically sensitive rivers like the Loire—was compelled to buy power from neighboring countries on the European spot market. The state-owned EDF, which normally is a major power exporter, ended up paying ten times the price of domestic power, incurring a financial cost of €300 million.[99] Although the 2006 European heat wave was less intense, water and heat problems forced Germany, Spain, and France to take some nuclear power plants offline and reduce operations at others. In 2006, plant operators in Western Europe also secured exemptions from regulations preventing them from discharging overheated water into natural ecosystems.[100]

The very conditions in 2003 and 2006 that made it impossible for the nuclear industry in Europe to deliver full power created peak demand for electricity, especially due to the increased use of air conditioning. Such heat waves could become more common as the climate changes. Indeed, in another sign of trouble for nuclear power in a warming climate, two of the three reactors at the Browns Ferry nuclear plant along the Tennessee River in the United States reduced electricity production during a 2007 drought, and the third had to be temporarily shut down due to diminished river flows and high water-discharge temperature.[101]

Nuclear power plants located by the sea do not face such problems in hot conditions, because ocean waters do not heat up anywhere near as rapidly as rivers and lakes. And because such plants rely largely on seawater, they do not contribute to freshwater scarcity. But as Fukushima showed, coastal nuclear plants are prone to serious dangers. Moreover, with nearly two-fifths of the world's population living within one hundred kilometers of a coastline, finding suitable seaside sites for new nuclear plants is no longer easy.[102] Coastal areas are often not only heavily populated but also constitute prime real estate. For example, India, despite a 6,000-kilometer coastline, has seen its plans for a huge expansion of nuclear power through seaside plants run into stiff grassroots objections.

France likes to showcase its nuclear power industry, which supplies almost 78 percent of the country's electricity. But its nuclear industry withdraws up to 19 billion cubic meters of water per year from rivers and lakes, or roughly half of France's total freshwater withdrawals, with nuclear-power-related water abstractions actually surpassing those by the agricultural industry in a country that prides itself as a major food exporter.[103] Indeed, no industry has wreaked a greater impact on France's freshwater ecosystems than nuclear power.

Nuclear power's early promise as a cheap, clean form of energy has simply not materialized. Decades after Lewis L. Strauss, the chairman of the U.S. Atomic Energy Agency, claimed that nuclear energy would become "too cheap to meter"—a hope that led to the U.S. "Atoms for Peace" program—the nuclear power industry everywhere still subsists on huge amounts of taxpayer subsidies, the burden of which leads the major nuclear-power-generating nations to push aggressively for reactor exports.[104] The subsidies, also universal in importing countries, help to create the illusion that nuclear power is cost effective.

Over the years, state support for the nuclear power industry has only become more generous—from loan guarantees and tax credits to major limits on accident liability and saddling taxpayers with nuclear-waste-related and plant-decommissioning expenses. A federal loan guarantee is like a parent cosigning a child's car loan but with a much greater risk—if the nuclear-reactor builders default, as happened on some U.S. projects in the 1980s, taxpayers could be saddled with multibillion-dollar liabilities. Instead of the cost of nuclear-generated power declining with the technology's maturation, the cost has actually escalated, making munificent taxpayer subsidies critical to help overcome unfavorable reactor economics and sustain the industry.[105]

It is illegal to drive a car without sufficient insurance but perfectly lawful to operate nuclear power plants without disaster insurance coverage. This is because liability laws usually shield the commercial nuclear industry by shifting most of the costs of a catastrophic accident to the public. Yet a single nuclear accident, if it results in significant release of radioactive materials, carries incalculable, long-lasting human and environmental consequences.

Nuclear power was supposed to open the door to an energy-secure future, but Fukushima—the world's third major nuclear accident in just about three decades—closed the door to the future for many local communities in Japan. It foisted a massive bill on a nation already trapped in low growth and deflation and with the highest debt level among developed nations.

Thirsty nuclear power had actually gained international credence as a cleaner energy alternative before Fukushima because of mounting concern about the environmental and public-health tolls of fossil fuels. Such acceptance helped obscure the fact that nuclear power has a major environmental downside that extends beyond its impact on water resources and the carbon intensity of its fuel cycle: nuclear power's front end may be clean, but its back end is remarkably dirty, with the safe disposal of radioactive wastes posing serious technical and environmental challenges. Although the United States operates the largest number of nuclear reactors in the world, it does not have any dedicated waste repository; in 1987, Congress designated Yucca Mountain—a desolate volcanic ridge in Nevada—as a national disposal site, but in 2009 the Obama administration scrapped that project.

Fuel discharged from reactors does not lose its radioactive potency for hundreds of thousands of years. Fukushima showed that when power and cooling-water supplies get disrupted, nuclear wastes stored in the form of spent-fuel rods are highly vulnerable to catching fire. The earthquake and tsunami at Fukushima cut off power and caused three reactor cores to melt, with the melting fuel releasing hydrogen gas that then exploded, throwing debris into the spent-fuel pools, which in turn released radioactive gases into the atmosphere. To allay concerns over extended storage of spent-fuel rods, nuclear wastes can be significantly reduced in size by reprocessing the spent fuel to extract reusable plutonium and uranium.

Whereas the appeal of nuclear power has declined considerably in the West, it has grown among the so-called nuclear newcomers, a development that throws up new challenges related to water, nuclear safety, and nuclear-weapons proliferation. Yet the global nuclear power industry—a powerful cartel of less than a dozen major state-owned or state-guided firms that had been trumpeting a global "nuclear renaissance"—sought to aggressively market reactors in water-stressed Asia and the Middle East in the decade before Fukushima, a period during which the share of nuclear energy in the world's total electric power supply remained in steady decline.[106] In fact, highlighting nuclear power's declining share and the rise of renewables, the worldwide aggregate installed capacity of just three renewable sources—wind power, solar power, and biomass—overtook installed nuclear-generating capacity by 2010.[107]

Yet the taxpayer-fattened nuclear power industry—while contributing more than $4.6 million to U.S. lawmakers during the 2000s, including then senator Barack Obama—ingeniously took advantage of growing international concerns over global warming to present nuclear energy as a "clean" alternative to carbon-producing fuels.[108] This industry has spent more than $10 million yearly since 2005 on broader U.S. lobbying.[109]

The Fukushima disaster, however, has reopened old safety concerns, stalling the nuclear power industry's momentum and casting doubt on nuclear power becoming a linchpin of efforts to slow down climate change. Just when nuclear power had ceased to be a hobgoblin to mainstream environmentalist groups, Fukushima dealt it a severe blow similar to the 1979 Three Mile Island plant accident in Pennsylvania and the 1986 Chernobyl meltdown in the now-independent Ukraine. The International Energy Agency has acknowledged that the nuclear power outlook has dimmed after Fukushima, with even France joining other countries that are reconsidering commercial nuclear plans.[110] But if the fallout from Chernobyl and Three Mile Island is a reliable guide, the advocates of nuclear power will eventually be back.

The use of helium, lead-bismuth, sodium, or fluoride salt as reactor coolant is being explored in the next generation of nuclear-energy technologies now under research and development by the thirteen-nation Generation IV International Forum. However, of all the important sources of electricity other than large-scale hydropower, nuclear power at present has the highest water and capital intensity and the longest plant-construction time frame (which, with licensing approval, averages a decade). Yet it enjoys the largest taxpayer subsidies, carries the greatest safety risks, and generates the most dangerous wastes whose safe disposal saddles future generations. More important, without a breakthrough in fusion energy or greater commercial advances in breeder (and thorium) reactors, nuclear power is in no position to lead the world out of the fossil-fuel age.

Addressing Water-Related Energy Challenges

At a time when national economies crave additional water and energy supplies to sustain growth, freshwater shortages are accentuating the energy challenges, and vice versa. As water becomes scarcer, its energy intensity amplifies, as more energy is needed to pump groundwater up, transport surface water across long distances, chemically filter water, and tap nontraditional sources of supply. Water shortages, however, act as a bottleneck for generating more power. Economies need water for growth, but growth is not possible without power, underscoring the importance of carefully calibrated trade-offs between water needs and electricity needs.

The water/energy nexus, of course, holds implications beyond power-generation challenges. Higher energy prices, for example, have an impact on food prices. Energy, according to one report, makes up 15 to 30 percent of the cost of crop production, 25 to 40 percent of the cost of steel, about 70 percent of the cost of groundwater, and 50 to 75 percent of the cost of desalination.[111] A business-as-usual approach toward the water/energy nexus thus threatens to impose long-term socioeconomic costs.

Constraints on water availability are increasingly shaping decisions about energy facilities, cooling technologies, and plant sites. In several regions, water paucity has already turned into a key limiting factor on a major expansion of the energy infrastructure. And in some other regions, freshwater availability is set to come under pressure as more plants are built to meet the fast-rising energy demand. The conflict between water and energy needs is likely to worsen in regions where energy demand is increasing rapidly but water resources are already constrained. Water stress, by impeding further expansion of thermoelectric generating capacity, threatens to foster major electricity-supply shortages and thereby trigger knock-on economic effects.

Global energy consumption has been increasing much faster than the population growth rate, a trend set to further accelerate. After all, per capita energy consumption in the developing world remains very low by Western standards. Indeed, almost 1.3 billion people, mainly in Africa and Asia, still lack access to electricity, while 2.7 billion continue to depend on traditional biomass like firewood and dung for cooking food.[112] Add to the picture the intensifying geopolitical competition over hydrocarbon resources, especially along the ancient Silk Road stretching from East Asia to southern Europe, through energy-resource-rich Central Asia, the Caspian Sea Basin, and the Middle East.

Global water use for energy production would need to more than double over the next two decades to meet both increasing electricity demand and national targets for production of biofuels and other alternative fuels. The International Energy Agency has projected that the water requirements for energy production will indeed grow at twice the pace of energy demand between 2012 and 2035.[113] Water shortages, however, are already accentuating the energy dilemma. Depletion of watercourses is necessitating the energy-intensive pumping and transport of water from greater depths and distances, while deterioration of water quality is dictating increased reliance on energy-intensive treatment and purification processes. Even efforts to save water use in agriculture demand greater energy consumption, such as by shifting from gravity-driven irrigation to drip-feed irrigation (which directs water flow to the root zone of plants) and computerized fertigation (the technique of supplying dissolved fertilizers to crops through drip irrigation).

Until technological innovations help decrease the water intensity of the energy sector and offer environmentally sustainable options, there is little alternative to better management of the energy/water interrelationship, including by choosing lower-water-intensity designs in energy infrastructure expansion. Given that operational water withdrawal and consumption factors in thermoelectric generation vary significantly across fuel and cooling technologies, the greater use of natural gas over coal or uranium, for example, will bring significant savings in water consumed per megawatt-hour of electricity produced.[114] Combined cycle gas turbine (CCGT) facilities, in comparison with nuclear and coal-fired power plants, are exceptionally economical in water withdrawal and consumption.

If alternative cooling systems in the coming years cease to impose either a cost or efficiency penalty, the relative share of water consumed by the electric power sector would shrink even as more power plants come online. The widespread use of dry or hybrid cooling, or the advent of more water-efficient wet cooling, will yield enormous water savings. At present, however, governments and companies face difficult choices on cooling technologies, which determine the amount of water withdrawn and consumed per unit of electricity output. A shift from once-through cooling to closed-loop cooling, for example, entails much lower water withdrawals but greater evaporative losses.

The water intensity of energy technologies can be moderated by supplementing freshwater withdrawals with seawater, brackish coastal water, saline groundwater, or degraded surface water. Minimal freshwater quality is important for growing the corn or soy feedstock for biofuels but not for thermoelectric cooling.

The power sector's impact on freshwater availability must be reduced by utilizing non-freshwater sources for cooling. Where it is possible, the larger use of seawater, or impaired groundwater or surface water, in energy extraction, processing, and production will bring important freshwater savings. Saline-water use in the U.S. energy sector already accounts for 30 percent of total water use, and in California nearly all the water used for thermoelectric cooling is saline.[115] Saline groundwater underlies several parts of the United States.

Most power plants in the world drawing seawater employ the once-through cooling process because of their access to an essentially limitless resource. The wider use of seawater as coolant in the thermoelectric power sector, however, is constrained by difficulties in finding coastal sites for new power plants. Such difficulties, already apparent in the United States, are particularly acute in China, India, Japan, South Korea, Vietnam, and other Asian states, where coastal areas are densely populated and serve as economic-boom zones.

Wastewater recycled after cleansing treatment offers another important source of supply to reduce the energy sector's freshwater withdrawals. Such reclaimed water must be used for resource extraction, energy conversion and transportation, and power generation. Two-fifths of all water used in Arizona in thermoelectric power production is reclaimed water.[116] In fact, the largest U.S. nuclear power plant, Palo Verde in the Arizona desert—the world's only nuclear-power-generating facility not situated next to a large body of water—uses water recycled from sewage from Phoenix, Scottsdale, Glendale, and other surrounding cities. An important drawback of recycled water, however, is that the cleansing treatment is highly energy intensive.

To save water, it is essential to save energy. For example, improved efficiency standards for lighting, heating, appliances, and transportation systems will free considerable freshwater resources for uses other than energy production. Reduced water consumption, for its part, will save energy used in water treatment and supply. Combined water and energy conservation is thus critical for relieving pressure on water resources. This calls for greater state support and tax incentives for the development of commercially viable technologies that slash water use in energy production and cut energy consumption by different sectors.

More fundamentally, energy and water must be managed not separately but jointly so as to promote synergistic approaches and technologies and to ensure the long-term supply reliability and sustainability of both resources. Absence of integrated planning and management can cause adverse impacts on basins and energy development. But smartly integrated water and energy planning can help identify and manage the water impacts of the various types of energy facilities and their vulnerabilities to water-resource changes. Indeed, biogas from wastewater-recycling plants can be channeled to generate power, while the waste heat from electric power stations has potential applications in desalination.

It is important that economies, in seeking to address the energy crunch, do not exacerbate challenges related to water, a more critical resource. Through integrated planning, they must promote access to alternative water resources.

Power plants should be located so that they are able to employ local brackish or degraded water, or reclaimed water, including from oil, gas, or coal-bed natural-gas production. Industries, to the extent it is technically feasible, must be made to recycle their wastewater. For example, at a time when the shale oil and gas industries are expanding rapidly, controlling the contamination risk they pose to local water resources demands increased treatment and reuse of flowback water containing low total dissolved solids, or TDS.[117]

To keep freshwater-supply stress in check, making water-smart energy choices is crucial, including selecting the right plant types and cooling technologies at the planning stage. After all, power plants have a long lifespan, and retrofitting to low-water cooling can be expensive. The appeal of non-thermal renewable-energy technologies that demand little water and cause no evaporative losses, such as solar photovoltaic and wind power, is certain to grow, especially in water-constrained regions, even though such sources of energy may offer intermittent power output. The only credible alternative to greater water stress and mounting energy costs is significant improvements in energy and water efficiency and strong conservation measures.

CAN "VIRTUAL WATER" TRADE BE THE ANSWER?

Direct bulk-water trade between water-rich and water-deficient countries is seriously hindered by distance and cost. As a local product, freshwater is expensive to ship overseas. Given that water is one resource that cannot be directly secured in sufficient quantities through long-distance international trade deals, attention has focused on the indirect trade role of "virtual water," or freshwater used in production processes and thus embedded in traded goods and services. Virtual-water trade is particularly important in agriculture, helping to keep much of the world fed and moving.[118]

For an understanding of the sustainability and opportunity costs, the utility of this concept in agricultural trade can be enhanced by identifying the source of virtual water—precipitation (green water) or irrigation water (blue water). Globally, green water dominates virtual-water crop exports, with blue water (including nonrenewable fossil water) making up only 16 percent of such exports.[119] Several of the leading wheat or corn sellers such as Canada, Russia, the United States, the European Union, Argentina, Australia, and Ukraine practice largely rainfed agriculture and thus export mainly green virtual water. By contrast, major rice exporters, such as the extensively irrigated Thailand, Vietnam, Pakistan, India, and China, mainly trade blue virtual water.

Unlike blue water, which can be moved and used for industrial and household purposes, green water generally has to be used in situ for agricultural purposes. This means that the export of green virtual water represents an efficient harnessing of precipitation. And nonexport may actually bring little water savings because natural vegetation instead of crops on the same land would consume similar or higher amounts of green water. But as one study has highlighted, most countries consume more green and blue waters on their own territories than from abroad and therefore nations with high levels of per capita water consumption mainly upset the water-resource situation within their own frontiers.[120]

The important point is that the exporters of green virtual water control the dominant share of the global food trade. The major food-exporting countries also boast much higher water-productivity levels than the major importing countries. This means that much of the virtual water that flows internationally originates in countries that have achieved high water productivity through improved farm varieties, mechanized technologies, and heavy inputs of chemical fertilizers and pesticides and by taking advantage of more favorable climate and soil. The international rice trade, however, illustrates an exception to this pattern.

Export of goods from countries with high water productivity to nations with low water productivity means net global water savings because the same products otherwise would have to be produced with greater water withdrawals. Higher water productivity actually translates into lower virtual-water content in a product, just as low water productivity increases the virtual-water content.

Virtual water, which is economically invisible, has been largely explored in relation to crop and livestock production and trade because agriculture accounts for more than two-thirds of global water withdrawals. Most studies indeed have focused on the cereal or meat trade.[121] Yet the concept must also extend to industrial products and services for a holistic analysis.[122] Only by understanding the amount of water embedded in manufactured goods, extending from cars to clothes, can the true water intensity of any economy be gauged. According to United Nations estimates, it takes about 400,000 liters to build a car and 11,000 liters to produce a pair of jeans.[123]

Conceptually, water-scarce countries are likely to find it attractive to import rather than domestically produce water-intensive products. A high import dependency, however, could make them vulnerable to international political and financial pressures as well as to long-term risks. Such imports could also act as a disincentive to addressing continued water profligacy at home.

In this context, three fundamental but oft-ignored questions arise. One, can any water-scarce economy's water needs for growth be addressed through virtual-water trade? Two, to what extent does such trade currently help alleviate water-stressed conditions in the world? And three, considering that the majority of the world population lives in water distress, do water-rich countries have enough renewable water resources to help alleviate water stress elsewhere through virtual-water trade with water-deficient states?

These questions, in turn, demand examination of various countries' water footprints, or the total freshwater resources used to produce the goods and services for a country's population. The costs and benefits of prevailing patterns of water use, however, can be understood only by assessing both the true value of how much water is being utilized to produce any product or service and the net benefits accruing from such usage. Moreover, developing a long-term national water policy requires a better understanding of not just an economy's water footprint but also its ecological footprint so as to factor in the broader ecosystem implications of production and consumption activities as well as the essential support provided by water flows to multiple ecosystem services that sustain social and ecological systems.[124]

A Nascent, Evolving Field

Few can fault the logic that in a world increasingly tied by economic interdependence, water-constrained nations must open the path to greater water efficiency and productivity by importing commodities with high virtual-water content and exporting those with low virtual-water content. Yet even the proponents of greater reliance on virtual-water trade acknowledge that "few governments explicitly consider options to save water through import of water-intensive products" or, in the case of water-rich nations, "to make use of relative water abundance to produce water-intensive commodities for export."[125]

Water demand/supply issues are looked at by governments strictly within a national territorial perspective, whether considering supply-side approaches to augment availability or demand-side options to emphasize water conservation and quality. No effort is made by governments to quantify exports and imports of virtual water in order to develop realistic data on national water consumption.

In truth, the study of national water footprints and virtual-water trade is still nascent and evolving, with a wide divergence visible in the scope, conceptual basis, methodology, hydrological models, assumptions, and sources of data.[126] As two scholars have pointed out, "The methodologies and databases of the studies are often crude, affecting the robustness and reliability of the results."[127]

Water is just one of several key factors in determining optimal production and trading strategies. By focusing solely on water-resource endowment, the notion of virtual water, according to another scholar, can hardly serve as "a valid policy prescriptive tool."[128] Because it is not easy to quantify all the factors involved—which range from political and financial to developmental and environmental—there has been little analysis of the potential trade-offs a water-stressed state will have to make to implement a strategy to rely increasingly on imports of virtual water.

In fact, actual water consumed in the production of goods and services varies considerably between and within countries due to a host of factors, including climate, technology, agronomic practices, and field management. Advantageous climatic, soil, and management conditions, for example, can lower virtual water content in a crop product. Although high water productivity usually translates into low virtual-water content in products, water productivity itself is variable in space and time. There are disagreements among scholars even on the methodology to evaluate the virtual-water content of crop products and on the real value of the embedded water.

In this light, employing aggregate ratios or narrowly based simulation models can scarcely provide an accurate picture. Quantification of the water embedded in international trade flows thus constitutes, at best, rough estimates. This is underscored by considerable variation in different studies on the probable virtual-water content in such flows. The direct policy relevance of such estimates thus remains a moot question.

Indeed, virtual-water trade, as a term, has come to be used so loosely as to become synonymous with international food trade.[129] Proffered estimates of national water savings resulting from food imports should be treated as little more than ballpark indicators open to further debate and refinement.[130] In fact, few scholarly attempts have been made to assess whether the blue water left unutilized due to food imports represents a real saving in the sense that it was put to more beneficial uses.[131]

Still, it is true that the virtual-water and water-footprint literature has helped improve international understanding of the link between water and food security and the greater role that food trade can potentially play in addressing water distress. The literature can also help encourage governments to more prudently utilize scarce resources and improve water-use efficiency by correcting underlying market and policy failures.

Striking Realities

Logically, water-rich states should serve as virtual-water suppliers to water-poor nations. International trade, however, does not represent a straightforward flow of virtual water from states well endowed with water resources to nations in water distress. One indicator that global virtual-water flows are not aligned with national water conditions is the fact that barely a quarter of the international cereal exports are from water-rich to water-scarce nations.[132]

In fact, a study that examined the impact of full liberalization of global agricultural trade on virtual-water flows in the future, including potential shifts in meat and cereal trade, found that virtual-water flows are "independent of water-resource endowments."[133] This may explain why countries with fairly similar-sized water resources have very different virtual-water trade patterns.

The patterns of virtual-water trade are actually complex and even perplexing. For example, some of the world's most water-scarce nations, ranging from South Africa to Pakistan, have been growing food for export on such a scale as to rank among the important exporters in the world. The world's three largest gross virtual-water exporters—the United States, China, and India—are partially or wholly under water stress.[134] Israel, which is two-thirds arid, is an important food and cotton exporter despite a very low freshwater availability of 252.4 cubic meters per person per year, which is well below the global water-poverty line. It is, however, a net beneficiary of virtual-water flows because it imports more water-intensive agricultural products than it exports; yet it earns significant trade income from exporting high-value crops of fruits, flowers, and vegetables.

Even when various countries' imports of virtual water are subtracted from their exports, some of the world's biggest net virtual-water exporters are water-constrained economies. In fact, some of the world's largest cotton exporters, including Uzbekistan, Turkmenistan, Pakistan, Egypt, Kazakhstan, Tanzania, and Syria, are water-stressed states, underlining the fact that buyers elsewhere are at least partly to blame for the water scarcity and other adverse environmental impacts in such exporting countries.[135] The water-surplus European Union is a net virtual-water importer. Although a major food exporter to the Middle East and North Africa, it is, for example, the world's fourth-largest importer of the thirstiest food product, rice.[136]

No less significant is the fact that the world's top five blue virtual-water exporters—the United States, India, Thailand, China, and Pakistan[137]—are water stressed at least in part. Because extensive or intensive irrigation often leads to salinization and waterlogging problems as well as to the overexploitation of surface and subterranean waters, these countries' blue virtual-water exports do not reflect the true resource and environmental costs of the traded products. Continued reliance on food exports by a water-distressed country as its main foreign-exchange earner can only compound its water challenges, as illustrated by the case of Pakistan, where per capita water availability has fallen sharply owing to its export-oriented agriculture and explosive population growth.

The attraction of earning valuable foreign exchange by exporting virtual-water-rich products tends to trump better management of water resources in some water-deficient nations. When water is supplied free or at heavily subsidized prices to farmers, as still happens in a number of water-scarce countries, the real value of the water embedded in food exports goes unappreciated. State subsidies on water, seeds, and fertilizers in developing economies are designed to keep domestic food prices affordable for the poor and the low middle class, yet in several cases they end up financing consumption in richer countries. Food exports by water-stressed nations will become uneconomical only if the true value is attached to the water utilized to produce those commodities.

To be sure, virtual-water exports of agricultural products—estimated to total 1,597 cubic kilometers yearly between 1996 and 2005, or nineteen times the average annual yield of the mighty Nile River, the world's longest—play a conflict-mitigating role.[138] Countries that lack water resources to grow enough food—from Kuwait and Djibouti to Libya and Malta—rely on virtual-water imports.[139] Virtual-water flows thus help keep people in water-scarce nations fed. But for this virtual-water trade, there probably would have been more conflicts in modern history. In the Arab world, subsidized resources, including virtual-water imports in the form of food, free or cheap water supply, and state-supported fuel prices, are integral to the rulers' political legitimacy. Yet, as the Arab revolts since 2011 have shown, inflation can undermine subsidies and the social contract, spurring popular unrest.

There are also a number of other countries that significantly, even if inadvertently, benefit from the virtual-water trade. Take Japan, which, despite state incentives designed to increase domestic cereal production, has witnessed a rise in grain imports. Its population may be aging and declining, yet a growing popular preference for meat, wheat-based instant noodles, and bakery goods has made Japan the world's top corn importer by far and a leading wheat and meat importer.

According to one study, Japan's pig and broiler meat imports alone make it a huge beneficiary of virtual water, with these imports embodying the virtual equivalent of 50 percent of the country's total arable land and releasing it from the imports' true production-related environmental costs.[140] Without its price-distorting policies that protect domestic production, Japan's agricultural imports would be even higher. Still, with a per capita freshwater availability of nearly half the global average of 6,380 cubic meters, Japan clearly gains from the fact that 64 percent of its total water footprint falls outside its frontiers.[141] Similarly, Egypt's wheat imports save it 3.6 billion cubic meters of water annually, according to one estimate.[142]

However, the virtual-water crop trade, far from being entirely market driven, is influenced by interstate political relationships, with importing countries careful not to rely for food on those they do not trust. There are plenty of examples of countries, even when faced with a severe drought, opting to import grains, sugar, oilseeds, and other food from distant states rather than from an available next-door supplier that they distrust or view as a rival.

Stable, cooperative ties are essential to forge an enduring food-trade partnership between two countries, while growing strains in an already-fraught relationship are very likely to obstruct virtual-water flows between them.

The level of food and energy consumption shapes national water footprints more than any other factor. For example, North America and Western Europe have much larger water footprints than the world's most populous nations, China and India, in keeping with the fact that wealthy nations are much bigger per capita consumers of meat and energy than developing states.

Admittedly, there are considerable national variations within both developed and developing worlds in water consumption. For example, America's annual per capita water footprint (which, as opposed to personal water use, is the average amount of water needed per head for the production of goods and services for national consumption) is the highest in the world. It is estimated at 2,842 cubic meters—greater than the holding capacity of an Olympic-size swimming pool—whereas Japan's footprint, at 1,379 cubic meters, is less than half of that.[143] A similar pattern, interestingly, prevails in their carbon footprint: the United States, with the world's highest per capita greenhouse-gas emissions, belches out twice as much carbon dioxide per head as Japan, although the two countries have fairly similar per capita incomes. Americans actually use water even more profligately than oil, in spite of the fact that thirty-six of the fifty U.S. states confront water shortages: per capita daily water use exceeds yearly oil consumption.[144]

Many developing countries' per capita water footprint is smaller than the global yearly average of 1,385 cubic meters. They include Bangladesh (769 cubic meters), Chile (1,155 cubic meters), China (1,071 cubic meters), the Democratic Republic of Congo (552 cubic meters), India (1,089 cubic meters), Indonesia (1,124 cubic meters), Nigeria (1,242 cubic meters), South Africa (931 cubic meters), and Vietnam (1,324 cubic meters). However, some—like Mexico (1,978 cubic meters) and Kazakhstan (2,376 cubic meters)—have a bigger footprint than the world average.[145]

The larger a water-stressed country's water footprint, the greater the challenge it likely faces to ameliorate its water situation through virtual-water imports. But if a water-deficient country has a small water footprint and yet lacks adequate financial resources—as is often the case in impoverished sub-Saharan Africa—it will not be possible for it to externally outsource water-intensive food and industrial production for its needs.

Many nations—usually not on purpose—have externalized significant portions of their water footprints as part of normal international trade in finished products and services. For an accurate comparative picture, countries' actual water consumption must reflect both internal and external footprints, just as national carbon footprints cannot be reliably measured without examining the trade elements, especially at a time when low-cost, carbon-intensive manufacturing has been shifting to the developing world while the advanced industrial economies are concentrating on high-value-added production. The true water footprint of a country is the sum of its national water withdrawals, plus its gross virtual-water imports, but minus its gross virtual-water exports.

Significantly, the extent to which a water footprint is externalized often bears little relation to the size of national water endowments. For example, water-rich Switzerland has externalized 79 percent of its water footprint, but water-stressed India barely 2 percent; similarly 19 percent of America's total water footprint falls outside its frontiers but only 7 percent of China's.[146]

The highest virtual-water import dependency is that of Kuwait and Malta at 87 percent each, followed by the Netherlands at 82 percent and Bahrain, Belgium, and Luxembourg at 80 percent. In contrast, Burma, the Central African Republic, Chad, Congo, Ethiopia, Haiti, Kazakhstan, Malawi, Mali, Sudan, and Zimbabwe have the lowest external water footprint, at just 1 percent. In the 2-percent external footprint category, besides India, is another large country, Nigeria, and also small states like Burkina Faso and Burundi.[147] In the case of countries with large external water footprints, such as Belgium, the Netherlands, and Japan, their consumption patterns have an important bearing on water withdrawals and pollution elsewhere.

These realities, if anything, underscore the restricted utility of the notions of virtual water, water footprint, and green/blue water as policy prescriptive tools by themselves. Even as these terms have brought much-needed attention to pressing water-related challenges, "none is a sufficient criterion for determining optimal policy decisions."[148] They do, however, help to highlight valuable aspects about international trade.

The Limits of Virtual-Water Trade Benefits

The idea of using virtual-water trade to ease national or substate water shortages is theoretically alluring. Water-distressed economies can mitigate their water stress, for example, by gradually slashing domestic production of water-intensive commodity crops to rely on greater imports. The water savings that accrue would be available for alternative, high-productivity uses. In practice, however, the appeal of virtual-water trade is blunted by several political, social, and economic factors. In fact, few states view it—or have consciously employed it—as a sustainable path to easing their water insecurity or reducing pressures on the natural environment.

Politically, given the power of concepts such as "food sovereignty" and "food security," a greater reliance on imports is a sensitive matter in transitioning and developing economies. In fact, as modern world history attests, rapid economic advancement generally has followed major advances in food and energy production and building resource security.

Few advanced industrial economies depend on other states to feed their populations; many of them, on the contrary, are important food exporters. Food aid indeed has been employed as an instrument to advance foreign-policy interests. A 1974 U.S. report prepared under the leadership of National Security Advisor Henry Kissinger even advocated that U.S. food aid "take account of what steps a country is taking in population control."[149] In more recent times, Western food aid has been used as leverage against the Stalinist regime in North Korea.

Food as a weapon can be as potent as the use of guns and bombs. This may explain why many countries attach strategic importance to food security, often equating that goal, imprudently, with food sovereignty. Countries are loath to lose their ability to substantially feed their own populations, or else they would be at the mercy of global commodity markets and other governments.[150] So, even though virtual-water imports conceptually make sense for any water-stressed economy, states, except when they confront a major drought, will likely move slowly—with a great deal of circumspection—toward relying more on imports to feed their populations.

Indeed, such is the reluctance to depend heavily on international markets for meeting domestic food requirements that Saudi Arabia—one of the world's poorest nations in water resources—embarked on the path to becoming self-sufficient in food by sponsoring an irrigated farming boom on its desert lands. The large-scale cultivation, drawing on the waters of its largest fossil aquifer, began in response to fears that the 1973–1974 Saudi-led Arab oil embargo—which amounted to the use of oil exports as a weapon—would provoke a retaliatory Western ban on food exports to Saudi Arabia. By the early 1990s, the riverless Saudi Arabia was exporting $500 million worth of wheat, dairy products, eggs, fish, poultry, vegetables, and flowers to other countries annually, thanks to an agricultural expansion propped up with heavy state subsidies, including artificial procurement prices.[151]

In other words, a country whose annual freshwater availability is abysmally less than 90 cubic meters per person became an important exporter of virtual water. It was only after the farming boom had depleted much of its fossil-aquifer resources, threatening to leave it waterless, that the absolute monarchy in this cash-rich, jihad-bankrolling state grudgingly decided in 2008 to phase out wheat cultivation by 2016. Yet Saudi Arabia plans to continue growing some other foods.

Socially, shifting food production to distant lands through a greater reliance on virtual-water imports is likely to carry a direct impact on the country's rural masses. It would promote rural unemployment and discontent and spur accelerated migration from the countryside to cities by undercutting the very sector that holds the key to poverty alleviation—agriculture. More than the overall growth of a country's gross domestic product, it is agricultural growth that holds the key to reducing rural poverty, according to studies by the United Nations Development Programme.[152]

Economically, one uncomfortable reality is that world food markets presently are not large enough to allow water-stressed nations to substantially shift to virtual-water imports. If just China, with its overflowing foreign-exchange coffers, were to become a major food importer, it would destabilize global food markets and reduce the supplies available for low-income countries, particularly in Africa but even for Asian states like Bangladesh and the Maldives. Rising international food prices also underscore supply constraints and the risks associated with greater reliance on food imports.

A more fundamental economic reality that water-constrained nations face is the centrality of water even in nonagricultural production and services. Many industrial processes are water intensive, with the indirect ("embodied") water use by industrial sectors usually much higher than the direct use. A study of all 428 sectors of the U.S. economy, for example, found that while agriculture and electric power generation account for the bulk of all direct water withdrawals, most sectors use more water indirectly in their supply chains through processing than by direct means, with indirect water use actually making up three-fifths of total water abstractions.[153] The water intensity of any sector thus cannot be measured merely in terms of direct water withdrawals.

Put simply, water scarcity and rapid economic advance cannot go hand in hand. If a water-stressed country were to rely on virtual-water imports to meet most of its needs for water-intensive products—ranging from manufactured goods to food products—it would sacrifice its own all-round economic development. Also, virtual-water trade can help ease shortages of food but not of electricity, whose production places significant demand on water resources.

Dependency on outsiders cannot be equated with comprehensive national development. When a country lacks adequate water resources to grow all its food, virtual-water imports, of course, become inevitable. If such imports are made as part of a farsighted water-management strategy aimed at reducing domestic production of low-value crops and tapping nonconventional water sources, they can help alleviate water distress to an important extent.

Had water-deficient states viewed a greater reliance on virtual-water imports as a politically sustainable and stable option, those among them that have made a beeline to Africa would have found little incentive to try to secure leases of underdeveloped farmland there to grow food for their home markets. The farmland leases indeed spotlight the determination of such countries to control their food production, even if on overseas lands, so as to cushion themselves from the political and financial vagaries of international food markets.

Just like the purchase of oil and gas fields in distant lands by some emerging powers is designed to lock up supplies, cropland leases in other countries represent an attempt to build assured, uninterrupted food supplies by effectively appropriating others' freshwater and land resources in order to overcome resource constraints at home. For example, Saudi Arabia, while deciding to phase out irrigated wheat cultivation at home, unveiled its ambitious King Abdullah Initiative for Agricultural Investment Overseas, under which the government has been providing billions of dollars of low-interest credit to private Saudi firms to farm land overseas. Saudi companies have initiated agricultural arrangements not just in Africa, but also in Asian nations like Pakistan, Cambodia, Vietnam, and the Philippines, where the aromatic and long-grain basmati rice is being grown for the Saudi market. Arab financial institutions, including the Jeddah-based Islamic Development Bank, have supported such cropland leases by other oil sheikhdoms as well.

Few states have calculatedly employed virtual-water trade as a policy tool to improve their water conditions—unless they have landed themselves in dire straits, like Saudi Arabia, with no other option. Some water-scarce nations indeed are locked in an odd situation from which they need to break out: instead of relying on imports of water-intensive food products, they are major exporters of virtual water in the form of food. As their capacity to increasingly grow water-intensive crops for export with irrigation subsidies falters in the face of intensifying water stress, a fundamental restructuring of such economies will become unavoidable. Their current production of crops for export with high water content, such as citrus fruit, or involving water-intensive cultivation, such as rice and cotton, is already at the expense of domestic socioeconomic needs or ecosystem services.

Still, virtual-water trade, even if it does not open the path to water security, offers any water-stressed state the means—however limited—to reap water savings and thereby ameliorate water-use efficiency and lessen pressures on the natural environment. Importing large quantities of cereals and meats from water-rich states allows water-constrained nations to channel greater water resources for other socioeconomic needs. Intrastate virtual-water trade, for its part, can help mitigate the spatial and seasonal variability of water resources within any nation.

Water-intensive crop cultivation and manufacturing should be concentrated in areas within a nation where water supplies are abundant, while water-scarce provinces ought to produce value-added, water-efficient products. Yet China, instead of growing most of its food in its water-rich south, has turned the semiarid northern plains into its breadbasket and is now completing the Great

South–North Water Diversion Project to transfer waters from the south to the north. The Huai, Yellow, and Hai-Luan river basins in the north have just 7 percent of China's total water resources yet produce 67 percent of the country's wheat, 44 percent of its corn, and 72 percent of its millet. Rather than correcting this paradox, authorities have designed the South–North Water Diversion Project to promote the environmentally unsound transfer of huge volumes of river waters from the south to the center of economic activity in the north.

This controversial, $62-billion project has drawn inspiration from America's diversion of 9.3 billion cubic meters of water annually from the lower Colorado River to California, Arizona, and New Mexico. But it is on a much larger scale: its Eastern and Central routes have been designed to jointly take 44.8 billion cubic meters of Yangtze Basin waters per year to the north, while the third and final leg, centered on the Tibetan Plateau, is intended to transfer another 200 billion cubic meters of river waters. The 1,156-kilometer, already-operational Eastern Route—the shortest and least challenging of the three legs—exemplifies how such transfers can prove expensive and cause water quality to deteriorate, including in the source basin.[154]

Globally, less water is likely to be available for food production in the years ahead, even though there will be many more mouths to feed and higher per capita consumption levels to meet. Many developing countries are set to become more dependent on grain imports. But if all water-deficient states were to use trade as a policy instrument to import products with high virtual-water content while perpetuating the mismanagement of their own water resources, the world would witness skyrocketing prices of water-intensive products.

The big question is whether the existing major food exporters will be able to produce the grain and meat surpluses to meet the greater international demand in the coming years. Generating larger surpluses for export will entail larger resource and environmental costs for the producing countries. Indeed, given rising international food prices, the poorest countries with the most acute food insecurity may be hard put to pay cash up front for imports, especially when they will have to compete with Asia's cash-laden emerging economies, which could emerge as important food importers.

On the positive side, the traditional grain exporters, such as North America, Argentina, and Australia, have been joined by a major new exporter, the European Union. The EU had been a net importer of 21 million tons of cereals a year in the mid-1970s, when it was known as the European Economic Community, and became a net exporter from 1997 onward. The EU now has emerged as the world's second largest wheat exporter, closely behind the United States. Ukraine and Russia have also become important suppliers. Russia, given its immense land and water wealth, has the potential to significantly step up its grain exports, including by raising yields.

However, the global cereal-import demand, as highlighted by spiraling prices, has grown faster than available export surpluses. Wheat demand in major importing nations like Egypt, Brazil, and Indonesia has continued to increase sharply. And in the United States—the world's leading food exporter—several factors, including water shortages and less profitability relative to other crops, suggest that U.S. wheat acreage will continue to decline.[155]

The world market for rice—which unlike wheat and corn is grown mostly for direct human consumption—is particularly tight, with much of the world's rice consumed in the same countries where it is produced. Barely 7 percent of the world's annual rice production is exported, as compared with more than 20 percent of the global wheat harvest.[156]

Yet the list of major importers seeking greater rice surpluses from the producing countries extends from the Philippines and Nigeria to the European Union and the Iran-Iraq belt (table 2.3). Several of the world's largest rice exporters, including the top three—Thailand, Vietnam, and Pakistan—are coming under greater water stress, at least in parts of their national territory, and, as their water woes intensify, they may find it difficult to sustain such virtual-water exports at the current rates of increase.

Paddy rice cultivation consumes large quantities of water. Rice fields extending from Sichuan to Texas remain waterlogged for up to four months, especially to produce the most fragrant, better-tasting rice varieties that are in the greatest demand in export markets. It is thus scarcely a surprise that the leading rice exporters are among the world's leading virtual-water exporters: Thailand's rice and other agricultural exports, for example, carry 47 billion cubic meters of embedded water annually.[157] Pakistan—the world's third-largest rice exporter and fourth-largest cotton producer—is sinking into water-related despair.

Table 2.3. World's Top Ten Rice Exporting and Importing Countries

Top Exporters	Tons (×1,000)	Top Importers	Tons (×1,000)
Thailand	9,047	Philippines	2,400
Vietnam	6,734	Nigeria	2,000
Pakistan	4,000	Indonesia	1,150
USA	3,856	Iraq	1,140
India	2,050	Saudi Arabia	1,069
Cambodia	1,000	Malaysia	907
Uruguay	808	Cote d'Ivoire	840
China	619	South Africa	733
Egypt	570	Bangladesh	700
Argentina	468	Japan	700

Source: Data for 2010 drawn from the Food and Agriculture Organization of the United Nations and the U.S. Department of Agriculture.

Maintaining continued increases in rice output to meet soaring global demand is becoming a major challenge, especially with yield growth slowing. If more rice is to be grown with the existing share of water resources, a necessary yet difficult task is to break free from the ponded rice culture, even though the waterlogged rice fields provide crucial habitat for waterfowl and wild fish.

Water shortages indeed suggest that less water may be available in the coming years for paddy rice cultivation. In this light, new paddy rice farming techniques and new rice varieties of good quality and taste that save irrigation water are urgently required. Genetically engineered rice varieties thus far have found little public acceptance.

The international food trade, which has risen dramatically since the 1960s, is expected to play an even bigger role in the future in the virtual reallocation of waters. At the same time, water deficits that act as a fundamental constraint on food security are likely to haunt several parts of the developing world, and it is far from certain that international trade will be able to fill the food gaps or that the exportable surpluses will be affordable for low-income importers.[158] After all, two present realities are unlikely to change in the foreseeable future—the international trading of a relatively modest proportion of total global food production, and the low purchasing power of some water- and food-scarce countries in an era of rising international food prices. In any event, virtual-water trade must not undercut prudent water management and environmental sustainability in both exporting and importing nations.

Objective and quantifiable standards could be developed to identify the water footprints of tradable goods and services. Carbon labeling of consumer goods has already started in some parts of the world. Similarly, to help build public awareness of the amount of water employed in the production chain, water footprinting of consumer goods could begin once objective measurement criteria evolve—a challenging task indeed. A report on water footprinting has advised companies to be ready to "engage outside their own fence line and traditional comfort zone to ensure the long-term viability" of a critical resource like water.[159] If governments and companies start to base their purchasing decisions on the water footprints of their suppliers, it would help catalyze improved water-efficiency standards.

Still, the utility of virtual-water trade as a policy instrument to balance national, regional, and global water supply/demand is likely to remain limited. Food- and water-insecure countries are expected to face tough choices. To raise their virtual-water purchasing power, such countries will have to accelerate their economic growth or invest a larger share of their gross domestic product in food and other imports. But as water shortages in these states worsen, virtual-water imports may become less effectual in addressing larger national needs, despite the continued critical importance of incoming food shipments.

Against this background, the virtual-water-import strategy of a water-distressed state is likely to yield better dividends if it is pursued in conjunction with other scarcity-mitigation measures, such as wastewater reclamation, improved water-use efficiency, and the channeling of irrigation water for high-value crops. Rather than being speciously pursued as a solution by itself, a virtual-water strategy ought to be a vital component of a country's integrated water resources management.

In such a holistic framework, reliance on virtual-water imports is likely to mesh well with other elements like watershed management, conservation, sustainable irrigation practices, and nontraditional supply-side measures. Even for exporting countries, virtual water must be part of integrated resource management so as to minimize resource and environmental costs.

Chapter Three

The Future of Water

Water, water everywhere, but not a drop to drink. This depiction in *The Rime of the Ancient Mariner*, published by English poet Samuel Taylor Coleridge in 1798, is about men on a ship running out of drinking water on a long voyage while surrounded by the waters of the ocean.[1] Too much water but too little to drink could become a reality in the coming world, given the rise in sea levels and the spreading freshwater crisis.

The Ancient Mariner scenario already haunts some seaside regions—freshwater scarcity against the backdrop of a shimmering ocean vista. Securing a clean, reliable water supply has become a pressing priority for many communities throughout the world. Yet such is the extent of the degradation of many watercourses that water is now seen as "blue gold," with businesses eager to commoditize freshwater in different ways.

Although water is more essential than anything other than breathable air, half of humanity lives in areas where the limits of sustainable water supply have already been breached, resulting in diminished flows and the decline of freshwater ecosystems and many fisheries. Other warning signs include rapidly falling water tables and rivers that peter out before reaching the sea, such as the Yellow River, known as "the Mother River of China" because it served as the cradle of the Chinese civilization. From the Aral Sea in Central Asia to Lake Chad in sub-Saharan Africa, even lakes have been shrinking at an alarming rate. Droughts, which are deceptive disasters because they knock down no buildings yet wreak high socioeconomic costs, have become more frequent. At least two-thirds of the world's population is likely to be living in water-stressed conditions by 2025.[2]

The rise of ocean levels threatens to not only inundate low-lying areas, but also to promote saltwater infiltration into arable lands along the coast. Overexploitation of coastal aquifers is also inviting seawater to intrude.

For example, saline water is jeopardizing the potability of supplies from the Potomac-Raritan-Magothy aquifer system, one of the primary sources of freshwater in the Coastal Plain of New Jersey, particularly in its heavily developed areas.

Groundwater depletion contributes to sea-level rise, although to a smaller extent than climate change; some of the waters extracted from aquifers ultimately end up in the oceans. One study estimates the contribution of groundwater depletion to such rise at 0.8 millimeters per year, or about a quarter of the total rise of ocean levels.[3] Water everywhere but not enough to drink is a plausible scenario in a world confronting resource depletion.

There is now better appreciation among governments, businesses, civil society, and international organizations of the scale and complexity of the world's water challenges and the long-lasting risks they carry for socioeconomic progress. Increasing water shortages and scarcity of other resources such as land, food, and energy are already contributing to intrastate conflicts and intercountry political tensions. Water shortages in several regions are set to turn into major scarcities, which could precipitate violent conflict, including civil strife and insurgencies, and spur refugee movements within and across national frontiers.

At the subnational level, the competition and conflict rooted in water despair hits the poorest people the hardest because they have little buffer against the socioeconomic impacts of scarcity. Acute water stress also erodes state capacity to ameliorate the conditions of the most vulnerable sections of society, thereby delivering a double blow to the poor. At the intercountry level, water distress often provokes silent resource wars over transnational rivers or aquifers, with the risk that rival appropriation moves could trigger diplomatic strong-arming or even armed conflict.

NEW CHALLENGES, NEW BEGINNINGS

An indication of what the future might hold was the chartering of ships by Spain's drought-seared Catalonia region in 2008 to bring freshwater from abroad to its principal port city, Barcelona. Over several months—at a cost of millions of euros monthly—the regional government chartered a fleet of six large tankers to import water from France, as well as from southern Spain, in order to ease the impact of the worst drought in Catalonia in decades. Compounding the drought-triggered water rationing in Barcelona was the fact that an expensive new pipeline to bring water to the city from the Ebro River Delta—one of the largest wetland areas in the western Mediterranean—only fueled a water dispute with the Spanish regions of Valencia and Murcia.

When the bills were added up, Catalonia's 2008 water imports proved so expensive—about three dollars a cubic meter—that regional authorities decided that, to meet future needs, desalinating local seawater would be a better and less costly option. As a result, new desalination plants were ordered. In contrast, the United States is paying Mexico 13.7 cents a cubic meter as part of a five-year agreement signed in late 2012, under which Mexico will surrender a total quantity of 153 million cubic meters of water from its share of the binational Colorado River in exchange for $21 million in U.S. infrastructure support.[4]

Water stress in Spain—which boasts Europe's greatest number of large dams (1,172)—has been accentuated by the hydrological impacts of anthropogenic land-cover changes and intensive irrigation. Important changes in the land cover in the mountainous catchment regions over the past six decades have contributed to decreased flows in El Jucar, El Segora, and other rivers.[5]

The situation has been aggravated by extensive water pollution and a sharp growth in irrigation-related water abstractions, with agriculture accounting for three-fifths of all water withdrawals in the country. The fact that Spain is Europe's most dammed country has only added to the anthropogenic impacts on water resources. The water crisis, in turn, has stoked water-sharing disputes between Spain's autonomous regions, as transfers to water-deficient areas have brought resources under pressure even in its relatively water-rich central provinces.

Another glimpse of the likely future was the drought that afflicted Australia from 1996 for more than a decade, extending in some areas up to 2012. The drought—the longest in more than a century—was paradoxically followed by serious flooding in 2010–2011. Australia's extremes of drought and flood, which seesaw with the cycles of the El Niño and La Niña events, have been compounded by natural variability in climate and the warming effect of greenhouse-gas emissions, raising the possibility of a "drier and warmer" future for its southeastern heartland.[6]

The record drought prompted the federal government to go from short-term contingencies like chartering tanker ships to unveiling a $13.8 billion, ten-year plan in 2008 to tap new sources of supply, including recycled and desalinated water. The plan also sought to buy back some irrigation-water rights from farmers and to upgrade and repair ailing rivers. "Water shortages are a serious threat to our economy and way of life," Climate Change Minister Penny Wong declared in 2008.

The water plan focused on the Darling and Murray rivers basin, which provides nearly three-quarters of the water consumed nationally and sustains Australia's food bowl. This southeastern basin, covering an area the size of Spain and France combined and boasting thousands of wetlands, bore the brunt of the long drought.

By producing two-fifths of Australia's farm output and almost four-fifths of its irrigated crops, the Murray-Darling Basin (which includes the Murrumbidgee River and covers much of New South Wales and Victoria, parts of Queensland and South Australia, and the entire National Capital Territory) generates 40 percent of the nation's agricultural income, including nearly $25 billion worth of food exports. But for the constant dredging, its overexploited rivers, which empty into the Murray, would silt up, and the Murray itself would run dry before reaching the sea (map 3.1). It was the imperative to combat salinity, especially in the vast lakes near the Murray mouth in the state of South Australia, that prompted the moves to buy back irrigation-water rights from farmers so that more water could flow in the river.

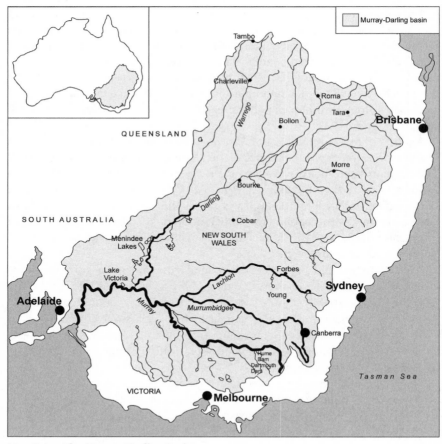

Map 3.1. The Murray-Darling Basin

Water squabbles between Australian states, especially those that share the Murray-Darling Basin, have grown louder. Despite an impressive per capita freshwater availability of 21,764 cubic meters annually—3.4 times larger than the global average and more than nine times greater than Spain's—Australia's heavy water withdrawals for agriculture (over 60 percent of the national total) have significantly contributed to resource depletion and degradation.[7]

Nine-tenths of Australia's irrigated area is in the three states of New South Wales, Victoria, and Queensland, where the irrigators' needs to grow cotton, rice, wheat, oilseeds, sugar, grapes, and other thirsty crops and to raise livestock, including pigs, poultry, goats, deer, ostriches, alpacas, and horses, have for more than a century taken precedence over environmental needs. Underscoring Australia's environmental challenges is the fact that its greenhouse-gas emissions virtually equal the combined emissions of France and Italy, which have six times more citizens than Australia.

If rich countries like Australia and Spain confront serious water challenges, the situation must be worse for many countries in Asia and Africa that have to make do with a much lower level of per capita freshwater availability. In the face of a relentless increase in water demand, a growing number of countries in the developing world are finding it difficult to mobilize sufficient water resources to meet the needs of their industry, agriculture, and municipalities. These nations are headed toward a grave water crisis that will have an important bearing on public health, economic growth, food security, migration flows, interriparian relations, and other components of national security. The most vulnerable regions stretch from the Maghreb and the Sahel to the North China Plain, covering the Middle East and the Iran–Central Asia–Indian subcontinent triangle.

This must not obscure the fact that more than half the club of rich countries, the Group of Eight (G8), has come under varying degrees of water stress. Britain, Germany, Italy, Japan, and the United States face water deficiency at least in some regions. The annual per capita freshwater availability in Germany is only 1,874 cubic meters, the lowest in the G8 and just 2.2 percent of Canada's.[8]

Even close allies in the developed world tend to suspect each other's intentions regarding freshwater resources. Canada is symbiotically tied to the United States, yet many Canadians have long believed that Americans covet their country's rich resources, especially its freshwater.[9] The majority of Canadian rivers flow north into the Hudson Bay and the Arctic Ocean.

Canada and the United States, however, share vast transnational water resources—best symbolized by the Great Lakes-St. Lawrence system—that are governed by robust institutional arrangements under a landmark 1909 water treaty. Bottled water is covered under the North American Free Trade

Agreement (NAFTA). Yet Canada has been reluctant to accept free trade in freshwater, rebuffing U.S. business proposals for bulk-water trade through pipelines or supertankers. Such proposals have only prompted calls in Canada for an outright ban on large-scale water exports.[10] Federal officials indeed have successfully leaned on provincial governments in the past to reverse the grant of licenses to export bulk shipments of water, prompting one American company, Sun Belt International, to file an arbitration claim of up to $10.5 billion against Ottawa under NAFTA's Chapter 11.

Canada is exceptionally well endowed with water resources, boasting one of the world's highest per capita freshwater availabilities and 6.85 percent of the total water resources on the planet. Yet, on environmental and political grounds, it contends that it has little water to spare. It thus opposes moves to make bulk freshwater a freely tradable commodity under NAFTA laws.

Although freshwater in Canada comes under provincial, not federal, jurisdiction, the Canadian Parliament in 1999 unanimously approved legislation aimed at enforcing a national moratorium on bulk-water removals from major basins, whether for export or otherwise. Some Canadian provinces, including Alberta, British Columbia, and Ontario, have enacted their own bans on bulk-water exports. Canadian opposition to bulk-water exports was actually stoked by U.S. proposals in past decades for colossal interriver and interbasin water transfers, including a stillborn western-corridor plan named the North American Water and Power Alliance to tap Alaskan water and reroute river flows from British Columbia to California, the Southwest, and the Great Plains.

A 2001 comment by U.S. president George W. Bush that he was "open" to discussing the possibility of bulk-water imports from Canada struck a raw nerve there, with newspapers accusing him of firing the first shot in a U.S.-Canadian diplomatic war over water. "George W. Bush is at the Door, and He Wants Canada's Water," screamed the *Globe and Mail*, a leading Canadian newspaper. Its editorial said Bush's remark "sent a chill up many a Canadian spine, as such talk has done for the past decade."[11] Tellingly, there is opposition within America's own water-rich Great Lakes states to taking water by pipeline from there to other U.S. states. In late 2005, the governors of the eight U.S. states in the Great Lakes region and the premiers of the adjacent Ontario and Quebec provinces in Canada reached agreement to ban any "new or increased diversions of water to areas outside of the Great Lakes-St. Lawrence River Basin."

The broader Canadian opposition to bulk-water trade with the United States, however, raises some interesting, policy-relevant questions: Can making bulk water an exportable commodity help to address water crises elsewhere that may be rooted in unsustainable practices? Or will bulk-water imports only serve as an incentive to perpetuate unsustainable practices? And is it justifiable to hoard a vital but renewable natural resource like water by invoking the principle of national or provincial sovereignty to bar exports?[12]

Significantly, there has been little opposition in Canada to the export by pipeline of another important natural resource: oil, extracted from the tar sands lying below Alberta Province's vast boreal forest, despite the environmental costs of dredging up oil sands from a depth of about 120 meters in a carbon- and water-intensive process that is virtually akin to strip mining.[13] As a result of the emissions from tar-sands mining, Canada is breaching the greenhouse-gas targets it agreed upon internationally.

Yet the Canadian public has largely supported oil-export pipeline projects, including TransCanada's 2,750-kilometer Keystone XL steel pipeline from Alberta's oil sands to the hungry Texan oil refineries along the Gulf of Mexico, notwithstanding the pipeline's potential threat to the Ogallala Aquifer—one of the world's largest underground sources of freshwater. In fact, Canada's public safety ministry in 2012 disparaged environmentalism's "grievances, real or imagined," against the tar-sands industry as a potential terrorism-inspiring cause, while a best-selling book branded the Canadian exports as "ethical oil," free from the human-rights-abusing taint of major oil producers like Saudi Arabia and Venezuela.[14]

Yet the debate in Canada has shied away from applying the same export logic to bulk water, although, unlike oil, freshwater is renewable and abundant across much of Canadian territory. U.S.-Canadian commerce in bulk water, if it were to happen, will mark a momentous new beginning in shipping freshwater as a major tradable commodity between nations.

THE COMMODITIZATION AND MARKETING OF WATER

As water becomes a prized commodity, there is a growing trend toward its commoditization, despite social protests in both the developing and developed worlds. Although water resources are still largely owned by governments and not by the private sector, the commercial push for water's commoditization is gaining traction, even before bulk water has become a globally traded commodity. Many investors hope the market for water becomes the next big investment opportunity. Yet water rights in an era of water stress have become an ever more contentious subject.

Businesses searching for the next big commodity to capitalize on have zeroed in on water—an underpriced yet increasingly scarce resource that continues to be heavily subsidized in most countries. For example, the bill for water consumption in homes usually reflects the extraction, treatment, distribution, and supply costs, but not the price of the water itself. Even the price of bottled water, more often than not, reflects the costs involved in processing, bottling, delivering, and marketing—not the cost of the water itself. In many places, there is no charge for withdrawing surface water or groundwater, as if water is a resource with no price.

This has encouraged an overexploitation of water resources. For instance, the absence of pricing or regulation on groundwater extraction, coupled with the provision in some places of free or subsidized electricity to farmers, has led to falling water tables in a number of nations. Even where farmers pay for water, the price is low, reflecting an unacknowledged subsidy. For instance, rice growers in central Texas pay at least twenty times less for their water than do city customers supplied by the Lower Colorado River Authority, a state organization that controls waters in two artificial lakes—Buchanan and Travis—and much of the Colorado River (a wholly Texan river not to be confused with the much longer Colorado River that flows from the Rocky Mountains to Mexico through five American states).[15]

In some regions, river or aquifer depletion has compelled authorities to impose limits on water abstraction. The limits often take the form of permits.

Not surprisingly, the push to commodify water and to market pumping rights and permits has come primarily from industry, which views water as a potential new oil. Corporations have convinced politicians in several countries to lend support to commodification as a way to control wasteful water use and attract investments to upgrade and maintain water infrastructure. Their contention is that, given the endemic political wrangling and bureaucratic foot-dragging over water issues, only a free marketplace can meet the challenges of prudently allocating scarce resources and financing improved infrastructure.

The marketing of water rights, as if they were akin to property rights, actually evolved in the arid American West, where there is an old axiom that whereas water flows downhill, money flows uphill. To help establish individual entitlements to use scarce water resources there, the concept of proprietary water rights for the owner of land contiguous to a river or stream developed more than a century ago.

In practice, water rights were allowed to be exercised, mortgaged, or transferred independently of the land on which the water originated or on which it was used. The doctrine of prior appropriation—now popularly known as the "Colorado Doctrine" of water law, after the U.S. Supreme Court case *Wyoming v. Colorado*—also evolved in the American West to create a "first in time, first in right" system that conferred a priority right on the first user or appropriator of river waters. Today, as rapidly expanding cities in the American West drive the value of water rights to new highs, there is a renewed gold rush for water rights there. By contrast, there is plenty of water in America's northern tier.

More broadly, with the number of companies specializing in water-related solutions growing, the international commoditization of water is being pursued through forceful arguments. One argument is that the marketplace, within a regulatory framework, can serve as the most efficient allocator of a finite resource vital to human existence and development. It is also contended that the marketplace, not the state, is the best instrument for setting the price of water. According to advocates of this argument, if water is priced to market rates, consumers will use it more carefully, avoiding wastage, while investors will look for more water to bring to the market, as with other resources. A public debate continues to rage, however, on a more fundamental issue: whether water is a public good and thus must remain in public hands or whether it may become a commodity for profit.

Social concerns have acted as a brake on any steep increase in water pricing. Yet the commodification push is getting an impetus from the decline not just in supplies but also in quality of water. Even in regions where quantity is still not a problem, quality can be a concern.

As contamination has become a growing problem, filter manufacturers and firms specializing in water treatment have thrived. Sophisticated water-filtration technology increasingly has come into everyday use, including in homes. In addition to making it fit for human consumption, water needs to be purified for medical, pharmacological, chemical, industrial, and other applications.

Water filters have developed over many centuries. Indeed, the earliest recorded references to water purification methods date back to 2000 BCE in Sanskrit texts in India. The methods in that period ranged from boiling water or placing hot metal instruments in it to employing sand or charcoal filters.[16] Centuries later, Hippocrates designed his own filter to purify the water he used for his patients—a cloth bag to trap sediments that became known as the "Hippocratic sleeve." Today's filters use sophisticated technologies, including reverse osmosis and nanotechnology. Yet contamination of drinking water remains a cause of major health problems globally.

In the coming years, degraded or scarce resources are likely to increasingly define water pricing. Nascent water markets, where people can hold investments in water, like another merchantable commodity, have already emerged within some countries. However, long-distance international water transfers are expected to remain uneconomical for the foreseeable future. Prices will have to increase to a point where long-distance water shipments become commercially viable. But even in that situation, desalination is expected to remain a more practicable and limitless source of direct supply than freshwater imports from distant lands. The choices for landlocked, water-scarce nations are more constricted, but even they may find importing desalinated water from a neighboring state to be a better option than importing freshwater by ship and then transporting it by pipeline from a port in an adjacent country.

The Rise of the Bottled-Water Industry

The most obvious symbol of the creeping commoditization of water has been the dramatic global rise of the bottled-water industry over the past two decades. Sales of bottled water have skyrocketed, promising to rise even more spectacularly in the years ahead.[17] The big bottlers, including Nestlé, Coca-Cola Company, PepsiCo, and Group Danone, see enormous potential for their water sales to continue growing not only in the developed world, but also in emerging markets, where a fast-increasing middle class recognizes the importance of good-quality drinking water. While the poor struggle to get basic access to water for their daily consumption and household chores, the prosperous everywhere now rely largely on bottled drinking water.

Mineral water first started being marketed a century ago in Europe as a curative. It was not until the late 1970s that Perrier, Evian, and other brands reached North America as high-end products aimed at those with sophisticated palates willing to accept a novel replacement beverage. By the late 1990s, the bottled-water industry took off in a major way, marketing plain-water products for the average consumer.

Today, the industry's commercial push has extended globally to marketing myriad types of water—going beyond sparkling, still, mineral, purified, and artesian—in attractive bottles, as if it were a new sort of wine. The slew of bottled waters on sale in many countries includes flavored still, flavored fizzy, and enriched with vitamins, minerals, electrolytes, or other avowedly health-promoting nutrients. In the marketplace, bottled noncarbonated drinking water competes with carbonated beverages, including carbonated water. Some bottled-water products, including fruit-flavored or enhanced waters, are actually soft drinks rather than plain, pure water.

High-end bottled water is traded across political frontiers in a big way. In fact, the price of "designer" bottled water is much steeper than the international spot price of crude oil (or "black gold," as it is called). Despite the major volatility in oil prices in recent years, one barrel (or 159 liters) of crude oil sells cheaper than an equivalent quantity of bottled water from, say, Alaska or Fiji. Unless the value of oil surges to an all-time high in the international spot market, the regular price of even ordinary bottled-water brands like Aquafina and Dasani in a U.S. convenience store will stay higher than "black gold."

Significantly, the bottled-water industry has been successfully convincing increasing numbers of citizens in wealthy states to buy its stuff in bottles when empirical tests have repeatedly shown that the local water out of the tap in many of these nations is as safe as the brands on sale, although occasionally both "can have quality problems."[18] Also, despite the commercial claims that bottled water tastes better, taste tests conducted blindly show that people usually are unable to differentiate between bottled and tap water.[19]

Yet the aggregate bottled-water purchases per year by Americans now virtually equal the national spending on the water-supply system.[20] The explosive growth in this industry has placed bottled water in nearly every American supermarket, convenience store, and vending machine from coast to coast, with scores of brands competing for consumers' dollars. The same is true in many other countries. In Germany, for example, customers can choose from five hundred domestic water brands, "all of them different in terms of taste and origin," in addition to imported mineral water.[21]

It is as if the era of safe, reliable tap water is over in the advanced industrial economies. Whereas many in the developing world still have no access to piped water, increasing numbers of citizens in the developed world distrust the potable water they can comfortably and limitlessly access from their taps at a tiny fraction of the cost of bottled water.

The source of some of the water that is bottled and sold is the runoff from glaciers in the Alps, Andes, Arctic, Cascades, Himalayas, Patagonia, Rockies, and elsewhere. The capture of increasing amounts of glacial runoff by the bottled-water industry is clearly at the expense of ecosystem services. If not siphoned off, this runoff would normally flow for ecosystem services extending from the recharging of wetlands to the sustenance of biodiversity.

Yet big bottlers and other investors have aggressively sought to buy glacier-water rights. For example, Canadian hedge fund Sextant Capital Management has bought 95-year water rights to three northern European glaciers that are close to ports. Meltwater from Alaska's Eklutna and other glaciers is bottled and shipped to East Asia and elsewhere. China's thriving mineral-water industry, by tapping into Himalayan glaciers, is compounding the anthropogenic impacts on Tibet's natural ecosystems.

Bottlers are also increasingly tapping waters of aquifers that feed rivers and lakes. The bulk of bottled water sold globally is actually drawn from the subterranean water reserves of aquifers and springs. The commandeering of these resources by the industry for shipment elsewhere is lowering the water table in some regions. A liter of bottled water requires 1.6 liters of water to produce, making the industry a major water consumer and wastewater generator.

Much of the bottled water sold across the world is not glacier or natural spring water but—as epitomized by two well-known brands, Dasani and Aquafina—processed water, which is municipal water or directly extracted groundwater subjected to reverse osmosis or other purification treatment. Large-scale water abstractions have entangled bottlers in disputes with local authorities and citizens' groups in some water-stressed regions over their contribution to resource depletion and even pollution.

Bottlers may own or lease land with an underground water source, or need fixed-term, renewable licenses to draw water. If they require permits, it may be only if they extract water beyond a set quantity per day. In some states, the bottlers are still not subject to any extraction-related levies or regulations. Yet disputes between them and local groups or authorities have flared in many areas around the world, leading at times to changes in water extraction and licensing rules.

Processed water in bottles, because of chemical treatment, often does not contain fluoride, which is naturally present in most groundwater or is added in tiny amounts to municipal water supplies in some regions to promote dental health. Whereas excessive fluoride content—found mostly in calcium-deficient groundwater—poses a health hazard, the absence of fluoride ions in drinking water can possibly increase tooth-decay risks, especially among children.[22] A small minority of bottlers do market fluoridated water. The fact is that the rising consumption of bottled water is reducing many people's daily exposure to the mineral that is believed to help prevent or slow tooth decay, although the use of fluoride toothpastes and mouth rinses may compensate for nonfluoridated drinking water.

The marketing of packaged ordinary water as a mass-produced commodity also generates colossal plastic-bottle waste as well as packing-carton waste. By selling the planet's most basic product at lucrative prices, the bottled-water industry reaps huge profits while burdening taxpayers with the costs of clearing the mountains of waste it produces. Thanks to a low recycling rate, tens of billions of plastic water bottles end up as garbage every year, taking up increasing space in landfills. Some cities have used up most of their landfill capacity and are experiencing difficulty finding acceptable new landfill sites.

Besides the direct environmental impacts of bottled water, other issues include the carbon footprint of water shipped across long distances and health concerns over the potential leaching of chemical compounds from water bottles, which are commonly made from polyethylene terephthalate (PET)—the chemical name for polyester plastic whose raw materials are derived from crude oil and natural gas.

It was the introduction of single-serve PET bottles in the 1990s that helped move bottled water from the niche market to the mainstream. It turned water into a portable, lightweight, convenience product, and also prompted producers of branded fruit juices, sports drinks, teas, and carbonated soft drinks to switch from glass to PET bottles.

As a result, there has been a phenomenal global growth in the PET packaging market. The market's focus has shifted to innovative bottle designs and caps, which now include push-pull sports caps, flip-top caps, and even "child-safe" caps. Considerable energy resources are being consumed to meet the growing international demand for PET plastic, which has also become popular for packaging salad dressings, cooking oils, peanut butter, and other food products.

Scientific studies on bottled water have spotlighted health concerns that suboptimal storage conditions—such as prolonged exposure to sunlight and high temperatures—can cause leaching of estrogenic compounds from PET bottles into their fluid contents.[23] The estrogenic activity in bottled water may thus expose its consumers to endocrine-disrupting chemicals, which alter the function of the endocrine system by mimicking the role of the body's natural hormones.[24] Hormones are secreted through endocrine glands and serve different functions throughout the body. In hot places in the world, as in Asia and Africa, exposure of water bottles to heat is often unavoidable, especially during the scorching summer.

The large, reusable polycarbonate containers of up to nineteen-liter (five-gallon) size in which bottlers deliver water to homes and offices also pose a potential health risk by releasing bisphenol A (BPA), another endocrine disruptor that has been shown in animal studies to affect reproduction and brain development. BPA is used by industry to harden polycarbonate plastics. One group of researchers found that the amount of BPA released from both new and used polycarbonate drinking bottles into cool or temperate water was the same, but that the level went up drastically once the bottles were briefly exposed to boiling water.[25]

The National Institute of Environmental Health Sciences in the United States has also reported that the degree to which BPA leaches from polycarbonate containers into any liquid depends more on the temperature of the liquid or bottle than the bottle's age.[26] The U.S. Food and Drug Administration (FDA) and the National Toxicology Program at the U.S. National Institutes of Health have both underlined concern about "the potential effects of BPA on the brain, behavior, and prostate gland in fetuses, infants, and young children."[27]

Although the science around endocrine disrupters is far from settled, concerns that BPA from plastics can leach into water and food have led to bans on BPA-containing baby bottles and infant feeding cups in several markets, including the European Union and North America. Bottlers reuse the large, translucent polycarbonate water containers several dozen times. While lingering doubts about the suitability of BPA in water containers have prompted the development of BPA-free plastic alternatives, the large bottles that go onto water coolers remain predominantly made of polycarbonate.

Developing international safety standards for bottled water remains a work in progress under the authority of the Codex Alimentarius Commission. Founded jointly by the Food and Agriculture Organization and the World Health Organization in 1963, the commission seeks to establish internationally recognized standards, codes of practice, guidelines, and other recommendations on food production and food safety.[28] Some bottled-water-related codes of practice (which essentially are recommended codes of hygienic practice) form part of Codex Alimentarius, but the codex system has still to address all concerns.[29]

While bottlers genuinely worry about groundwater pollution, given that pollutants can seep into their water sources from surrounding areas, their production process itself has on occasion caused contamination of bottled water. For example, the Coca-Cola Company—just weeks after it introduced Dasani into Britain—was forced to recall its entire processed-water range there in 2004 after excess levels of bromate were found in its products, sourced from local municipal water. The firm admitted that the contamination resulted from its regular practice of adding calcium to Dasani—a practice that led to bromate's formation in excess of legal levels.[30] Bromate, a derivative of bromine, has been linked to increased cancer risk through long-term exposure.

There have been scores of other recalls of bottled water across the world, with the discovered contaminants in the products ranging from benzene and coliform to algae and arsenic. For example, algae forced the 2005 withdrawal of Coles Farmland mineral water from supermarket shelves in Australia, while high levels of arsenic in mineral water imported from Armenia prompted the FDA to issue a U.S. recall on the Jermuk brand of products in 2007. Bacterial contamination led to the 2002 recall of Cotswold spring water in Britain, and paraffin discovery to the 2006 withdrawal of Sao Bac Dau bottled water in Vietnam.

Although PET can be commercially recycled by remelting or chemically breaking it down to its component materials to make new PET resin, the majority of water bottles sold worldwide find their way to landfills. In the United States alone, according to Pacific Institute founder Peter Gleick, 85 million water bottles a day—or more than 30 billion bottles a year—are discarded, with booming bottled-water sales reflecting "the long-term decay of our public water systems, inequitable access to safe water around the world, our susceptibility to advertising and marketing, and a society trained from birth to buy, consume, and throw away."[31]

Like the oil industry, the bottled-water industry has responded to a popular backlash over its growing impact on the environment by trumpeting its own environmental bona fides. Just as Exxon Mobil, BP, Shell, and other energy behemoths that were once the lightning rod in the debate about climate change have ostensibly gone "green," water bottlers have sought to deflect criticism by environmental groups and others by touting their eco-friendliness and promoting green initiatives. From funding water projects and schools in the developing world to seeking to reduce their carbon footprint and the wastewater generated by their bottling facilities, Nestlé, PepsiCo, Coca-Cola Company, and other leading bottlers are trying to wrap themselves in green and, in the process, reap new profit opportunities.

Interestingly, greater environmental consciousness among consumers, as in Europe, does not spell trouble for the bottled-water industry. Europeans, in fact, remain the largest per capita consumers of bottled water in the world. The exceptional rise of this industry globally in a relatively short period of time is a reminder that the world must move to more sustainable practices to address its resource challenges, including the provision of piped water that consumers can trust as safe and reliable to drink.

Water Trading within and between Nations

Water is both fungible (water from one source can be substituted with water from another source) and liquid (it can be exchanged for cash)—the two elements that conceptually define a tradable commodity. Water, however, is a unique resource, and the market cannot by itself assess all the values society places on water. After all, some values associated with water—extending from social and environmental to recreational—elude market assessment. Valuing water is thus complex and challenging. Still, as part of the commoditization and marketing of water, some national markets have been developed or are under development to let water be priced and traded like any other commodity.

Water scarcity or prolonged drought has typically acted as the main driver for the formation of water markets, as in the American West and Australia. A formal marketing system can be built only on the basis of clearly delineated water-related property rights, without which the sale or purchase of water rights will be legally hampered. Water rights are usually considered mineral interests in land. After water is captured by diversion or extraction from a watercourse, it becomes a merchantable good. Property rights apply to such ownership or water-use entitlement.

In the water markets in California, New Mexico, and Wyoming, for example, private investors and public authorities can sell or buy water rights, while farmers are entitled to purchase other landowners' water rights so as to irrigate their fields. Some other parched U.S. regions, such as West Texas, also have a history of water-rights trading. By contrast, a state relatively well endowed in water resources like Wisconsin has constitutionally embraced the

Chapter Three

public trust doctrine and defined the state as holding property rights over all natural and navigable waterways. This shows how water laws and rights vary widely within the United States itself.[32]

Australia has sought to be one of the global leaders in developing a free market in water, making its first tentative moves in the 1980s to encourage farmers to explore trade in water rights or annual allocations. In 1993, state governments in Australia agreed to establish a water market underpinned eventually by a national register of water entitlements—a process aimed at creating transparent water transactions, availability of up-to-date market information, and high-performing state and territory water registers leading to a standardized national registry system. Local-level water registers are designed to accurately record water rights and support water accounting and resource management.

The Australian water market began to progress only this century after a decadelong drought threatened to drastically cut the country's food exports from the agriculturally rich Murray-Darling Basin. At the height of the drought in 2004, states and territories in Australia reached agreement on initiating major water reforms, including a nationally compatible market and a regulatory system. And in 2008, the Australian Competition and Consumer Commission was tasked with developing rules for pricing and trading water, while the Australian government decided to invest billions of dollars in the development of the so-called National Water Market System.

Now the buying and selling of permanent and temporary water rights across the major irrigation regions in Australia has been growing at 20 percent a year, according to national water broker Waterfind. A legislative and regulatory framework, by removing bureaucratic hurdles and decoupling water rights from land ownership, has allowed such trading to thrive. The federal government's $3 billion scheme to buy back water rights from farmers in the Murray-Darling Basin, however, did little to correct spot-market price volatility.

A lot of the trading is still done between farmers, although investment groups (including mutual fund companies and savvy offshore players) are moving in to buy water entitlements and lease them back to irrigators for a three- to five-year period. Investors can also sell water entitlements into the temporary transfer market or own entitlements in a joint venture with an irrigation enterprise. The development of a full and mature national market, however, has been hobbled by water-allocation discord between regions and the complexity of the various state-level water-rights regimes.

Australia—a world leader in the international trade of primary commodities—has effectively established itself as the leading example of intracountry trade in a more essential commodity: water. Its water investors and brokers consequently are well positioned to seize potential international opportunities in water rights elsewhere. Its total market turnover in water rights is now well over $3 billion a year. In volume, such trades surpass 20 percent of the yearly surface-water abstractions in the Murray-Darling Basin; in comparison, water transactions in Arizona, California, Colorado, Nevada, and Texas total between 5 to 15 percent of all freshwater diversions annually in those states.[33] Australia's legislation, by guaranteeing public access to water and setting aside allocations for environmental protection, has sought to address moral and ethical issues about allowing the most vital resource to be traded on an open market where it can be bought and sold independent of land rights.

It is thus possible for a government to create a common national water market by amalgamating provincial-level water registers or banks, with water being sold on a cost-plus basis in a clearinghouse of buyers and sellers, or even in a spot market, where buyers and sellers set the price by posting their trading offers online or on bulletin boards of local irrigation offices. If national water markets were to develop in a number of countries and become integrated, futures markets and other derivative water-based financial instruments could also emerge. But what is possible may not necessarily happen.

When water rights are marketed, any transaction means that the buying party secures the water and the selling party gains a monetary reward. This may promote economic efficiency of water use. But it could promote economically inefficient use of land. For example, some farmers may find it more profitable to simply sell their water rights than to cultivate their fields. This explains why some in California prefer to leave their fields fallow. Some thirsty metropolises in the American West have been buying up farmland for water rights.

Also, when water has a market price, water resources may get allocated for high-value uses at the expense of other important uses. Water resources, along with the land on which they originate, may also be grabbed by foreign investors, who could put them to use in a way detrimental to the social and environmental interests of local communities. Water rights, in other words, may go to the highest bidder, thereby possibly disadvantaging other users and undercutting the principle of equity associated with a public good.

Setting up formal water markets in nations or provinces where water is publicly owned and provided is often politically taxing and divisive because it entails jettisoning long-established traditions. Yet despite the difficult challenges, moves to establish a formal water marketing system have been made even in the non-Western world on the grounds that water must seek its own best use based on its value.

Chile is an example of the conversion of water from a public resource to a tradable commodity, with water rights held like private property. Designed to allocate water to the highest economic use, the Chilean water-rights trading system was set up way back in 1981 during the brutal dictatorship of General Augusto Pinochet, who became a notorious symbol of human-rights abuses and corruption in his nearly two-decade rule. In practice, the Chilean water-trading system—largely because of little regulatory oversight or environmental safeguards—has permitted big businesses to muscle out smaller users, besides promoting speculation and environmental degradation.[34]

Chile thus offers some sobering lessons on what poor regulation and unbridled water trading can bring. If water rights are sold to the highest bidder, the losers may not only be the poor but also natural ecosystems. By contrast, Australia's establishment of water rights and trading mechanisms in the Murray-Darling Basin has helped boost agricultural productivity by creating the needed water-price signals to incentivize greater agricultural output or a shift to high-value crops, even though prices for seasonal water have remained far more volatile than those for permanent water entitlements.[35] Chile, however, has shown that the channeling of water for high-value economic activity can force other operations, such as mines, to rely on seawater, thus saving freshwater.

Bolivia is an interesting example of how a popular backlash against the local commercialization of the water supply can rapidly gain momentum, especially in countries where protests cannot be muzzled by harsh autocratic rule. A violent popular reaction in the Andean city of Cochabamba in 1999–2000 over price spikes resulting from the privatization of local water services helped sweep Evo Morales to power as Bolivia's president and led to the eventual ouster of two major multinational companies, Bechtel and Suez, from water services in Cochabamba and the Bolivian capital of La Paz, respectively.

Cochabamba became the symbol of people's power to regain control of a water-supply system from a transnational corporation.[36] The backlash—which included a refusal by residents to pay the much-higher water charges to Bechtel's local subsidiary—as well as Morales's vow that "water cannot be turned over to private business" stymied any hope of creating a national water market. Such water wars in Bolivia inspired the 2008 James Bond movie *Quantum of Solace.*

Some other countries have also witnessed a grassroots backlash on the water issue. Whereas local water-privatization plans in countries ranging from Tanzania to Ecuador have been halted, new laws were passed in the Netherlands and Uruguay explicitly prohibiting the transfer of water systems and services to private companies. These legal actions implicitly drew a contrast between a private good like bottled water and a public good such as natural water resources that must be held by the state in trust for the public.

A 2004 popular referendum in Uruguay gave the go-ahead to the government to enact a constitutional amendment recognizing access to water as a fundamental human right and guaranteeing public ownership of the water system and services. This was the first instance in the world of a national referendum specially devised to win popular support to make water privatization illegal. As illustrated by the separate case of Senegal, where a French consortium took over water services, privatization can actually accentuate local water woes.

The Netherlands, for its part, enacted legislation in 2004 blocking any privately held company from providing drinking-water services to the public. In doing so, it showed that the European Union rules are no barrier to such a patently antimarket action in the water sector. In fact, Italian voters in a mid-2011 referendum delivered a stinging rebuke to their controversial prime minister, Silvio Berlusconi, by overturning laws passed by his government that, among other things, would have privatized the water supply.

Still, as water stress accentuates, moves to privatize and set up, consolidate, or expand intracountry water markets are likely to gain traction in some parts of the world. After all, when finite supply meets fast-growing demand, commoditization of the resource—along with higher prices—may seem inevitable.

Just as bottled water is marketed in myriad types, bulk water, theoretically, could be internationally sold in different grades. Transcontinental bulk shipments of high-end water from glaciers are already happening. For example, companies are transporting bulk glacier water by tankers from Alaskan, Canadian, and Nordic sources to third countries like China so that it can be bottled at coastal plants, using local cheap labor. But prospects of bulk ordinary water becoming an internationally tradable commodity remain bleak at present because of the problems associated with the methods and costs of long-distance shipment. Tankered bulk water must first be able to compete with locally recycled or desalinated water in any destination country.

The commodification process and increasing economic value of water, however, have influenced the issue of piped-water trade between neighboring countries, as illustrated by the Canada-U.S. and Malaysia-Singapore cases. Canada has clamped a virtual blanket prohibition on foreign or domestic investment in bulk-water exports. In fact, its protectionist ban seeks to preempt NAFTA provisions that permit commoditizing water and empower private citizens and corporations to sue a state party for taking actions that impair their investments.

Ottawa failed to secure an explicit exception for water in the NAFTA text. The 1993 Canada-U.S.-Mexico Joint Statement following the conclusion of the NAFTA negotiations declared that water "in its natural state" in watercourses is not covered by the treaty's provisions, "unless water, in any form, has entered commerce and become a good or product."[37] Yet it left ambiguity—which a

state party can exploit—by not clarifying at what point in the continuum between being a resource in its natural state and a fully merchantable good, water
would become subject to the treaty obligations.[38]

Canada has sought to prevent bulk water from its basins entering North
American commerce because of the larger legal implications of exporting
water as a "good." It is concerned that if it permits even limited bulk-water
exports to the United States, NAFTA obligations would kick in and it could
face lawsuits before NAFTA Chapter 11 tribunals for restrictive trade practices in the form of export limits. After all, quantitative export restrictions
are not permitted under the rules of NAFTA and the international General
Agreement on Tariffs and Trade.

Whereas American commercial interest in the transfer of surplus Canadian
waters via pipeline to water-stressed U.S. regions has only heightened public
opposition in Canada to bulk-water exports, Malaysia has sought unsuccess-
fully to revise the price at which it has been selling piped water to Singapore
for about half a century. The price—three Malaysian cents (just under one
U.S. cent) per thousand imperial gallons (4,546 liters)—was set in the 1960s,
when Singapore seceded after being part of the newly independent Malaysia
for barely two years.

The Singaporean government has refused to pay a higher charge for the
water it imports, describing the Malaysian demand to match the price at
which mainland China sells water to Hong Kong as an unacceptable attempt
to reopen an issue settled by the 1965 secession agreement. Instead Singapore has embarked on an ambitious, high-cost domestic program to slash its
water dependence on Malaysia by tapping nontraditional sources, including
captured rainwater, recycled water, and desalinated water.

The increasing tendency to value freshwater as an economic good, however, prompted America's water deal with Mexico in late 2012. Under the
five-year agreement, the United States will "contribute a total amount of
$21 million" for Mexican water projects, and in return for the U.S. "infra-
structure investments," Mexico is to surrender 153 million cubic meters of
its share of the Colorado River waters for American use.[39] In other words,
the water deal prices the forfeited Mexican share at 13.72 cents per 1,000
liters (1 cubic meter).

The deal has opened the way for water agencies in southern California,
Arizona, and Nevada to purchase additional water rights. The Metropolitan
Water District of Southern California will receive 58.6 million cubic meters
of water from Mexico's share for $5 million, while the Central Arizona Water
Conservation District and the Southern Nevada Water Authority, which sup-
plies Las Vegas, will each get 29.3 million cubic meters for $2.5 million.[40]

These regional water agencies will thus be paying only 8.6 cents per cubic meter for the Mexican water purchased by the federal government.

Israel agreed in principle in 2004 to import 50 million cubic meters of water annually over twenty years from next-door Turkey, but the deal never went through for economic and strategic reasons. It was actually devised as a water-for-arms deal, with water-rich Turkey agreeing in return to buy Israeli weapons systems, including tanks and air force technology. The two countries—both dominant riparian powers in their respective basins—do not have contiguous borders. The consequent necessity of using converted oil tankers or giant floating polyurethane bags to transport water supplies from Turkey's Manavgat River to Israel across the Mediterranean Sea would have driven up the costs, possibly making the imports dearer than locally desalinated seawater in Israel.

The aborted deal was crucial to Turkey's larger ambitions to emerge as an unmatched freshwater-exporting power supplying not just Israel but also Malta, Crete, both parts of divided Cyprus, and Jordan. To underpin its water-export ambitions, Turkey even invested more than $100 million in building necessary infrastructure, including channels to divert Manavgat waters to a treatment facility and a ten-kilometer pipeline to take the processed water to two custom-built sea terminals for tankers.

Israel concluded after long reflection that, although the proposed water imports only amounted to 3 percent of its then needs, the strategic costs of creating water dependency on Turkey were too high. Rather than arm Turkey with hydro-leverage against it, Israel decided to cost-effectively invest in desalination technology at home in a major way. After all, Turkey—the starting point of the Tigris and Euphrates Rivers—had in the past not been shy to flaunt its riparian leverage against downstream Syria and Iraq, with a senior Turkish official once boasting that his country could stop transboundary flows "in order to regulate the Arabs' political behavior."[41] Large upstream hydroengineering projects in Turkey since the 1970s have indeed diminished Tigris-Euphrates cross-border flows.

The desire of water-rich but otherwise resource-poor Kyrgyzstan and Tajikistan to fashion natural transboundary water flows into commodity exports akin to the hydrocarbon exports of their downstream neighbors—Kazakhstan, Uzbekistan, and Turkmenistan—is more of a dream because it is fraught with serious, larger implications. A line, after all, must be drawn on the commoditization of water. If commoditization were to extend to natural cross-border flows, the vulnerability of downriver countries would swell considerably. Unlike piped-water exports, which can be made only after major investments in infrastructure, natural transboundary flows defy pricing.

Globally, the push toward greater privatization, commoditization, and securitization of water reflects the increasing application of business principles to a resource that had escaped such pressure in the past. This trend could promote efficient, market-based distribution and supply as well as more efficient use of water. However, if a resource that was once available without charge gets increasingly traded at market prices, it could possibly affect low-value production and exacerbate social and income disparities.[42] Maude Barlow, a Canadian antiglobalization activist and leader of the so-called Water Justice Movement, contends that to "deny the right to water is to deny the right to life."[43] Yet the commodification push makes it less likely that access to clean water for basic needs will ever become a fundamental human right.

When a resource becomes scarce, a key issue that must be addressed is how best to price or ration it. The goal of fair, efficient allocation and distribution of water demands consideration of scarcity and opportunity costs. Water shortages necessitate trade-offs, while trade-offs are likely to impose opportunity costs. Reconciling efficiency with equity is never easy. Efficiency and equity also must be reconciled with environmental sustainability and financial viability.

Food, like water, is essential for life, yet governments usually dole out subsidized food, not free food, to the poor. Differential water pricing certainly is one way to shield poor households while charging industrial-commercial users the full supply costs. A broader challenge is to ensure, through strong regulatory oversight, that market-based provision of water does not foster corrupt practices or take away water rights from vulnerable communities.

From municipalities to markets, new efforts are now under way to establish the right price signals on water. There has been an upswing in investments by companies seeking to profit from the demand for clean, regular water supplies. As this investment trend accelerates, more firms will seek to deliver water to the marketplace. Water promises to become a new profit engine for private investors and speculators, as soaring demand and global warming strain its availability. A possible global market in water futures could set the stage for water to eventually be traded like oil and gold.

Privatization and commoditization are also expected to help catalyze the development of new clean-water technologies. The advent of advanced clean-water technologies would represent a major leap forward by potentially offering affordable and sustainable ways to ease the global water crisis. New technologies, however, are likely to only accelerate the commodification push. After all, development of new clean-water technologies—by permitting energy-efficient treatment of polluted freshwater, wastewater, brackish water, and seawater—is expected to dramatically increase the turnover of the global water business.

Global sales of water-related equipment, services, and products, including irrigation gear, already total up to half a trillion dollars a year. Such sales are likely to more than double as new technologies come online to more efficiently treat, distribute, and use water, which is unlikely to be available in the future for all applications at the same low cost as it is today.[44]

However, a globally integrated market for physical water remains a far cry from a market for water rights or water equipment, services, and products. Water is innately local and thus hard and expensive to transport in bulk across seas. Water also tends to degrade when it is taken from a natural body of water and transported across long distances.

For the foreseeable future, long-distance bulk transport of water is unlikely to be economically attractive, other than to deal with a temporary water crisis in a coastal city or region or to supply a water-distressed island that has no other workable option. Some Caribbean and Mediterranean islands, for example, cannot make do without intermittent import of freshwater by tankers or barges to meet local demand, especially from a growing tourism industry.

NEW THREATS TO WATER: ENVIRONMENTAL AND CLIMATE CHANGE

There is an intimate relationship between water and climate: water's movement by gravity, along with the processes of evaporation and condensation, helps to propel the earth's biogeochemical cycles and regulate its climate.[45] Human-induced changes in the hydrological cycle affect regional climate, and climate variability in turn carries significant impacts on water resources.

Although the international community accepts that protecting the atmosphere, hydrosphere, lithosphere, biosphere, and cyberspace—the "global commons"—is the common responsibility of all, it is proving difficult to develop and enforce supporting norms. The natural systems, including the land, water, marine, and atmospheric resources utilized by the human race, must be managed in such a way, according to the World Charter for Nature, as to "achieve and maintain optimum sustainable productivity."[46] Yet human-induced alterations of natural systems have become so profound that the earth has entered what is being called a new age of geological time, the Anthropocene, in which human civilization—not nature—is the dominant force driving transformations.[47]

Human activities are altering freshwater ecosystems even more profoundly than terrestrial ecosystems. Climate stability is becoming a casualty of various anthropogenic transformations, whose net effect is to raise regional temperatures. For example, intensive irrigation fosters atmospheric humidity in semiarid areas, while land degradation and other land-use changes perturb the climate, including by affecting surface albedo (or the percentage of incoming solar energy that is reflected back into space and not absorbed by the planet).[48] A warming climate, by reducing the highly reflective snow cover, only allows more radiation from the sun to get absorbed by the ground and water, further increasing temperatures.

Tensions between the hydrosphere and the human overexploitation of natural resources, including growing demands for freshwater, exemplify changing threats to fragile ecosystems. They signal that, from a resource perspective, human civilization is approaching a tipping point. If the integrity of essential ecological processes and life-support systems is to be preserved, the link between economic progress and the degradation of water and other resources must be broken.[49] As the 2002 Earth Summit in Johannesburg proclaimed in its implementation plan, "Fundamental changes in the way societies produce and consume are indispensable for achieving global sustainable development."[50]

Besides freshwater, nowhere is the difficulty greater in shielding the common good than in controlling the buildup of planet-warming greenhouse gases. Effectively combating such buildup in the atmosphere demands fundamental shifts in national policies and approaches, as well as lifestyle changes. It is easier to visualize than to devise carbon standards capable of protecting the long-term material and social benefits of continued economic growth.

As the international experience since the 1992 United Nations Framework Convention on Climate Change (UNFCCC) bears out, it is often easier to set international goals than to implement them. For example, how many industrialized countries that voluntarily became party to the Kyoto Protocol faithfully implemented their varying obligations? The binding targets set under the UNFCCC-linked Kyoto Protocol, which took effect in 2005, averaged a cut of 5 percent in greenhouse-gas emissions below the industrialized countries' 1990 levels over a five-year period from 2008 to 2012.

With global emissions continuing to grow at over 1 percent a year, the world dumps some 90 million tons of greenhouse gases every twenty-four hours into the atmosphere, as if it were an open sewer.[51] This situation has primarily been created by the burning of fossil fuels, in which carbon was locked up over millions of years.

Water scarcity and global warming are two of the most worrying problems in this millennium. Global warming is a long-term trend that can be stemmed, at best, but not reversed. Rapid economic growth is bringing two key resources linked to climate change under increasing strain. One resource is energy, the main contributor to the buildup of greenhouse gases. And the other is water, whose availability will be seriously affected by global warming, thereby increasing the likelihood of water-related conflicts, as the Inter-governmental Panel on Climate Change (IPCC) has warned.[52]

Such is the interconnectedness of climate and freshwater systems that changes in one system are likely to bring about changes in the other. For example, elimination of wetlands can decrease regional precipitation by altering moisture recycling, but intensive irrigation in semiarid areas tends to boost precipitation and thunderstorm frequency.[53] Climate change can bring on increased flooding and more frequent droughts. The already visible effects of climate change range from a retreating polar sea icecap and coastal erosion to glacial recession and thawing of permafrost (permanently frozen ground) in parts of North and South America, Tibet, and Siberia.

If global warming accelerates, the primary sources of rivers—mountain snows, glaciers, and springs—will be directly affected. Glaciers store so much freshwater that if they totally melted, the seas would rise as much as 80 meters, inundating many coastal regions.[54] Even a rise of a few meters in sea levels holds serious implications for the 37 percent of the global population that lives within one hundred kilometers of coasts.[55]

In fact, reflecting the large-scale migration of people to coastlines in modern times, most megacities, and two-fifths of all cities with more than 1 million residents, are located in coastal zones.[56] In Asia, where the economic-boom zones are concentrated along the coasts, such vulnerability is greater because it has not only the world's longest coastline, but also a larger proportion of its population living in coastal areas.

In a world increasingly affected by climate change, accelerated snowmelt in mountain ranges and faster glacier thaw would initially trigger serious flooding in the warm months before river flows began to deplete irreversibly. A scenario of this type would promote recurrent drought and drive large numbers of subsistence farmers into cities and other areas with relatively better availability of water.

Assessing the Growing Anthropogenic Impacts

The anthropogenic impacts on the earth's diverse and sensitive ecosystems carry long-term implications for human health, well-being, and productivity. Technologies and practices that were once thought to be sustainable have become unsustainable in the face of population, consumption, and developmental pressures.

Coal, for instance, was embraced by eighteenth-century Europe as an excellent substitute for dwindling timber supplies. But today, coal burning is the leading problem on the climate front. Kerosene—the first petroleum product to be refined—was developed in the nineteenth century as a cleaner-burning alternative to whale oil. Now kerosene is the dominant aviation fuel. But kerosene combustion emits almost as much carbon dioxide as gasoline combustion.

After the remarkable improvements in infant mortality and life expectancy in the past century, the continuing human alterations of earth threaten to raise morbidity and mortality rates. Heavy use of nitrogen fertilizers, for instance, has dramatically boosted food production and helped lower mortality levels through better diets, but only to promote land, water, and air pollution, which is raising morbidity and mortality levels. Fossil-fuel combustion helped power the industrial revolution and boosted living standards. But now heavy reliance on fossil fuels is a major contributor to global warming, which is likely to increase heat-, flood-, and drought-related morbidity and mortality. The spread of antibiotic-resistant bacteria is yet another example of human impacts on the environment. That is why, as illustrated by the freshwater challenge, the world needs sustainability on all key fronts, including population, resources, consumption, and technology.

To be sure, there are continuing gaps in the scientific understanding of the phenomenon of climate change. The world now knows more about the anthropogenic factors contributing to global warming than about the earth's natural climatic variations. What, for example, caused the Little Ice Age from about the fifteenth to the nineteenth centuries? The end of the Little Ice Age coincided with the start of the industrial era, which resulted in higher emissions of greenhouse gases.[57] Since then, mean global temperatures have increased steadily, while the level of carbon dioxide in the atmosphere has risen from 280 parts per million by volume (ppmv) to more than 370 ppmv.[58] Climate, however, has been changing since ancient times, with a study published in 2012 reporting that Central America's Maya civilization likely collapsed due to climate change.[59] Further scientific findings are necessary to understand to what extent current global warming is attributable to human activities or to natural climate shifts that usually extend over several centuries.

In addition, there are genuine differences among experts on the long-term impacts of climate shifts. The British Treasury–commissioned Stern Report, for instance, expressed greater alarm about the impacts of climate change than the IPCC scientific reports, published periodically since the 1990s.[60] Such differences among experts are understandable, given that the science of global warming is still young. However, "Climategate," as the 2009 publication of damaging e-mails and other documents from the Climate Research Unit at Britain's University of East Anglia became known, exposed disturbingly politicized scientific research in the form of manipulated or suppressed data on human-driven climate change.[61]

The IPCC's own 2010 "Glaciergate" scandal over one of its key claims in a 2007 report helped dent public trust in the independence and accuracy of research on climate change—a subject that increasingly has been politicized, with tensions running high between competing interests over when and how, and at what cost, to shape a low-carbon future for the world.[62] The IPCC—the supposed gold standard in climate science—admitted that its published claim that the Himalayan glaciers were "very likely" to disappear "by 2035 and perhaps sooner" if earth "keeps warming at the current rate" was based not on peer-reviewed scientific research but on a 1999 magazine interview with one glaciologist, Syed Hasnain—a claim which had been recycled in a 2005 report by the environmental campaign group World Wide Fund for Nature (WWF) and then enthusiastically picked up by the IPCC without any investigation.[63] To IPCC's acute discomfiture, the concerned glaciologist later went public to say that he had been misquoted in the magazine interview.[64]

To make matters worse, Murari Lal, the coordinating lead author of the portion of the IPCC report where the claim appeared, said the bogus assertion had been intentionally incorporated to help put political pressure on regional leaders in Asia. "It related to several countries in this region and their water sources. We thought that if we can highlight it, it will impact policymakers and politicians and encourage them to take some concrete action," a British newspaper quoted Lal as saying in an astonishing admission.[65]

The IPCC apologized for the "Glaciergate" scandal, saying, "The Chair, Vice-Chairs, and Co-chairs of the IPCC regret the poor application of well-established IPCC procedures in this instance. This episode demonstrates that the quality of the assessment depends on absolute adherence to the IPCC standards, including thorough review of 'the quality and validity of each source before incorporating results from the source into an IPCC Report.'"[66]

In yet another embarrassment for the IPCC, it emerged that its headline-grabbing claim that global warming could damage up to 40 percent of the Amazon rainforests in Latin America was devoid of any independent scientific investigation.[67] The Amazon River Basin, roughly the size of the continental United States, is home to the largest rain forest on earth. In fact, it is one of only three areas of very extensive natural "frontier forest" that still remain, the other two being in Russia and Canada/Alaska. The Amazon, the most voluminous river on earth, with 12.6 times the average carrying volume of North America's largest river, the Mississippi, is made up of more than 1,100 tributaries, two of which (the Negro and the Madeira) boast a discharge at mouth greater than that of the mighty Congo River.

The IPCC asserted in its 2007 report, "Up to 40 percent of the Amazonian forests could react drastically to even a slight reduction in precipitation; this means that the tropical vegetation, hydrology, and climate system in South America could change very rapidly to another steady state, not necessarily producing gradual changes between the current and the future situation."[68]

The IPCC was also forced to acknowledge that the very same report incorrectly stated that 55 percent of the Netherlands is under sea level. In truth, that figure signified only the total area at risk of flooding.[69]

Separately, as if to highlight the evolving nature of climate-change science, three scientists were forced to retract their 2009 study on a projected climate-induced rise in sea levels after acknowledging fundamental mistakes that undermined their original findings.[70] Their study, published in a leading journal, *Nature Geoscience*, had used fossil coral data and temperature records derived from ice-core measurements to broadly support the 2007 IPCC projections of an eighteen- to seventy-six-centimeter rise in ocean levels by the year 2100.

These incidents helped highlight the need for accurate, objective research uninfluenced by the geopolitics and ideology surrounding the climate-change debate. To understand the planet's long-term climatic trajectory, climate variations must be measured over decades and perhaps centuries, not months or years. No individual episode of severe weather can directly be attributed to climate trends, although scientific research indicates that extreme weather events are likely to become more frequent as global temperatures rise.

Although it is easy to exaggerate or underestimate the impact of climate change on water resources, the effects are likely to be serious even in the most conservative of scenarios. Two water-related implications of global warming indeed are beyond dispute.

First, water stress will intensify and spread to new areas as accelerated glacial thaw, degradation of watercourses, and cycles of flooding and drought result in the diminished quality and quantity of available freshwater. And second, shifts in precipitation and runoff patterns will mean greater hydrological variability, negatively affecting food production in some regions unless more water efficient and drought- and flood-resistant crop varieties emerge from agricultural research.

In addition, global warming is expected to make river flows and levels unpredictable, encourage saltwater intrusion into coastal aquifers and lands, reduce aquifer recharge through increased evapotranspiration, and promote smaller snowpacks and thus a significant loss of water storage. More evapotranspiration would result in drier vegetation and soils, while earlier spring and summer runoff in glaciated regions would mean reduced dry-season flows in streams that rely on snow and glacial melt. Higher concentrations of carbon dioxide in the atmosphere, however, could promote two contradictory trends: it could reduce evaporation by depressing the passing of water vapor through plant pores, yet intensify aggregate evapotranspiration through increased plant growth.

Overall, the losses in freshwater storage and supplies could create water shortages even in areas that today take abundant freshwater availability for granted. And shifts in precipitation and melt characteristics are expected to lead to greater interannual climate variations and to heighten the risks of droughts, floods, and other unusual events. This suggests that even with improved water management, negative impacts on sustainable development may become unavoidable.

Through political, economic, or technological means, constraints on the availability of water or other resources can be managed only up to a point, beyond which conflict may ensue.[71] Many of the historical responses to water scarcity will not work in a climate-driven situation. Indeed, there is no single silver bullet "for adapting to the stress that climate change will put on water resources."[72] A new water-ethics process—centered on doing away with profligate water-use practices and embracing conservation, recycling, and environmentally sound allocation and utilization—would have to be developed at national, provincial, and local levels to deal with water stress.

The Hydrological Impact of Environmental Change

Environmental change is distinct from climate change, although there is a popular tendency to blur the difference and even to link all extreme weather events to global warming. Environmental degradation unrelated to the effects of the buildup of greenhouse gases and aerosol concentrations in the atmosphere must not be passed off as global warming. What has climate change to do with the impacts of reckless land use, overgrazing, contamination of surface-water resources, groundwater depletion, environmentally unsustainable irrigation, degradation of coastal ecosystems, waste mismanagement, or the destruction of forests, mangroves, and other natural habitats?

Climate change cannot be turned into a convenient, blame-all phenomenon.[73] The widespread devastation from Hurricane Katrina that hit America's central Gulf Coast states in 2005, for example, was more likely facilitated by human-made environmental degradation. Similarly, the increased monsoonal flooding in Bangladesh is apparently the consequence of upstream deforestation and hydroengineering works, riverbank erosion, and other developmental and population pressures. No single extreme weather event, including the 2012 Hurricane Sandy that lashed the U.S. East Coast, can conclusively be attributed to global warming. Human-caused environmental change, however, does contribute to climate variation over time.

Anthropogenic alterations to ecosystems indeed carry significant hydrological impacts. For example, elimination of natural forest cover, which helps regulate streamflows and serves as a carbon sink, disrupts the water-runoff regime and increases regional temperatures. Along with deforestation, the depletion of swamps (nature's water storage and absorption cover) contributes to a cycle of chronic flooding and drought, besides allowing deserts to advance and swallow up grasslands. Removal of vegetation in catchment areas eliminates a natural instrument to trap sediments and stabilize riverbanks. This can degrade water quality and diminish groundwater recharge. Reckless exploitation of mineral resources, for its part, contributes to contamination of water resources through tailings and land erosion.

Many coastal wetlands, which bridge the transition between freshwater and saltwater ecosystems, have been irremediably altered, especially by cutting down native vegetation. Whereas rain forests, by filtering the air, serve as the earth's "lungs," coastal wetlands act as part of the world's "kidneys" by filtering water before it gets into aquifers and rivers.[74] The vegetation of these wetlands also provides valuable flood protection against storms and helps prevent coastal erosion and saltwater intrusion.

Mangrove swamps are coastal wetlands found in tropical and subtropical regions. Disturbingly, about half of all mangrove forests have been destroyed and 10 percent of coral reefs degraded. Mangroves host unique species and act as a bulwark against hurricanes and tsunamis.

Coral reefs are a rich source of biodiversity, with Australia's Great Barrier Reef—one of the seven wonders of the natural world—alone having more than 700 species of coral, 1,500 species of fish, and 4,000 species of mollusk. Its coral cover, however, has been reduced by about half just since 1985, according to an Australian study published in 2012.[75] Reefs are among the most threatened habitats in the world, even though they yield up to a quarter of all the fish catch of developing nations and serve as a major tourist attraction in areas dependent on income from foreign tourism.

Saltwater incursion in coastal regions is being facilitated by the removal of vegetation and the overexploitation of coastal aquifers. The increase in toxic algal blooms in coastal areas is a reminder of how human activity is endangering marine food chains. Organic carbon from rocks, organisms, and soils continues to be released into the atmosphere as carbon dioxide at an alarming rate as a result of the reduction of forests and grasslands, mining operations, fossil-fuel combustion, and other human activities.

Take forests, which help to keep the planet cool through their ecological-service role in the water and carbon cycles. Each hectare of rainforest stores about five hundred tons of carbon dioxide. Forests of all types also filter pollution to yield clean water and help generate rainfall, even if indirectly. Forests are the largest storehouse of biological diversity, containing more than half the world's plant and animal species—a fact that led to the 1992

Convention on Biological Diversity. The convention grants countries additional rights to the ownership of their forest-based genetic resources, valuable to pharmaceuticals and new farm varieties.

Yet about half of the world's natural forests have disappeared, largely in the past century. During the 1990s alone, 15.2 million hectares of forests were lost on average each year.[76] Hundreds of millions of indigenous, tribal, and poor people still rely on forest resources for their livelihoods. Fortunately, most of the forests in sparsely populated regions remain intact, including the major equatorial rainforests of Central Africa, the Amazon Basin, and the Southeast Asian islands of Sumatra, Borneo, and New Guinea, as well as the boreal forests of Siberia and North America.

Environmental change often serves as a stepping-stone to climate change. For example, fossil-fuel burning and chemical-fertilizer application has altered the natural cycles of carbon and nitrogen, contributing to climate variation. Anthropogenic impacts have thinned the ozone layer that shields life on earth from harmful ultraviolet radiation. Land transformation directly affects freshwater ecosystems and climate. Land transformation actually contributes

up to 20 percent to current anthropogenic CO_2 emissions, and more substantially to the increasing concentrations of the greenhouse gases methane and nitrous oxide; fires associated with it alter the reactive chemistry of the troposphere, bringing elevated carbon-monoxide concentrations and episodes of urban-like photochemical air pollution to remote tropical areas of Africa and South America; and it causes runoff of sediments and nutrients that drive substantial changes in stream, lake, estuarine, and coral-reef ecosystems.[77]

Environmental change indeed is the main threat to the integrity of freshwater reserves.

To lessen the hydrological impacts of environmental and climate change, countries must strategically invest in ecological restoration—growing and preserving rainforests, conserving wetlands, shielding species critical to ecosystems, and restoring rivers and other natural heritage. Ecological-restoration programs, by assisting in the recovery of damaged or degraded ecosystems, can help bring wider benefits in regulating regional climate, slowing soil and coastal erosion, augmenting freshwater storage and supply, and controlling droughts and flooding.

Some people illusorily believe that environmental and climate change would change the relative strategic weight of nations, with those in the colder climes gaining, such as Canada, the Nordic bloc, and Russia, but many others suffering an erosion of security and status.[78] Spurred by such thinking, Russia, Denmark, Canada, the United States, Norway, and Iceland are in a race to claim the potentially vast energy and other resources of the Arctic by asserting their territorial claims to the polar region, where melting ice, they believe, could open water for drilling or create the long-sought Northwest Passage for shipping.[79]

Some other persons, like Czech president Vaclav Klaus, have even mocked international efforts to control anthropogenic impacts on eco-systems. Just as Austrian-trained economist Friedrich von Hayek, during World War II, asserted in *The Road to Serfdom* that fascism and communism shared the same fundamental roots in that they both suppress human freedom, Klaus in his controversial 2007 book, *Blue Planet in Green Shackle—What Is Endangered: Climate or Freedom?*, has contended that communism and environmentalism are two sides of the same coin as they stifle freedom and innovation.

Klaus is thus more than just a climate-change contrarian. "The largest threat to freedom, democracy, the market economy, and prosperity at the end of the twentieth and at the beginning of the twenty-first century is no longer socialism," Klaus wrote in his book. "It is instead the ambitious, arrogant, unscrupulous ideology of environmentalism."[80] While it is unconscionable that global warming is offered by some as the explanation for any natural disaster, Klaus stands out as a doom-monger by continuing to rail against what he sees as a mortal threat to human freedom and the market economy posed by "fear-arousing" environmentalism.

Despite both climate-change enthusiasts and skeptics falling to the lure of hyperbole, few scientists doubt the serious, long-term implications that climate and environmental change holds for water resources. Such change, given its effects on resource security and socioeconomic stability, is clearly a threat multiplier. Yet the international agenda on global warming has become politically loaded because important actors have tagged onto it all sorts of competing interests, economic and otherwise. Climate change must not remain a convenient peg on which to hang assorted national interests. Yet this trend serves as a reminder that climate change is not just a matter of science but also a matter of geopolitics. Without improved global geopolitics, there can be no real fight against climate change.

In an increasingly interconnected and interdependent world, the blunt fact is that there will be no winners from environmental and climate change, only losers. The effects of global warming will be universal, including making weather patterns more unpredictable in higher latitudes, where agriculture, public health, and ecosystems could be adversely affected. The Arctic and the Antarctic are warming faster than the rest of the world, except the Tibetan Plateau, with mean annual temperatures at the two poles having risen by about two to three degrees Celsius since the 1950s.[81] The projected greater impacts of climate change in the developing world are likely to carry such international consequences as to offset any potential benefits to the rich nations in the colder climes.

Water-Related Challenges in an Era of Global Warming

As global warming accelerates, the water-related challenges will extend from the existential threat to low-lying nations from rising ocean levels to the permanently displaced climate migrants potentially bringing freshwater resources under strain in their new habitats. The challenges will test many countries' disaster-response and emergency-management capabilities, exposing national institutional vulnerabilities. Dealing with the challenges will demand greater intranational, regional, and international cooperation.

Low-lying states' water challenges will extend beyond freshwater shortages to potential inundation by more frequent storms and loss of arable land from saltwater incursion. The South Pacific Islands, for instance, are likely to be battered both by an increased frequency of tropical storms and by rising sea levels[82]—a scenario that could force the migration of many residents to Australia and New Zealand. The same specter could also haunt some major coastal cities in the world.

What is common between the Maldives—the flattest state in the world and the smallest country in Asia in terms of population—and Kiribati, a Pacific nation also made up of low-lying islands? In addition to these countries boasting among the world's highest per capita fish-consumption levels, the presidents of both the Maldives and Kiribati proposed the establishment of a fund to buy a new homeland for their citizens if global warming raised sea levels dangerously.[83]

Take another case, Bangladesh, which has almost 170 million citizens. In addition to the millions of Bangladeshis that already have illegally settled in India, many Bangladeshis have moved internally from rural areas to the capital city, Dhaka, as environmental refugees, driven out by floods, cyclones, and saltwater incursion from the Bay of Bengal.[84] Infiltration of saltwater is swallowing up valuable arable land in one of the world's most densely populated countries, whose land area is less than half the size of Germany but with a population more than double. Excluding microstates, Bangladesh boasts the greatest population density in the world.

Much of Bangladesh is made up of the massive estuary deltas of the Brahmaputra, Ganges, and Meghna rivers. Such is the extent of low-lying floodplains and deltas that about one-third of the country is situated less than six meters above sea level. With the IPCC warning that Bangladesh risks losing 17 percent of its land and 30 percent of its food production by 2050 due to increasing saltwater incursion, many more Bangladeshis could be compelled to relocate to India, already home to more than 15 million illegally settled Bangladeshis.[85] The estimated number of Bangladeshi refugees resident in India is actually greater than the number of Mexican immigrants—legal and illegal—living in the United States.[86]

Bangladesh's vulnerability has been heightened by the fact that sediments disgorged by rivers have shaped the seafloor near the coast in such a manner as to facilitate the funneling of seawater inland during storms and hurricanes.[87] Growing salinity in the lower part of the country's deltaic region threatens a UNESCO World Heritage site—the Sunderbans, the world's largest contiguous mangrove forests, located along the Bay of Bengal and extending into India. Bangladesh's low-lying topography also facilitates river flooding during the monsoon season, when most of the country's 230 rivers and streams are in spate.

Another major dimension of climate change relates to shifts in rainfall patterns. In many parts of the world, farmers are greatly dependent on rainfall for crop cultivation, and any major variability in rainfall patterns would be detrimental to agriculture. Several studies on climate change actually point to a potential silver lining—that global warming is likely to intensify the hydrological cycle and bring increased rainfall in the wet season, especially in the tropics. For example, simulations of changes in precipitation patterns in tropical Asia indicate greater rainfall during the monsoons—the Southwest Monsoon and Southeast Monsoon in the summer and the Northeast Monsoon in the winter—with the greatest shifts in summer rainfall patterns projected in southern and southeastern Asia, excluding Indonesia.[88] The studies, however, have found it difficult to gauge the likely changes in dry-season rainfall.

The IPCC indeed has projected that mean annual precipitation, and heavy-precipitation events, will increase globally throughout the twenty-first century.[89] This is in keeping with the trend that global precipitation levels increased during the twentieth century and, consequently, that days with heavy rain and snow have become more frequent in the world.[90] The projected further increase in intense-precipitation events could mean record-setting rainstorms and snowstorms.

As average temperatures rise, the atmosphere's capacity to hold moisture will increase, because warmer air carries more moisture. For each one-degree Celsius rise in temperature, the water-holding capacity of the atmosphere is expected to rise 7 percent.[91] So, when it rains, it may pour. But regions subject to extreme climatic stress are expected to face more extensive flooding and/or more prolonged droughts, which could make wildfires more common. More frequent wildfires would contribute to land transformation and erosion, thereby adversely affecting water resources.

The likely greater precipitation, however, suggests that global warming is not going to be an unmitigated disaster, although the general rule of thumb is that wet regions will get wetter and dry regions will become drier. As river-flow seasonality heightens in the rain-dominated catchments, there will be greater flows in the peak-flow season and lower flows during the low-flow season or extended dry periods.[92] Adaptation to the shifts in precipitation and river-flow patterns would demand, among other things, rainwater harvesting, much greater reservoir storage capacity, water conservation, and recycling of water.

The greater precipitation, after all, will likely be offset by unfavorable trends, including increased evapotranspiration, larger variability in flows, and more frequent and intense droughts, flooding, and hurricanes. Greater rainfall variability will negatively affect agriculture, which does best with a relatively stable water supply. A study by the U.S.-based National Center for Atmospheric Research has warned that dry spells could significantly lengthen in several parts of the world.[93] The reason is simple: whereas rising temperatures will allow more water to evaporate from oceans, rivers, lakes, and reservoirs, thereby creating extra moisture in the air for rain or snow, the warmer weather will also draw more moisture out of the ground, thus worsening drought conditions when it has not rained for an extended period.

According to another study, published in the American Meteorological Society's *Journal of Climate*, flows of the rivers that serve as the lifeblood for the most populous countries in the world are already falling, partly due to global warming. The study, led by scientists at the National Center for Atmospheric Research, examined river flows from 1948 to 2004. It looked at the historical monthly streamflows at the farthest-downriver stations for the world's 925 largest ocean-reaching rivers, blending the raw data with computer-based stream-flow models to plug data gaps. It found significant shifts in the discharge of about a third of the rivers (including the Columbia, Congo, Ganges, Mississippi, Niger, Paraná, Uruguay, and Yenisey), with the number of rivers experiencing reduced flow exceeding those with increased flow by a ratio of almost 2.5 to 1.[94]

The diminished ocean discharge of rivers flowing through densely populated regions highlights troubling global shifts in continental runoff. Rivers with reduced discharge at the mouth include, for example, China's Yellow, India's Ganges, and West Africa's Niger. The runoff shifts, indicative of overexploitation of water resources, mean that many rivers are also disgorging into the oceans reduced nutrients and minerals, which are vital to marine life.

However, Europe and the United States, by and large, have experienced an opposite trend—greater river runoff. A notable exception is the binational

Colorado River. Globally, the National Center for Atmospheric Research study found that in the fifty-six-year period it examined, the average annual river waters discharged into the Indian Ocean fell by about 3 percent and into the Pacific Ocean by 6 percent but rose by about 10 percent into the Arctic Ocean.

Yet another trend that portends greater water insecurity in the coming years is the accelerated thawing of glacier ice, the earth's largest reservoir of stored freshwater. As a result of the increasing levels of greenhouse gases and aerosols in the atmosphere and other factors, earth is now said to be absorbing 0.85 ± 0.15 watts more energy from the sun per square meter than it is reflecting into space, with this imbalance expected to promote "acceleration of ice-sheet disintegration and sea-level rise."[95] Mountain snowpacks serve as major surface-water reservoirs, with seasonal snowpack melt a key and essential source of freshwater supply. But warmer climate is likely to result in smaller snowpacks, earlier snow and glacial melt, and decreased dry-season river flows.

Accelerated thawing of snow and ice layers will cause peak river runoff to advance to late spring and early summer from the midsummer and autumn months when water demand crests, especially for agriculture.[96] Maximum river flows in many regions could be advanced by more than a month. Such peak-runoff cycle shifts, the effects of which would extend to the operation of reservoirs and hydropower plants, have led one scientific study to gloomily conclude that the "current demands for water in many parts of the world will not be met under plausible future climate conditions, much less the demands of a larger population and a larger economy."[97]

Many glaciers in the world are already melting at an accelerating clip, a trend that would, in the near to midterm, likely increase flows of the rivers draining glaciated regions, but with the glacier-melt contribution to the flows decreasing over time as glaciers gradually become smaller. In several parts of the world, snowpacks have declined in recent decades owing to global warming and natural climate swings. One study has warned that average snow accumulation will further decrease, compared to historical patterns, within the next three decades across large parts of the Northern Hemisphere.[98] For the people and wildlife relying on the seasonal snowpack melt for their water supply, the consequences of the changes in their hydrological "insurance" are likely to be serious.

Snowpacks, for example, provide more than three-quarters of the water supply in the American West. Yet, with snowpack reductions across the northern Rocky Mountains in recent decades almost unprecedented in scale, the increasing contribution of springtime warming to large-scale snowpack variability, according to one study, presages "fundamental impacts on streamflow and water supplies across the western United States."[99] A long-term decline in the Rockies snowpack could mean the loss of the American West's natural freshwater storehouse.

Declining snowpacks, as table 3.1 shows, have also been documented elsewhere in the world, including in South America's Andean region and on Mount Kilimanjaro and Mount Kenya in Eastern Africa. The attrition rate of Kilimanjaro's icecap has significantly increased since the end of the 1980s, and such loss is contemporaneous with glacier retreat elsewhere in mid to low latitudes.[100]

One of the world's greatest glacial and snowpack declines is centered in the Great Himalayan Watershed and the adjacent Karakoram, Kunlun, Hindu Kush, Pamir, Alay, and Tian Shan ranges. This extended Himalayan region is home to thousands of glaciers and the source of some of the world's greatest river systems, ranging from China's two main rivers, the Yangtze and the Yellow, to the principal rivers of Southeast Asia, northern India, the Afghanistan-Pakistan belt, and Central Asia. Compared to the European Alps—the water tower of Europe—Himalayan Asia, which boasts the world's tallest mountains, is a water tower of unmatched size that supports nearly half the global population living in the basins of the rivers flowing down from its mountains.

Asia's great river systems rely on the constant flux of glaciers. With the monsoon domination bringing most of the year's rain in a three- to four-month period to large parts of the continent, the rivers are sustained by melting snow and ice in the warmer months. The accelerated thawing of glaciers in Himalayan Asia will have adverse impacts on irrigation and hydropower by increasing the seasonality of runoff. It thus holds major long-term implications for Asia's most heavily populated regions, which find themselves on the front lines of global warming. Glacial shrinkage will generate increased meltwaters in the next few decades, helping to meet dry-season human needs, before the situation morphs from plenty to paucity of resources.

Climate-change effects are already being felt in Tibet, the world's largest and highest plateau, which is called the "Third Pole" because it has the largest perennial ice mass on the planet after the Arctic and Antarctica. In terms of accessible freshwater, the Tibetan Plateau is actually the world's largest water repository. Its vast snowfields, thousands of glaciers, hundreds of lakes, and huge underground springs have made it the source of many of the world's greatest river systems. Its mountain springs in the winter and snowpack in the summer ensure dependable year-round river flows.

The effects of global warming indeed are more visible on the Tibetan Plateau than in the North and South poles. The plateau—nearly two-thirds the size of the European continent—is warming at a rate faster than the global average, in part because of its exceptionally high elevation, which makes it

Table 3.1. Accelerated Thawing of Snow and Ice Layers and Its Significance for Freshwater Supply

Rocky Mountains and Pacific Northwest snowpacks	Western U.S.	From the Rockies to the Cascades, springtime snowpacks have declined. The Rockies snowpack is central to freshwater supply in much of the irrigation-dependent American West. About 85 percent of the flow of the Colorado River, the primary source of irrigation in the American Southwest, comes from Rockies snow and ice melt. Meltwaters from the Sierra Nevada snowpack help irrigate California's Central Valley, the world's reputed fruit basket. The flows of the runoff-fed Columbia and Missouri rivers are also at risk of diminishing.
Himalayan glaciers	Great Himalayan Watershed	The more than 5,000 large glaciers in the Himalayan Range—among the biggest and most spectacular in the world—serve as massive storehouses of freshwater for the world's largest concentration of population. Glaciers cover 17 percent of the Himalayas and store an estimated 3,870 cubic kilometers of water. These glaciers extend from Burma's northern tip to Afghanistan. The snout of several of the leading glaciers in the Indian Himalayas has retreated, including the Gangotri, a source of meltwaters to the Ganges.
Tibetan glaciers	Tibetan Plateau	These glaciers, many of them part of the Great Himalayan Watershed, supply meltwaters to the Yangtze, Yellow, Indus, Brahmaputra, Mekong, Salween, Irrawaddy, and other rivers that are the lifeblood for hundreds of millions of Asians. According to the Institute of Tibetan Plateau Research of the Chinese Academy of Sciences, these glaciers have reduced by 7 percent since the 1960s. The Himalayan portion in Tibet has at least three times more glaciated region than the part in India.
Karakoram, Kunlun, Hindu-Kush, Pamir, Alay, and Tian Shan glaciers	Himalayan orogenic belt extending into Central Asia	These glaciers are significant contributors of meltwaters to Central Asia's principal rivers—the Tarim, the Syr Darya, and the Amu Darya—as well as to a couple of South Asian rivers. Two of the tributaries of the Tarim originate in parts of the disputed Kashmir region controlled by China and Pakistan. Global warming and land-cover changes have apparently aided glacier recession in this belt.
Mount Kilimanjaro and Mount Kenya icecaps	East Africa	Ice loss on Africa's highest mountain, Kilimanjaro in Tanzania, has been dramatic, with 85 percent of the ice cover present in 1912 having vanished. There is also evidence of significant icecap recession on Africa's second-highest peak, Mount Kenya, with the result that there is a decrease in the flow of rivers that are the lifeline for people in and around Nairobi.
Andean glaciers	South America	The accelerated thawing of Andean glaciers other than in southern Patagonia threatens long-term water supplies to major cities and puts at risk populations and food production in Colombia, Peru, Chile, Venezuela, Ecuador, Argentina, and Bolivia. The Quelccaya icecap, covering forty-four square kilometers in the Peruvian Andes, is the world's largest tropical ice mass. Its biggest glacier, Qori Kalis, has alarmingly receded. In Bolivia, the Chacaltaya glacier—the source of freshwater for La Paz and El Alto cities—is retreating. So too the O'Higgins glacier in Chile and Argentina's Upsala glacier.

Sources: United Nations Environment Program, 2007; Food and Agriculture Organization, 2011; and Intergovernmental Panel on Climate Change, 2007.

"the Roof of the World." This warming, extending to the eastern Himalayas, is leading to the accelerated melting of snow and ice fields across the wider Himalayan region. The faster warming also holds significant implications for Asian precipitation patterns because the plateau, by serving as a high-elevation heat pump, draws into the Asian hinterland the monsoonal currents from the Arabian Sea, the Bay of Bengal, and the East and South China seas.[101]

This trend actually poses a potential threat to climate stability beyond Asia. The towering Tibetan Plateau, in addition to playing a unique role in Asian hydrology and climate, influences the Northern Hemisphere's atmospheric general circulation, or the system of winds that helps transport warm air from the equator toward the higher latitudes, producing different climate zones. So the warming trend, accentuated by reckless environmental degradation on the plateau, also has a bearing on climatic patterns in Europe and North America.

Several studies have highlighted glacier erosion in Himalayan Asia. A 2012 report by the U.S. National Academy of Sciences said glaciers in the eastern and central Himalayan regions are retreating at accelerating rates but glaciers toward the western Himalayan rim seem more stable, and some may even be growing.[102] This largely mirrored the finding of another study, which reported that most Himalayan glaciers other than those in the Karakoram Range in the west are losing mass at rates similar to glaciers elsewhere in the world.[103] The Chinese Academy of Sciences has also recorded a decrease in the area and mass of Himalayan glaciers in Tibet.[104]

Another study specifically found a significant increase in the Tarim River runoff in southern Xinjiang owing to the accelerated glacier thawing.[105] According to yet another study, a combination of rising greenhouse-gas levels in the atmosphere and the clouds of aerosol particles over the Indian Ocean and Asia from biomass burning and fossil-fuel combustion "may be sufficient to account for the observed retreat of the Himalayan glaciers."[106] Although aerosol particles play a cooling role by reflecting sunlight back into space and by promoting increased light scattering and cloud cover, they absorb solar radiation and contribute, like greenhouse gases, to lower-atmospheric warming trends.

The magnitude of the glacial attrition is smallest in Tibet's arid interior and greatest along the plateau's southern and eastern rim, which boasts the world's supreme combination of glaciers and riverheads and is part of the Great Himalayan Watershed.[107] Besides relying on runoff from rainfall, several of the great Asian river systems have particularly high dependency on snow and glacier melt to sustain their flows from late spring and are therefore very vulnerable to climate-driven shifts in melt characteristics.[108] Table 3.2 indicates the vulnerability of the major Asian rivers relative to their level of dependency on meltwaters.

Table 3.2. Dependence of the Major Rivers on Glacial and Snow Meltwaters That Originate in or around the Great Himalayan Range

River	Basin Area (thousand km²)	Cropland (percent)	Dependence on Meltwaters
Amu Darya	535	22	Very high
Brahmaputra	651	29	Very high
Ganges	1,016	72	High
Indus	1,082	30	Very high
Irrawaddy	414	31	Low
Mekong	806	38	Moderate
Salween	272	6	Moderate
Syr Darya	783	22	Very high
Tarim	1,152	2	Very high
Yangtze	1,722	48	Moderate
Yellow	945	30	Moderate

Sources: United Nations Environment Program, *Global Outlook for Ice and Snow* (Nairobi, Kenya: UNEP, 2007); Walter Immerzeel et al., "Climate Change Will Affect the Asian Water Towers," *Science* 328, no. 5983 (June 11, 2010); M. L. Parry et al. (eds.), *Climate Change 2007: Impacts, Adaptation, and Vulnerability* (Cambridge, UK: Cambridge University Press, 2007); C. Revenga et al., *Watersheds of the World: Ecological Value and Vulnerability* (Washington, DC: World Resources Institute, 1998).

More broadly, climate stress, with the attendant freshwater scarcity and cropland degradation, threatens to sharpen competition over scarce resources and engender civil strife. The political effects of climate change could extend from internal destabilization to even possible state failure. By overwhelming some nations' adaptive capacities, climate shifts could foster or intensify conditions leading to failed states—the breeding grounds for extremism, fundamentalism, and terrorism. The most fragile states, with little institutional capacity or financial resources to manage the impacts, would be at serious risk of collapsing.[109] There may also be an escalation of low-intensity military threats that today's conventional forces are already finding difficult to defeat—transnational terrorism, guerrilla campaigns, and insurgencies.

Large-scale internal strife or state failure, in turn, could invite foreign intervention—overt or covert. In any event, because of the projected higher frequency and intensity of extreme weather events, the armed forces of the major powers are likely to be increasingly called upon to respond to natural disasters elsewhere. There are already American calls that the Pentagon integrate such response capabilities into its regional military commands across the world.[110] Given the specter of greater water conflicts, dysfunctional states, refugee flows, and more frequent hurricanes, flooding, and droughts, global warming has been correctly characterized as a threat multiplier. In some cases, it may even serve like "Mother Nature's weapon of mass destruction," even if a slow-acting one.[111]

To be sure, more scientific research is needed to help improve existing climate-change simulation models and build a better understanding and quantitative estimation of the likely impacts on freshwater resources. So that appropriate adaptation measures can be developed across multiple, water-dependent sectors, extending from energy production to public health, existing gaps in research must be plugged, including by reliably downscaling global climate models to the level at which management is most needed—catchment areas.[112]

With plain old water on its way to becoming "blue gold," the world must begin to find ways to manage the geopolitical and human-security risks of climate change. At a time when climate change threatens to become an engine of destabilization, meeting the water-related challenges above all holds the key to continued socioeconomic stability and peace.[113] Strategies for adaptation to climate impacts must be central to any preventive policy.[114]

Climate change, more fundamentally, needs to be elevated to a strategic challenge in order to help build multitiered international, regional, and intranational cooperation, as well as a climate information system database and public-private partnerships. Promotion of sustainable uses of common-pool resources demands forward-looking policies, institutional diversity, and multilateral cooperation.[115]

The worst choice would be for the technologically advanced states to take matters into their own hands and try to regionally modify climate through the tools of geoengineering, including technically possible options such as stratospheric aerosol injection and boundary-layer marine cloud seeding.[116] For example, China, flaunting its new technological prowess, has openly experimented with weather modification in Beijing and elsewhere, including injecting silver iodide crystals into clouds to trigger more precipitation in the upper catchment of the dying Yellow River. In what was trumpeted as a big weather-modification success, the Chinese Air Force cleared the skies over Beijing's Tiananmen Square just in time for the 2009 National Day parade. However, another operation in 2009 misfired, bringing on so much snow that China had to close twelve highways around Beijing.

National experiments to geoengineer climate, including improving local weather or inducing more precipitation, open a new human-interventionist frontier with unpredictable, long-range consequences for nature. They raise serious concern about the adverse impacts that weather modification may cause in other regions owing to the climate system's global interconnections. Cloud seeding, for instance, could help suck in moisture from another region. Indeed, geoengineering technology ominously risks becoming a military instrument more powerful than any nuclear weapon.

FORESTALLING A THIRSTY FUTURE

Respect for the environment and sustainable management of common-pool resources are notions that are still not actively embraced across much of the world. Water holds the strategic key to peace, public health, and prosperity, yet it is the resource under the greatest pressure because of overexploitation. Worse, contamination is increasingly limiting public access to safe drinking water. The quality and quantity of available water resources is becoming a critical component of national security, helping to highlight the need for better management of a resource increasingly in short supply.

One stark reality is nature's unequal distribution of water resources in the world. Some of the world's largest rivers run through sparsely populated regions. They include the Amazon (which, with 15 percent of the world's runoff, supports just 0.4 percent of the world's population); the Congo, which empties into the Atlantic Ocean; and the major rivers draining into the Arctic Ocean from northern Canada and Siberia.[117] In contrast, very little of the waters of several great river systems such as the Yellow, the Nile, and the Indus reach the sea because of heavy abstractions in the densely populated regions through which they flow.

African water resources are concentrated in Central Africa and, to a lesser extent, West Africa, with the result that 300 million Africans lack adequate access to a water supply and sanitation. Asia's richest water resources are concentrated in its mountainous central regions that are far from the plains where most Asians live. The impressive aggregate size and per capita availability of water resources in Latin America obscures the local scarcity conditions in several of its heavily inhabited areas, such as Chile's Valle Central, the Peruvian and South Ecuadorian coast, sections of Argentina, Colombia's Cauca and Magdalena valleys, the Bolivian portion of the Antiplano—the world's largest mountain plateau after Tibet—the Brazilian northeast, the Pacific Coast of Central America, and large parts of Mexico.[118]

A large number of countries depend significantly on natural inflows from across their borders. Such reliance on neighbors for freshwater supplies can engender water insecurity. Countries with a very high dependency ratio (equal to the part of renewable water resources originating outside national frontiers) include Kuwait, 100 percent; Turkmenistan, 97.1 percent; Egypt, 96.9 percent; Mauritania, 96.5 percent; Hungary, 94.2 percent; Bangladesh, 91.4 percent; Niger, 89.6 percent; and the Netherlands, 87.9 percent.[119] States located at the mouth of major international river systems—like Bangladesh, Egypt, Iraq, the Netherlands, and Vietnam—usually have high dependency on transboundary inflows and are most vulnerable to the impacts of diminished flows.

At the other end of the spectrum are countries with no reliance on cross-border inflows. Angola, Ecuador, and Ethiopia have a zero dependency ratio, while just 1.8 percent of Canada's water resources flow in from across its borders. China and Turkey are most happily placed: they receive little water from across their frontiers yet are considered hydro-hegemons because each is the principal source of transboundary river flows in its region.[120]

Turkey—the origin of waters flowing to Syria, Iraq, and Georgia—is rich in wetlands, rivers, and lakes. Rivers, in fact, define one-fifth of its total borders with neighboring countries. But just as much of China's water resources are concentrated in the homelands of ethnic minorities, including the sprawling Tibetan Plateau, Turkey's southeastern and eastern Kurdish regions are the source of the Tigris, the Euphrates, and smaller rivers like the Murat, the Ceyhan, and the Buhtān.

The overexploitation of resources, however, is the principal cause of water shortages and insecurity. In most arid and semiarid regions, extending from the North China Plain to the American Southwest, water withdrawals are already greater than precipitation, which means that the strategic water reserves in aquifers are being used faster than they can be recharged through rain, snowmelt, and other sources. Such use of subterranean water reserves due to inadequate availability of surface resources increases the likelihood that vulnerable areas will experience major water shocks in the future.

Water Refugees and Water Warriors

When sources of freshwater begin to dry up, water refugees will stream across provincial and international frontiers, exacerbating ethnic, sectarian, and political tensions. The Stern Report presented a frightening scenario of 200 million climate migrants by the middle of this century. Such refugees, displaced by water and food crises or recurrent droughts, hurricanes, and flooding, are likely to be the poorest, most vulnerable people.[121] They would create new socioeconomic and cultural divides in the places they migrate to, besides potentially provoking a backlash against their influx.

For example, Yemen, an impoverished failing state that has become an incubator of Islamic terrorism, is headed toward a freshwater catastrophe with serious international implications. Heavy abstractions and global warming are set to worsen the water situation and further destabilize Yemen, where one of the world's fastest-growing populations threatens to completely drain aquifers by midcentury or earlier. Even the deeper fossil aquifers there are being rapidly depleted. When Sanaa and other cities in the Yemeni highlands run out of water, millions of thirsty water refugees are likely to flood coastal plains, with a number of them seeking to migrate to other countries. Yemen is emerging as a hydrological basket case.

However, Yemen is not the only country in danger of producing parched refugees who relocate far from their native villages or towns. Water refugees from elsewhere are also expected to stream across ethnic or national frontiers to escape acute water distress. Large-scale, cross-border migration driven by any factor is likely to compound internal or regional conflicts by stoking new ethnic and sectarian violence, as underscored by the bloody clashes in 2012 between tribal people and Muslim immigrants from Bangladesh in the northeastern Indian state of Assam and the subsequent flight of northeastern Indians from southern India to escape feared Muslim reprisals. Those clashes—like the recurrent violence in Burma between Buddhists and a Muslim ethnic group, the Rohingya, whose members the Burmese government does not recognize as citizens, referring to them as Bengalis from Bangladesh—were sparked by ongoing conflict over natural resources. Internal conflicts over water and other resources fester in many countries.

Persistent water scarcity tends to encourage local social strife and a resort to strong-arm tactics, with the poorest at the receiving end. It can also create water warriors, ready to spill blood to control or shield sources of supply. Villagers in some parched areas in South and Central Asia now protect their wells, tanks, and ponds from water thieves by engaging private security guards. In the parts of East Africa that are racked by recurrent drought, many wells are controlled by criminal gangs or warlords.

Water in these dry areas has become a precious resource worth fighting for and dying over. The warlords of water hold sway over local wells by employing water warriors, with sporadic deadly clashes leaving a trail of water widows.

For a holistic picture, water conflicts must be assessed at all levels—interstate, intranational, and local—along with the cross-linkages. Internal instability, for example, can intensify international water disputes, and vice versa.[122] Countries already troubled by domestic unrest or rising popular discontent tend to be more vulnerable to violent internal water disputes. If two such internally torn countries neighbor each other, share resources from a transboundary basin, suffer from serious water stress, and are locked in an adversarial relationship, the potential for water tensions or conflict between them is high.

Within water-stressed nations, intense competition for water occurs between different provinces, communities, and classes, as well as between farms and cities and between industry and rural residents. When such competition assumes tribal, ethnic, or sectarian dimensions, it poses significant internal-security challenges. If a water situation goes from bad to worse, water-related struggles can fan interprovincial or intercountry hostility, leading to greater water insecurity or possible overt conflict "caused by one party disrupting the water supply of another."[123]

Indeed, the threat of civil strife, loss of life, and humanitarian crisis is generally greater in intrastate than in interstate conflicts over water. There are, however, no well-developed principles of international humanitarian law directly relevant to internal conflicts. The history of domestic water conflicts over the past half a century highlights that water more often serves as a source of competition than cooperation—a trend likely to intensify as water distress accentuates.

Some domestic water conflicts arise from deep-seated sociopolitical grievances that breed violence. This is best exemplified by the deadly conflict in the western Sudanese province of Darfur that has its roots, according to a study by the United Nations Environment Program, in water and environmental issues.[124] A significant decline in rainfall since the 1960s has spurred acute water scarcity and turned millions of hectares of marginal arid grazing land into desert, exacerbating competition between sedentary farmers and seminomadic herders along racial fault lines.

This competition over access to grazing land and water in Darfur took the form of low-level conflict following droughts in the 1980s. It then escalated to a bloody conflict in 2003, when the Sudan Liberation Army (SLA) and Justice and Equality Movement (JEM) launched armed attacks, accusing the government in Khartoum of oppressing black Africans to favor Arabs. The Sudanese government's counterinsurgency campaign has enlisted proxy Arab militias that have terrorized the mostly black African population of the embattled, economically marginalized region, which is near Sudan's border with the new nation of South Sudan. Since 2003, as many as 300,000 people have died in the Darfur conflict—mostly from disease—and 2.7 million have been displaced, with some fleeing across the western border to Chad.[125]

A study has found a link between economic shock caused by poor rainfall and greater civil strife and refugee flows in sub-Saharan Africa. Using rainfall changes as an instrumental variable for economic growth in forty of the sub-Saharan countries from 1983 to 1999, the study reported that a negative growth shock of five percentage points resulting from deficient rainfall increased the likelihood of civil conflict by nearly one-half.[126] That such economic shocks have a dramatic causal relationship with civil war was illustrated by the example of Sierra Leone, which plunged into conflict in 1991, barely a year after a sharp decline in rainfall resulted in drought and economic hardship.

Scarcity of a resource as essential for human survival as water creates socioeconomic instability, which in turn engenders a political environment conducive to civil strife. The linkages between resource scarcity, environmental degradation, and conflict often spur a vicious cycle that chains the poor to interminable penury. Poverty is the enemy of the environment because it promotes greater degradation.

Lack of water can foster underdevelopment. And when lack of development encourages people to migrate in search of jobs and resources, it builds more pressures on the environment. In situations where water shortages have cut farm production, conflict over irrigable land and water resources can further limit food supplies. Water conflicts between groups, communities, and provinces are not only endemic in several parts of the world, but attacks on water facilities have also repeatedly occurred in civil wars.

Although violent water struggles occur mainly in developing countries, intrastate water disputes are not unknown in the rich world, including, for example, Spain and the two countries where European settlements triggered water conflicts with indigenous peoples—Australia and the United States. The Texas–New Mexico rivalry over water rights dates back to the nineteenth century, while Arizona in 1934 sought to "protect" its interests against water diversion by dispatching a militia to a border dam being built by California on the Colorado River.[127] Given that water struggles are common in the American Southwest, Andrew Wice's novel, *To the Last Drop*, with its story of Texas and New Mexico going to war over water, is more than just a fictional tale.[128]

In fact, water in the United States has become so contentious an issue that disputes have spread from the West to the East, and more than thirty states nationwide are now "fighting with their neighbors over water."[129] The bitter fight between Alabama, Florida, and Georgia since 1990 over the right to water from Lake Lanier is one example, with the U.S. Supreme Court in 2012 declining to intervene in the feud.

In a dispute between South Carolina and North Carolina over the shared Catawba River, the former sued its neighbor in 2007, claiming that North Carolina was moving to unlawfully divert water. In a 5–4 interim ruling in 2010, the U.S. Supreme Court handed South Carolina a partial victory by barring the city of Charlotte—the largest single user of the Catawba waters—from joining the legal challenge. An out-of-court agreement later ended the dispute but left the door open to future lawsuits in the event of shifts in water withdrawals. The water dispute between the state of Mississippi and the city of Memphis also went to the U.S. Supreme Court.

In Arizona, where Native American tribes are spread out over 28 percent of the state's arid land, water rights for these communities remain a more contentious issue than in other states.[130] In one case in Arizona, the upstream diversion of the Gila River waters by farmers plunged a downstream Native

American community into poverty and obesity. Deprived of the waters of a river on which it had depended for many centuries, the tribe lost its farming traditions and had to subsist on government rations of highly refined food, leaving the reservation with among the world's highest rates of diabetes.[131] After decades of struggle and litigation, the tribe won a water-rights settlement to get some of its waters back, starting in 2008.

It is in the developing world, however, that water-conflict risks are particularly high, along with the specter of water refugees. Water refugees can seek to either temporarily escape from a drought-seared region or permanently migrate when their sources of supply begin to dry up—a scenario more likely to emerge from the gross overexploitation of resources than from any natural cause. In conditions of scarcity, efforts by socially or economically dominant groups to control sources of water can help drive out powerless residents. Interethnic or intersectarian water conflict, compounded by recurrent drought, can also force the weaker groups to relocate.

If communities are to manage competition and conflict over scarce water resources, they need water warriors not for usurping or controlling water sources on behalf of the powerful but to help create greater public awareness of ways to manage the water problem through cooperative local-level actions. Such warriors of peace could play the traditional role of warriors—safeguarding the interests of entire communities to help keep the peace.

The Dangers of a Parched Future

Although freshwater is fundamental to all human activities, scarcity of this resource threatens to become the defining crisis of the twenty-first century— a situation that would imperil the stability and security of many nations and engender more conflicts.[132] To help mitigate the conflict potential of water scarcity, it is important to face up to the growing challenges by adopting more sustainable and cooperative practices. Lessons learned on how to manage water resources holistically must be applied to the volatile and vulnerable regions of the world.

Water scarcity arising from recurring drought or resource overexploitation holds far-reaching implications. The overextraction of groundwater from basins linked with surface watercourses is reducing river flows, drying up wetlands, and shrinking lakes, while overexploitation of coastal aquifers is inviting seawater to intrude and replace the lost freshwater. Excessive withdrawals from rivers are seriously contributing to degraded water quality and altering fluvial ecosystems.

But whereas reductions in river flows or lake levels are noticeable to the eye, the fall in underground water tables is invisible, emboldening users to blithely continue pumping groundwater at unsustainable rates. What is out of sight tends to be out of mind. The water in many aquifers has slowly accumulated over hundreds and even thousands of years, constituting a sort of savings deposit for communities to tap in hard times, such as a severe drought. Fortunately for many of those living in riverless regions, some of the world's largest aquifers are located under desert sands. But these aquifers are particularly difficult to replenish by rain.

So, when their sole source of freshwater dries up, the people dependent on groundwater will have little choice but to relocate. This is a potent threat in the riverless Arabian Peninsula, as well as in four of the five Maghreb Region countries that lack any perennial surface watercourse. There are, of course, other countries or dependencies that also lack any perennial river or stream, including Anguilla, Bermuda, Cyprus, Gibraltar, Kiribati, the Marshall Islands, the Maldives, Monaco, Nauru, Tonga, and Tuvalu, with aquifer depletion a serious problem in many of these places. In addition, some arid or semiarid areas in large countries, including the United States, rely solely on groundwater resources.

Water tables continue to fall as millions of electric and diesel-fuel pumps extract groundwater, including for the production of wheat, barley, oats, grapes, olives, citrus fruits, and meat on arid lands. With the rate of withdrawal often surpassing the rate of recharge, water must be pumped from ever-greater depths, with the risk that the aquifers will get fully depleted or yield only brackish water. Nowhere is this threat more real than in Saudi Arabia, which, by deciding to phase out all wheat cultivation by 2016, became probably the first country in the world to acknowledge that it made a mistake by seriously depleting its aquifer resources. It has since turned to the water resources of others to produce food for its market, encouraging its firms to lease farmland in poor countries, including the world's hungriest states, Ethiopia, Somalia, and Sudan.

The Aral Sea in former Soviet Central Asia stands out as a sordid example of a human-made ecological disaster, wreaked by unbridled water abstractions from its principal sources, the Amu Darya and Syr Darya rivers. Heavy withdrawals for irrigation from these rivers have cut water inflows to a trickle. Once the world's fourth largest lake, the Aral Sea has shrunk 74 percent in area and 90 percent in volume, and its salinity has grown ninefold.[133] The Soviet introduction of large-scale irrigated cotton cultivation in arid Central Asia wrought havoc, with the Aral Sea level falling by sixteen meters just between 1981 and 1990 and windstorms blowing toxic dust salt from the dry seabed into surrounding regions.

The Aral Sea's desiccation, symbolized by its shriveling into essentially three lakes, has wrought five grave ecological consequences.[134] The first is the exposure of its salt-laden bottom, which has become a major source of wind-blown dust. The exposed seabed is thick with salts and agricultural chemical residue that threaten human health and natural vegetation. The airborne particles, sickening local people and killing crops, are at times carried as far as five hundred kilometers.

Compounding the toxicity problem is the now-deserted Vozrozhdeniya Island (or "Renaissance Island" in Russian), a Soviet biological-weapon testing site that ceased to be an island and connected to the mainland just this century owing to the Aral's receding waters. Shared by the former Soviet republics of Uzbekistan and Kazakhstan, it is the world's largest anthrax burial ground.[135]

A second ecological consequence is that most of the thirty-two fish species have become extinct and a once-thriving fishing industry has collapsed, eliminating thousands of jobs. Another consequence is the broader loss of native biota and the destruction of the Aral Sea's unique ecosystem, including deltas. Yet another outcome has been the serious impacts on the people living around the lake. High local levels of respiratory illnesses, kidney ailments, and throat and esophageal cancer have been linked with the salt-laden air and contamination of freshwater resources. Many local residents, deprived of their farming and fishing livelihoods, have left as environmental refugees. And a fifth consequence is a drier, more continental climate in the region, along with an intensifying desertification.

The impaired, increasingly salty Salton Sea in southeastern California also illustrates the impacts of wrongheaded irrigation practices. Created between 1905 and 1907 when the Colorado River broke through irrigation diversion canals, the Salton Sea became one of the most productive ecosystems in North America, hosting many species of fish and birds and serving as a critical stop for migratory birds along the Pacific Flyway. Now, with its water about 50 percent saltier than the Pacific Ocean, this shrinking lake (once considered to be the Riviera of the West) is at risk of turning into a wasteland.[136]

Dependent on agricultural drainage from the Whitewater, New, and Alamo rivers, this accidental lake has increasingly been damaged by receding inflows, leading to loss of aquatic and wetland habitat. A sudden drop in oxygen content in the shallow lake's waters during heat waves often kills large numbers of fish. Nearly 8 million fish perished on a single day in 1999.

A more alarming example of how humans can wreck the natural environment is the slow death of Lake Chad, once a massive lake that served as the third-largest source of freshwater in Africa. It has been shrinking so rapidly in recent decades, chiefly because of irrigation-related withdrawals, that its future seems unpromising. Lake Chad was larger in size than Israel when that Jewish state was founded. But the lake's dry-season surface area has since shrunk from more than 23,000 square kilometers to barely 1,000 square kilometers, increasing tensions between herders, farmers, and fishermen.

Lake Chad is located in the middle of the arid Sahel belt, with the Sahara Desert to the north and the savannah to the south. Land that was once perennially covered with lake waters now is parched dry or reduced to a series of ponds and islands, even as the Sahara menacingly edges southward toward this lake and its associated wetlands.[137] The lake, whose maximum depth of barely seven meters heightens its vulnerability, borders Cameroon, Chad, Niger, and Nigeria.[138] The lake, however, has shriveled to such an extent that its waters now are almost entirely concentrated in one country, Chad.

This ecological disaster has been tied to uncontrolled water withdrawals from Lake Chad for irrigation, the upstream damming of the rivers feeding the lake, a decline in the mean annual rainfall, increasing desertification in the surrounding Sahel region, and a warming climate. The lake's rapid shrinking, however, has itself contributed to desertification and the changing climate.

A NASA-funded study found that rising water use by communities accounted for about half of the observed decrease in the lake area since the 1960s, when newly independent Nigeria launched the ambitious South Chad Irrigation Project to tap the lake waters for large-scale cultivation in the surrounding desert. Just between 1983 and 1994, according to the study, water abstractions for irrigation, compared with the 1953–1979 period, jumped fourfold.[139]

Lake Chad and the Chari/Logone river system, which transports much of the runoff generated in the combined basin, are critical sources of freshwater for more than 20 million people. But the lake's receding shoreline has reduced fisheries by about three-fifths, wiped out many pasturelands, and forced farmers relying on the lake to either move closer to its retreating waters or give up farming. The basin has also witnessed a falling groundwater table, shrinking floodplains, disappearing plant and animal species, and increasing soil erosion that has reduced cultivated land. A serious shortage of animal feed has resulted in cattle deaths and plummeting livestock production, which, together with declining fish stock, has aggravated malnutrition among children.

The specter of increasing conflicts over scarce water resources looms large in this basin. If there is any positive news, it is the start of multilateral efforts, however belatedly, to help save the disappearing lake. The 2009-launched Lake Chad Sustainable Development Support Program, partly funded by the African Development Bank, aims to clear the choked water channels entering the freshwater lake, implement antierosion measures, and explore the diversion of additional water to the lake.

The Lake Chad Basin Commission has drawn up an ambitious plan to replenish the lake by diverting water from the Oubangui River, in the Republic of Congo, to the Chari River that feeds the lake.[140] The Oubangui is the biggest tributary of the mighty Congo River. Concerns, however, linger that this major diversion by damming the Oubangui and constructing nearly one hundred kilometers of canals could adversely affect the Congo River Basin without succeeding in fully resuscitating Lake Chad.

The Dead Sea, an iconic tourist attraction hidden in the world's deepest valley and shielded by desert mountains, has also been rapidly shrinking because of the irrigation-related diversion of inflowing waters from the Jordan River system, shared by Israel, the West Bank, Jordan, Syria, and Lebanon. The Dead Sea, the lowest point on earth with salinity 8.6 times greater than the oceans, has seen its surface area shrink more than one-third and its depth drop by twenty-five meters in the past half century, even though it is also fed by seasonal flash floods flowing out of desert canyons.[141] Such desiccation holds serious implications for the surrounding wetlands that support endangered species—such as the Arabian wolf, ibex, hyrax, Griffon vulture, and Egyptian vulture—and serve as an important resting and breeding place for millions of birds migrating between Eurasia and Africa each year.

With most of the Jordan River's waters drained off for irrigation and other human needs, this once-vibrant and fast-flowing biblical river now delivers only a contaminated trickle to the dying Dead Sea—a reminder that the impressive success of Israel and Jordan in growing field crops in irrigated desert lands has come with a heavy environmental price. These two countries consume the bulk of this river system's waters, in which Christians believe Jesus was baptized. Take Israel: with the inflow of diaspora Jews swelling its population and the expanding housing complexes encroaching on farmland in the densely inhabited central region, agricultural production has expanded in the southern desert area between Be'er Sheva and Eilat (the Arava and the Negev)—a region that now accounts for 45 percent of the country's vegetables and field crops and most of its exported melons.

The increasing diversion of the waters of both Lake Tiberias (known also as Lake Kinneret, or the Sea of Galilee) and the Yarmouk (a Jordan River tributary) has impeded inflows into the Dead Sea. Israel's 1964 launch of its National Water Carrier, which draws off Lake Tiberias waters, set in motion a sharp reduction in flows to the Dead Sea. Now Lake Tiberias is at risk of becoming irremediably salinized by the saltwater springs below it, while much of the waters of the Yarmouk—which has the second-largest annual discharge of any river in the subregion—are siphoned off by Israel, Jordan, and Syria before it drains into the Jordan River, from whose southern section, according to the Bible, the people of Israel crossed into the Promised Land.

This is a water-scarce subregion, and these countries' per capita freshwater availability is well below the international water-poverty line, with Jordan's availability the lowest. Even Israel, despite its control over the subregion's freshwater resources, faces a water crisis: overexploitation of aquifers to meet rising domestic demand, according to one international report, "has led to the infiltration of seawater and salinity, the impoundment of springs has dried up perennial and ephemeral streams, and domestic, industrial, and agricultural practices have contaminated water sources."[142]

The Dead Sea—linked to Abraham, Moses, and Jesus and currently located about 416 meters below sea level—faces the threat of drying up within decades. It was Moses who in Genesis recounted how the Dead Sea was formed as part of the legend of Sodom and Gomorrah, probably the most infamous cities in the Bible. Lot was seduced by his two daughters. God asked Lot and his family to flee and not look back. But Lot's wife did not heed this advice, with the result that she was instantly turned into a pillar of salt, thereby creating the Dead Sea, a giant lake that is still praised for the healing powers of its minerals and waters.

As the Dead Sea drops by more than one meter a year and its water becomes saltier, an ecological catastrophe looms large, including for the faunal and floral species on its shores.[143] Reinforcing this specter is the reckless commercial extraction of potash and magnesium from the Dead Sea brine by evaporating the water. Salts from the Dead Sea have been famous for centuries for their supposed therapeutic properties. Unlike regular sea salts that are made up of more than 90 percent sodium, the Dead Sea salts (marketed across the world) contain only about 10 percent sodium and are rich in magnesium and potassium. They also contain traces of other minerals, such as bromine and calcium.

Being the lowest point on earth, the salt-carrying inflows into the Dead Sea cannot drain out, and the natural evaporative process helps to accentuate the lake's extremely high salinity. This creates such high water density and buoyancy that it is easy for swimmers to stay afloat. Tourists like to bathe in the Dead Sea's rich mineral mud and its shoreside hot sulfur springs. But how much longer will tourists be able to do that? The Dead Sea's shallow southern basin dried up by the 1980s, and sinkholes have caved into the former seabed; its deep northern basin is also receding.[144]

Jordan has resurrected a controversial plan with the support of Israel and the Palestinian Authority to replenish the Dead Sea by building a 180-kilometer-long canal system to bring water from the Red Sea—a financially costly and environmentally risky proposition, just like the separate Israeli interest in a Mediterranean-to-Dead Sea canal. Either project to link two seas would cost billions of dollars and likely have unintended consequences, including irremediably altering the Dead Sea's chemical makeup and threatening its native biota. The transferred water, being less salty and chemically different, could float atop the Dead Sea's own water, with the mismatch wreaking serious environmental impacts.

To be sure, the Red-to-Dead Sea link would yield some real benefits, including hydroelectricity and desalinated water for Jordanian and Israeli use (see map 1.1). The more-than-four-hundred-meter difference in elevation between the Red Sea (or the Mediterranean) and the Dead Sea will allow the harnessing of the forces of water flow and gravity to generate hydroelectricity to power desalination plants and run pumps for water transfer. More than one billion cubic meters of water would be removed annually from the Red Sea, with the diverted seawater being desalinated in plants along the canal route and the untreated portion draining into the Dead Sea. The transfers would help ease the water situation in Jordan and Israel and strengthen peaceful cooperation and coexistence between these former adversaries.

The canal system's length would be slightly longer than the Suez Canal, located next to the Red Sea. Because of a hilly elevation along the mainly Jordanian route of this canal, the rerouting plan would necessitate extensive tunneling and continuous pumping of the transferred water. In comparison, the Mediterranean–Dead Sea canal—for which two competing routes have been touted—would be shorter and thus less expensive. There is also a proposal to save $1 billion in costs by building a pipeline instead of a canal between the Red and Dead seas.

The canal paths will pass through a seismically active region and carry environmental and geological risks that cannot be easily discounted. The Dead Sea is located on the northern part of the Great Rift Valley fault line, which extends southward to the Red Sea and East Africa. The diversion of waters from the Red Sea or the Mediterranean through an earthquake-prone region close to the Syrian-African Rift will carry the danger of causing seawater contamination of freshwater aquifers that lie beneath the identified canal routes.

The more prudent way to save the Dead Sea is for the upstream users of the Jordan River system—Israel and Jordan, in particular—to reach agreement to increase natural flows into the Dead Sea. Only by gradually abandoning unsustainable regional practices, including large-scale irrigation of field crops on desert lands, can the natural runoff entering the Dead Sea be securely boosted. Switching to less thirsty crops and recycling water must form part of this initiative.

Israel is an important exporter of cotton and other water-rich agricultural products to water-surplus European countries, although it also imports water-intensive food products from elsewhere. Rather than producing grains, beef, and oilseeds at home, Israel and Jordan need to shift to greater fruit and vegetable production, and import water-intensive products from countries where they can be grown more efficiently and sustainably.

To be sure, Israel has perfected techniques like drip-feed irrigation—a highly efficient way of irrigating crops—and now exports about $1.5 billion worth of water-related technologies, equipment, and services a year. It recycles three-quarters of its wastewater, mostly for agriculture. The booming farm sector and food exports are an understandable source of pride for a largely arid country.

The issue is not whether Israel uses water efficiently but whether channeling slightly more than half of its total freshwater withdrawals to agriculture (which makes up just 2.5 percent of its GDP) and maintaining generous irrigation subsidies is an efficient utilization of scarce resources.[145] With its population projected to swell to 8.5 million by 2020, it must reduce freshwater allocations to agriculture to meet fast-rising urban demand—or find substantial quantities of additional water supply. Yet, hewing to the ideological mission of the earliest Zionist pioneers to make the desert bloom, Israel is literally making the Negev Desert bloom, at the expense of the Dead Sea and other ecosystems.

In Jordan, desert farming poses an even a bigger problem because water resources are more limited. Yet Jordan grows not only wheat and barley for its domestic market but also grapes, olives, figs, almonds, bananas, and other farm products for export. Its largest irrigation network is the East Ghor Canal, which was largely completed in 1966 and, after heavy damage from Israeli airstrikes, reconstructed and extended to 110.5 kilometers from the early 1970s onward. Subsequently renamed the King Abdullah Canal, it siphons water from the Yarmouk and, to a smaller extent, the Zarqa River to help irrigate crops in the Jordan Valley.

Jordan's agricultural water withdrawals as a percentage of its actual renewable water resources (ARWR)—65.23—are much above the internationally recognized danger point of 40 percent and among the highest in the world.[146] Yet the widening demand-supply gap for water has only accelerated the overexploitation of groundwater in a country heavily dependent on foreign aid.

If the world is to avert a thirsty future and contain the risks of greater intrastate and interstate conflict, it must, through sustainable development practices, protect freshwater ecosystems, which harbor the greatest concentration of species. The cases of dying lakes illustrate the dangers of a parched future. In addition to the examples discussed above, there are several other impaired lakes, including Yamdrok Tso, a sacred, 4,440-meter-high freshwater lake in Tibet blighted by a Chinese hydropower project, and the Horn of Africa's Lake Turkana, which has shrunk to such an extent as to have largely disappeared from Ethiopian territory, retreating south into Kenya. A number of important rivers are also at risk of dying. The world's leading endangered rivers range from the Nile and the Mekong to the La Plata and the Rio Grande, North America's fifth-longest river, which forms the entire border between Texas and Mexico.[147] Several major rivers, including the Yellow and the Colorado, run dry before reaching the sea in years when rainfall is below normal.

The continued shrinkage, degradation, or destruction of freshwater habitats is accelerating biodiversity loss. More and more wetlands, for example, are disappearing, although the Convention on Wetlands of International Importance—known as the Ramsar Convention—provides a framework for concerted multilateral action to conserve marshes, fens, peatlands, and other wetlands, both coastal and inland.[148] This is the only global environmental convention that deals with a specific ecosystem, which plays a valuable role in the biosphere, including by improving water quality and providing important habitat for plant and animal species.

Technology can serve as an important tool to protect freshwater ecosystems and improve water efficiency and management. How scientific techniques can help in the better management of water resources is illustrated by a nuclear method called isotope hydrology. It can help identify the size, origin, flow, and age of any water source, including an aquifer. The Vienna-based International Atomic Energy Agency has run scores of projects in Africa, Asia, Europe, and Latin America to help map aquifers, manage surface and underground resources, monitor dam leakage, and control water pollution.[149] A number of countries are employing isotope hydrology for similar purposes.

Technology, however, cannot offer all the solutions to the deepening problems springing from unsustainable human practices. Many technologies, in fact, are still far from mature. For example, less energy intensive and more cost effective technologies need to be found for desalinating seawater, reclaiming wastewater, and cleansing contaminated water, as well as for safely disposing of the chemical residues from such treatment processes.

Constraints on finding additional water supplies, however, make increased use of cleaned-up sewage—"reclaimed water," in industry parlance—inevitable. By irrigating farms, golf courses, and parks and being channeled for power-plant cooling, energy-resource extraction and refining, and outdoor fountains, recycled water can free up potable water for drinking and other human uses.

Reclaimed water can also be employed to recharge aquifers and other depleted freshwater sources. As nature cannot replenish groundwater to keep pace with its fast-rising extraction in several parts of the world, humans must intervene to pump recycled water regularly into aquifers, where the accumulating water will be pumped up again for use. Such artificial replenishment can also serve as an important aid in cleaning up polluted rivers and lakes.

Groundwater contamination is more difficult to tackle because of its cumulative effect and the hidden nature of the resources. In artificial aquifer recharge, however, as the injected water infiltrates the soil, natural biological, chemical, and physical processes usually help improve the quality of the new water entering the aquifer.

Greater investments and public-private partnerships have become imperative for finding cheaper, more environmentally friendly ways to convert seawater into drinking water and thereby fully realize the promise of desalination.[150] At present, desalination costs three to four times more than drawing water from conventional sources, besides using ten times more energy. However, water-scarce but energy-rich countries, like the oil sheikhdoms of the Persian Gulf, have little option other than to operate giant, energy-hogging desalination plants. Desalination is also the only option for several water-scarce coastal areas. The safe, cost-effective disposal of toxic residues from desalination and wastewater treatment poses a technological challenge by itself.

Water-efficient agricultural practices, including drip irrigation, must also be adopted, given the fact that the farm sector consumes more than two-thirds of all the water globally drawn from rivers, aquifers, and lakes. Drip irrigation is presently employed in just over 3 percent of all irrigated land in the world. Trapping rainwater with check dams to limit runoff can easily be practiced in hilly areas. But even in the plains, rainwater can economically be captured and stored on a large scale.

What is needed is a radical shift from the business-as-usual approach in industry and agriculture, along with lifestyle changes, to help conserve water. Without water-efficient farming, industrial production, and toilet technologies, global water woes will only worsen. Even as the search for new or improved technologies continues, human practices must change to stem the growing water-related challenges.

Chapter Four

Changing Water Cooperation, Competition, and Conflict

Cooperation or competition between upstream and downstream provinces or countries over the resources of shared rivers, lakes, and aquifers has principally focused on water allocations, distribution, diversion, quality, and management. Dam and other water projects, pollution, excessive river silting, perceived overexploitation of resources, and water sharing have often been at the center of interstate and intrastate disputes. Several regions now have become highly vulnerable to water-related flash points between provinces or countries. As United Nations secretary general Ban Ki-moon has cautioned, "Freshwater resources are stretched thin" and—as with oil—"problems that grow from the scarcity of a vital resource tend to spill over borders."[1]

Fast-growing demand has placed the world on the cusp of severe water shortages. Whereas engines stop without an ample supply of oil, life as we know it will be disrupted without minimally adequate supplies of water—not only will toilets not flush, but industrial production will grind to a halt, crops will shrivel up and die, fires will rage unchecked, epidemics will spread, and thirsty people will take the law into their own hands. A water crisis already haunts more than half the world, and as it accentuates and spreads to more regions, the risks of violent conflict or instability will grow.

Competition and conflict over transboundary groundwater resources poses a special challenge. Given the hidden nature of these resources and the absence of any international legal framework governing their extraction, there are no reliable estimates of the extent of such cross-border subterranean reserves or even the number of transnational aquifers.

This situation only emboldens countries that share groundwater basins to quietly get into a pumping race, spurred by "use it or lose it" thinking. The overpumping from transnational aquifers tends to raise political tensions, as between Israel and the occupied Palestinian territories and in the case of

al-Disi, a fossil sandstone aquifer largely located within Saudi Arabia's frontiers but which Jordan is seeking rapaciously to exploit.

There is thus a pressing need to develop interstate cooperation on transboundary aquifers so as to stem conflict, especially in basins where differences or disputes over mutual rights have created simmering tensions.[2] If
cross-border aquifers are to be managed in an environmentally sustainable
manner, rights must be balanced with responsibilities, and there has to be
greater political and scientific cooperation.

River-centered conflict tends to be even more potent. The construction
of upstream structures to augment water supply or storage capacity often
threatens to affect the interests of downstream basins, stoking political,
ethnic, or sectarian tensions. The interruption of natural flows by dams,
irrigation canals, interbasin transfers, or other diversions, while bringing
socioeconomic benefits to the intended communities, can cause resource
degradation, alter fluvial ecosystems, foster downstream scarcity, and breed
discord and violence.

The likelihood of river-basin conflict rises as the rate of change within
any basin surpasses the regional institutional capacity to absorb the change.[3]
Preventive hydrodiplomacy can deter conflicts only if it is anchored in realistic strategies geared toward anticipating and mediating disputes between
competing users and interests.[4]

WATER CONTROL AS POTENTIAL POLITICAL WEAPON

Current water-related political trends across large parts of the world are worrying: Shared water resources are often being siphoned off with little consideration for the interests of downstream users. Alterations of natural river
flows carry the seeds of interriparian conflict.[5] Damming rivers is the most
potent symbol of water impoundment that tends to spur discord and tensions
between co-riparian states. A chain of dams, reservoirs, barrages, and irrigation structures on a river system can actually cause river fragmentation, leading to the loss of riparian vegetation and disappearance of waterfalls, rapids,
and even wetlands, besides affecting the quantity and quality of downstream
flows, impeding organic-matter recycling, and blocking fish from migrating.

Even foreign direct investment in forest real estate has triggered lurking
suspicions that the motive may be to control streams and other water sources.
For example, in Japan—where 70 percent of the landmass is mountainous
forest—the flow of Chinese money to buy mountains and forests by taking
advantage of depressed woodland prices has struck a national nerve, raising

the specter of China gaining control of pristine forests and riverheads and serving as "a chilling reminder of the expanding shadow cast by China."[6]

A 2010 book by environmental scientist Hideki Hirano, *Japan's Forests under Siege: How Foreign Capital Threatens Our Water Sources*, highlighted national concerns over what has been portrayed as a stealthy Chinese land grab to control Japanese water resources.[7] Hirano, a Tokyo Foundation senior fellow, heads a land-conservation project examining how in Japan "interests backed by global capital have been making furtive moves to buy up forestry and water resources amid a prolonged slump in forestland prices."[8] Japanese law permits foreign ownership of forest real estate, yet the independent and widely respected Tokyo Foundation, in its own policy recommendations, has warned that the potential exploitation of Japan's natural resources by foreign investors threatens Japan's national security.[9]

Territorial claims can also spur water-related suspicions and intensify a water dispute. China's resurrection of its long-dormant claim to India's northeastern state of Arunachal Pradesh, bordering Tibet and Burma, has coincided with its eyeing of that region's rich water resources. China has also unveiled plans to build multiple dams on the Brahmaputra before the river enters Arunachal Pradesh. China's resource-driven claim to that Indian-controlled region parallels the way it became covetous of the Japanese-controlled Senkaku Islands—which it calls the Diaoyu Islands—only after the issue of developing hydrocarbon resources on the continental shelf of the East China Sea came up in the 1970s. The same story unfolded in the South China Sea, where the islands now at the center of intercountry tensions assumed importance for Beijing only after it became apparent in the early 1970s that they were located beside potentially vast oil and gas reserves.

The 2006 reactivation of its claim to Arunachal Pradesh raised wider concern that China was seeking to leverage its upper-riparian position to contend that the downstream effects of its damming activities are its own internal matter. The territorial claim, more ominously, signaled a desire to extend the Chinese control over the Tibetan Plateau to another water-rich region.

More broadly, hydrosupremacy can be fashioned at the basin level through water-resource capture and control, or perpetuation of inequitable utilization patterns.[10] Such domination often necessitates some degree of coercion or sustained pressure. It may involve continuation of unfair, preindependence arrangements, construction of hydroengineering structures, or assertive protection of "historical rights."

Extensive hydroengineering infrastructure centered on an international watercourse can arm the upstream state with a potential water weapon against a neighbor, which may effectively lose control over a key source of water supply. In such a situation, the upper riparian could wield the water weapon overtly in a war or subtly in peacetime to signal displeasure with a co-riparian state. Even if the upper riparian had no intention of turning water into a weapon, it is likely to gain significant leverage over a neighbor through a latent capability to control cross-border river flows—a leverage it could employ diplomatically to keep a co-riparian on good behavior, including to deter it from challenging its broader regional interests in any manner.

However, if the downriver state was militarily and economically dominant, as Egypt is in the Nile River Basin and Uzbekistan in the Amu Darya River Basin, an upper riparian would have little political leeway to fashion water into a potential weapon. Even to divert any significant volume of river waters for its own legitimate use, it would need the concurrence of the downstream power. In the Amu Darya Basin, an activity that was viewed positively during the Soviet times—to increase upstream water-storage capacity for the benefit of the regional economy—is in the postindependence era of sharpening intercountry rivalry considered a threat to the interests of the dominant user of water, the downriver power Uzbekistan.[11]

Uzbekistan, which has 45 percent of Central Asia's population and consumes more than half the region's river waters, has sought to perpetuate its riparian ascendancy by foiling the plans of upstream Tajikistan and Kyrgyzstan to build new dams to boost hydropower production. The weak, internally torn, and energy-poor Tajikistan and Kyrgyzstan contend that such dam building is imperative because Uzbekistan and the other two hydrocarbon-rich, downstream consumers of the bulk of the region's waters, Kazakhstan and Turkmenistan, are unwilling to supply them with oil and gas at concessional rates, as was the practice in the Soviet era. Uzbekistan, however, has threatened military reprisals if Tajikistan, for example, moved to resume work on unfinished Soviet-era hydropower projects, including one on the Vakhsh River that was intended to become the world's tallest dam.

Egypt, for its part, has held out the threat of unspecified reprisals if its use of the bulk of the Nile waters was endangered by upstream appropriations. As the preeminent power in the basin, Egypt has successfully lent legitimacy to colonial-era water allocations, linking their perpetuation to its national security and stymieing moves by Sudan, Ethiopia, and other upstream states for more equitable allocation of the basin resources. It has aggressively sought to protect what it calls its "historical rights" to the Nile waters, contending that it has various means—legal, political, and military—to shield them.

Although located farthest downstream, Egypt has fashioned a hydro-supremacy in this basin. But with upstream states itching to challenge its colonial-era monopoly on the Nile waters, it is unclear how long Egypt will be able to protect its dominant share from shrinking.

The 1959 Nile accord between Egypt and Sudan stated that Egypt had an "acquired right" to 48 billion cubic meters per year of the river's waters —as measured at Aswan—and Sudan 5 billion cubic meters, but did not quantify the share for Ethiopia, the underdeveloped country that generates the largest volume of the Nile flows and is the source of the Blue Nile. However, with Arab, Asian, and other agribusiness firms seeking to grow food for their home markets on arable land along the Upper Nile, and the upriver Nile states aiming to step up their own developmental activities, flows to Egypt could lessen even without upstream construction of major dams. Ethiopia, however, has laid out ambitious plans to emerge as Africa's leading power exporter by building an array of hydroelectric dams on the Nile and other rivers, even as it has complained that Egypt was pressuring donor countries and international lenders to withhold funding for such projects. Given that Egypt will remain the paramount power in the basin for the foreseeable future, the riparian advantage will continue to lie with it.

How the upstream diversion of an international river can precipitate an intercountry crisis was demonstrated in the Rio Lauca case in 1962 when Chile, ignoring downriver Bolivia's explicit warning not to commit "an act of aggression," opened the floodgates of a new project and began diverting the binational river's waters to irrigate its arid northern valleys of Sobraya and Azapa. Bolivia, claiming that the Chilean megaproject constituted a serious transgression of its rights as the lower riparian, responded by breaking diplomatic relations with Chile, prompting Chile to pay back in kind.

Violent anti-Chilean protests flared in the city of La Paz amid the tit-for-tat diplomatic actions, with police gunfire killing several demonstrators. The Council of the Organization of American States (OAS), convening specially to consider Bolivia's complaint about Chile's threat to Bolivian "territorial integrity," tried to mediate an end to the conflict.[12] However, with Chile presenting the diversion as a fait accompli, the OAS could do little more than call for a peaceful bilateral resolution. That eventually left Bolivia with no credible option. Today, China is similarly seeking to present a fait accompli to its riparian neighbors by quietly building giant new dams on the transnational Amur, Arun, Brahmaputra, Irtysh-Illy, Mekong, and Salween rivers, among others.

Make no mistake: power equations are central to interriparian relations. For example, in the Tigris-Euphrates Basin, Turkey, a longtime U.S. ally and member of the North Atlantic Treaty Organization, claims absolute rights over waters originating in its territory—a stance that runs counter to Iraq's assertion of historical rights dating back to the dawn of civilization and to Syria's riparian interests. But unlike Egypt, Iraq—also located farthest downstream—is militarily in no position to shield its avowed rights.

Turkey, a rising heavyweight in the Islamic world that aspires to a greater geopolitical role, has been unilaterally harnessing the waters of the twin rivers, unmindful of the potential downstream impacts. With its dynamic economy, vantage location, and increasing clout in world affairs under an assertive, Islamist-leaning leadership, Turkey is simply too important to Western interests to come under any sustained pressure.

Just 22 percent of the Tigris-Euphrates Basin falls in Turkey, which straddles two continents. Turkey, however, is the largest source of basin waters, contributing almost 90 percent of the Euphrates' mean annual flow and 38 percent of the Tigris' waters. It is also the source of 11 percent of the flow of the Tigris tributaries that merge with the Tigris in Iraq. Complicating the water picture is Syria's location midstream on the Euphrates—to which the Syrian contribution is 10 percent of the average annual flow—and the stateless situation of the Kurds, whose traditional homeland straddles the Tigris-Euphrates Basin. The fact is that conflicting approaches, unilateral projects, broken promises, and the absence of any water institution have long exposed the basin to recurrent water tensions and low-intensity conflict.[13]

In general, even if an upper riparian's actions pose no threat of water being turned into a political weapon yet risk affecting the quality or quantity of transboundary flows, the response of the downriver party—or the lack of it—will likely be shaped by prevailing power factors. If such upstream actions are undertaken by a power armed with superior military and economic capabilities and geopolitical influence, the lower riparian may be able to do little more than protest, unless a watercourse agreement between the two countries provides for international arbitration at the request of one side.

Overt conflict may become more probable if the downstream riparian is militarily strong and highly reliant on the international river whose cross-border flows are threatened by the upstream state's moves. In such a situation, the temptation to use military might to set right a perceived wrong could be strong.

Riparian dominance impervious to the interests of downstream communities and to international norms poses a special challenge. Backed by massive hydroengineering infrastructure, untrammeled riparian ascendancy can create a tense and conflict-laden situation where water allocations to co-riparian states become a function of upstream political fiat. This threat is exemplified by China's frenetic construction of upstream dams, barrages, reservoirs, and irrigation systems on international rivers flowing to Central, South, and Southeast Asia and to Russia, as a result of which transboundary flows are set to diminish and potentially even become a function of political concession by the upstream controller.

Even the denial or delayed transfer of hydrological data to a subjacent riparian at a critically important time like the floods-bringing monsoon season or the dry part of the year would mean the use of water as a political instrument. Supplying poor-quality hydrological data would amount to cynically playing adversarial hydropolitics. The politics of hydrological-data sharing indeed can become murky through a lack of transparency on information collection and dissemination. After all, like rice traded on the world market, hydrological data comes in different grades and qualities—from good, reliable data to inferior data and broken data.

In fact, an upper riparian can cause serious flooding in a downriver state by releasing massive quantities of water from dams and barrages in heavy rainstorm conditions in order to save those structures, or even as a means to wreak punishment. Such structures can be militarily used, however, only if they are located very close to the international border, or else the deliberately unleashed flooding would leave a trail of death and destruction on the wrong side of the frontier. Releases to save structures during torrential rains can also wreak havoc downstream. For example, the surprise flash floods that ravaged India's Himalayan states of Himachal Pradesh and Arunachal Pradesh between 2000 and 2005 were linked to abrupt discharges from China's projects in Tibet. The political furor over the flash floods, however, led Beijing to agree to sell hydrological data to India during the monsoon season.

Whereas it is usually the upper riparian that can cause harm to the downstream riparian in river basins, the situation with regard to transboundary aquifers is different. Either the up-flow or down-flow riparian can be the source of damage to the aquifer system and thus to the interests of another party. Such damage can result from contamination or overexploitation of resources. Many aquifers, however, are adjacent to and linked with surface watercourses, and overexploitation of one type of watercourse (be it an aquifer, river, or lake) holds consequences for another watercourse. Riverbeds, for example, can begin to dry out when aquifers are excessively drained.

The key point is that resource-harnessing projects on international watercourses cannot be treated just as an engineering issue because of their wider political and socioeconomic implications. Averting interriparian conflict demands the building of basin-level transparency on projects and joint cooperation for shared benefits. One riparian state's accumulation of hydrological leverage can only prompt another riparian over time to build up its military capabilities or acquire other levers to try and offset the strategic disadvantage or perceived discrimination.

WATER AS A DRIVER OF TERRITORIAL DISPUTES

Rich water resources in a disputed territory usually raise the stakes between two disputant nations, serving to rationalize political control over or claim to that region and fuelling tensions. In many cases, water and land disputes are interlocked. Breaking the nexus between water and territorial claims demands a resolution of land feuds and respect for political and hydrological boundaries—or, to lessen the water-related importance of a contested territory, an equitable apportionment of shared water resources or guarantees of uninterrupted natural flows.

A number of territorial disputes—varying in intensity from managed or dormant to violent or militarized—plague the 322 international land boundaries that separate geographically contiguous states in a world with 195 independent nations and 71 dependencies or special entities.[14] Such disputes, often constituting a dangerous legacy of arbitrary, colonially drawn borders or territorial usurpation by military means, are both a cause and a symptom of intercountry tensions.

Even though territorial feuds may be rooted in historical and cultural claims of the disputant nations, they are often driven by resource interest or resource competition. The resources at stake in territorial disputes range from water and minerals (including hydrocarbons) to fisheries and arable land. Not just the scarcity of a resource but even its abundance on a territory can distort the controlling state's extraction policies, fostering intercountry conflict and internal civil strife.[15] In a situation where the resource is exploited in a territory contested between two states and racked by separatist discontent, the overlapping intercountry and intracountry elements tend to fuel greater tensions and strife.

Resource-driven territorial feuds also instigate ethnic and cultural clashes, leading at times to the internal displacement of people or even the cross-border movement of refugees. Resource-related territorial feuds indeed serve as an important source of intercountry military rivalries, with the size of the land in dispute often secondary to the size of the potential resources at stake. For example, the disputed islands in the East China Sea at the center of China-Japan tensions occupy an area of only seven square kilometers, but their surrounding seas hold rich hydrocarbon reserves.

Similarly, the geopolitical and resource importance of an area controlling river headwaters usually surpasses the significance of its land size. Given the ravenous national and international demand for resources, it is no accident that territorial disputes and separatist struggles increasingly have a link with resource issues. Resources indeed have emerged as a major driver of conflict.

Against this background, water is a key factor behind several major territorial disputes and externally aided separatist movements. The roots of water conflicts range from border disputes and identity fights to the perceived threats from large hydroengineering projects to downstream flows and the environment. But when water is a driver of territorial disputes and separatist struggles, it also holds the key to resolving such problems.

The sharpening water competition between rival states threatens to exacerbate or reopen disputes over territories that are either the original source of water or through which important rivers flow, such as the Israeli-occupied Golan Heights and West Bank, Turkey's Kurdish-majority southeast (once part of Mesopotamia), Abyei in Sudan, the Tibetan Plateau, Central Asia's divided Fergana Valley, and Jammu and Kashmir, whose control is split among India, Pakistan, and China. In places as diverse as Mauritania, Mali, Somalia, Ethiopia, Sudan, Yemen, Afghanistan, and Pakistan-controlled Baluchistan, water stress has helped intensify separatist struggles.

Redrawing Political Maps Compounds Water Challenges

One key factor known to foster water conflict is change in the political or physical parameters of international basins. Such change usually flows from the territorial realignment of political frontiers—which has happened on a large scale in the post–World War II world, especially in the developing world—or the upstream construction of large dams and other water diversions, which can affect flow velocity, volume of water, currents, and other physical characteristics of a waterway.[16] Just between 1945 and 1960, the total number of nations in the world virtually doubled, as the end of World War II set in motion the process of decolonization. The number of river basins straddling international frontiers quadrupled in that same fifteen-year period.

Since 1960, the number of countries has almost doubled again. As a result, the classification of scores of additional rivers has changed from "internal" to "international" waterways. The number of transnational river and lake basins has grown to at least 276, with some rivers demarcating political boundaries. In addition, an undetermined number of aquifer systems straddle intercountry borders. No scientific effort has been made in some parts of the world to identify transnational aquifers.

The biggest changes in physical parameters of basins have also occurred in the post–World War II period. Most of the world's large dams and interbasin water transfer (IBWT) projects have come up since the late 1940s, with the reengineering of flows significantly altering the physical properties of waterways. In fact, the world has witnessed the greatest changes in political and physical parameters of basins in the most compressed time frame in history. These shifts have appreciably added to the complexity and challenges of sustainably managing water resources.

The rejiggering of political borders has in many cases led to intractable water disputes or other festering conflicts. The manner in which the new political frontiers were drawn by departing colonial powers—usually with little regard for natural contours or the national-security interests of the newly emerging states—have complicated issues related to water resources, bequeathing a troubling legacy of territorial and water discord.

For example, following World Wars I and II, European colonial powers and the United States sat around tables and redrew political frontiers in the Middle East, creating new nations—and still-persisting feuds.[17] In fact, other than Iran, Egypt, and Turkey, every other important nation in the Middle East is a modern construct created largely by the British and the French, who, according to former U.S. secretary of state Condoleezza Rice, "drew borders like lines on the back of an envelope."[18]

Many of the borders drawn at departing colonizers' tables or produced by postwar agreements among victors created countries whose boundaries coincided neither with a division of nationalities, ethnic groups, and cultures nor with geographic features, such as rivers and mountains. In picking winners and losers of their own, these powers drew new borders that lumped together antagonistic ethnic groups and sects or split nationalities with little regard for the aspirations of the local people.

Artificial states with no roots in history or preexisting identity were born. In fact, such was the arbitrariness of the boundary making that many straight-line frontiers were created. The bulk of political boundaries in Africa follow latitudinal and longitudinal lines, with the artificial borders contributing to hydrological stresses and even to political and economic failures.[19] In many cases, artificial frontiers have fostered revanchist and identity struggles at the expense of economic and political development and hydrologic integrity.

In Africa, the redrawing of political frontiers as part of decolonization changed the hydrological map so radically that fifty-four river basins became transnational. In a number of cases where river and lake basins got politically divided in Africa, conflict-ridden water situations emerged.

One such example is the territorial conflict between Malawi and Tanzania over Lake Malawi and their now-dormant dispute over the meandering Songwe River. The decades-old dispute over Lake Malawi—which boasts more freshwater fish than any other place in the world, although overfishing threatens to make many of its species extinct—was reignited in 2011 by Malawi's award of a hydrocarbon-exploration contract to a British firm. If significant oil or gas reserves are found in the Lake Malawi Basin, the border dispute could escalate and imperil subregional stability. Malawi cites the Anglo-German Heligoland Treaty of 1890 to claim ownership of the entire 570-kilometer-long lake, other than the portion that falls within Mozambique's boundaries. But Tanzania, relying on different provisions of the same colonial-era treaty, insists that its border with Malawi runs through the middle of the lake, which it calls Nyasa.

In some international basins elsewhere, water discord arose slowly, as water shortages became serious only as a result of changing regional needs. But in the cases where disputes were quick to occur between newly independent nations over the sanctity of colonially demarcated territorial frontiers, the feuding states had little incentive to try to jointly manage the resources of common rivers.

For example, the disputes in South Asia over the waters of the multiple-river Indus system arose almost immediately after Pakistan was carved out of India as an independent state for the region's Muslims in 1947. However, the premise on which Pakistan was created as a religion-based political entity—that Muslims constitute a separate nation—proved flawed almost immediately when Muslim Afghanistan vehemently opposed Pakistan's admission to the United Nations and later when Bangladesh broke away from Pakistan to become independent. Pakistan remains torn by internal conflict.

The India-Pakistan boundary was hastily drawn by an English lawyer, Cyril Radcliffe, on the map of British-administered India. The partition left the larger part of the Indus Basin in the newly created country but the headwaters and main catchment regions mainly in India. The bulk of the irrigation canals developed by the British to earn revenue in the form of local taxes on water use, as well as more than 80 percent of the 10.5 million hectares of farmland that those canals irrigated, fell in the part that became Pakistan. The new political borders split the basins of all the Indus-system rivers—the Beas, the Chenab, the Jhelum, the Ravi, the Sutlej, and the main stem Indus. The last two rivers originate in Tibet, but their largest water collections occur in India.

Political boundaries have changed even after decolonization as a result of internal conflicts and other developments linked with identity and nationality issues or structural problems. The breaking away of South Sudan, East Timor, and Eritrea and the disintegration of the Soviet Union and Yugoslavia since the 1990s showed that political maps are not carved in stone. In fact, the most profound global events in recent history have been the fragmentation of several countries.

Any basin's integrity is tied to natural hydrologic, topographic, physical, and other parameters. The political division of basins, in addition to carrying risks of water conflict, tends to make environmentally prudent water management more challenging.

The unraveling of the Soviet empire, while dramatically changing the international geopolitical landscape and spawning fifteen new nations, altered the political makeup of several basins, including of the Ob River and the Aral Sea. It internationalized important internal rivers like the Amu Darya, Dnieper, Don, Syr Darya, and Volga. Germany's reunification did the reverse by turning the intercountry Tiban and Weser basins into intrastate basins. Fragmentation of countries, however, has been far more common than mergers or reunification.

In Central Asia, the Soviet Union's sudden collapse turned what had been intracountry river basins into international basins, with the water sharing between the five newly independent "stans" being compounded by their frontiers bearing little resemblance, in most parts, to natural or ethnic fault lines. The water competition there pits the interests of the two smallest states that are the sources of the Amu Darya and the Syr Darya—Kyrgyzstan and Tajikistan—against those of the larger and militarily stronger countries that are the main water consumers: Uzbekistan, Kazakhstan, and Turkmenistan. The cotton monoculture in Uzbekistan and Turkmenistan from the Soviet era, meanwhile, continues to deplete and degrade local water resources.

Another case of water conflict arising from the Soviet collapse centers on the Kura-Araks, a twin river system that drains into the Caspian Sea. The Araks (also known as Aras, and historically seen as loosely defining the northern territorial limits of the Kurdish people) originates in Turkey and the Kura in Georgia, with the two rivers merging in Azerbaijan. Two-thirds of their combined basin—located south of the Caucasus Mountains—is spread across Azerbaijan, Armenia, Nagorno-Karabakh, and Georgia, and the remaining one-third is in northeastern Turkey and northwestern Iran.

The first war triggered by the Soviet empire's unraveling was between Armenians and Azerbaijanis over the mountainous region of Nagorno-Karabakh. It killed about 30,000 people before a 1994 cease-fire left Nagorno-Karabakh outside the control of Azerbaijan, the largest state in the Caucasus. Now an unrecognized, de facto republic in the hands of ethnic Armenians, Nagorno-Karabakh—with rich water resources of its own—remains a potential flash point for another war. It also serves as the main obstacle to agreements being reached to share or manage water resources in this politically unstable region, which is also roiled by ill will between Georgians and Abkhazians and between Georgians and Ossetians, who fought a brief renewed war in 2008. The volatility of this region indeed extends to the Russian control over Chechnya and other areas in the mostly Muslim North Caucasus.

With Europe and Russia locked in a new Great Game over the hydrocarbon resources in the Caspian Sea Basin, the West has sought to promote stability in the strategically important Caucasus by trying to build water cooperation among co-riparians Azerbaijan, Armenia, and Georgia, including through a river-monitoring and watershed-management project sponsored jointly by the North Atlantic Treaty Organization and the Organization for Security and Cooperation in Europe.[20] However, ethnic and nationalistic tensions—highlighted by saber-rattling rhetoric over Nagorno-Karabakh and sniper shootings across the cease-fire line—have blocked possible water treaties between Azerbaijan, Armenia, and Georgia, although the Kura-Araks is the lifeline of these states.[21] Amid the inflamed regional passions, the river system faces a growing problem of pollution of its resources, with the areas located farthest downstream in Azerbaijan also confronting water shortages.

Shifts in political control of international basins have also resulted from war. In fact, the desire for acquisition of water resources or consolidation of control over them has been a major driver of territorial claims, political disputes, and military actions. This was illustrated by the Israeli-initiated 1967 Six-Day War, which changed the subregional water map. A trigger of that war was water-related disputes and tensions, yet the militarily changed parameters of the subregion's river and groundwater basins have only made water an even more critical issue in finding an eventual peace settlement.

The Nexus between Territorial and Water Feuds

The link between territorial and water disputes is often glossed over. The fact is that festering territorial disputes can block water cooperation (as in the case of the Asi-Orontes River), compound water challenges (as in the Indus Basin), or do both (best exemplified by the Kura-Araks Basin). When serious land disputes rage, rival territorial interests usually trump genuine water cooperation, even if a water treaty is in force. Unsurprisingly, preventing a war by miscalculation and containing tensions in such situations tends to take precedence over joint water collaboration.

Put simply, festering territorial and other political disputes make meaningful intercountry cooperation on shared waters difficult. This is apparent from the case of Morocco and Algeria, embroiled in territorial feuds from the time the former gained independence in 1956. After Algeria freed itself from the French colonial yoke in 1962, it continued to press the territorial claims France had made against Morocco during 1956–1962, including in the Guir-Zousfana River Basin region.[22] In 1963, Algeria and Morocco fought a war over their border disputes when Morocco attempted to reclaim the Tindouf and Bechare regions, annexed to Algeria during French colonial times.

Algeria-Morocco relations have been weighed down by other issues as well, including the Western Sahara War of 1975–1991, the 1994 closing of the Algerian-Moroccan border following an Islamist attack in Marrakesh that was blamed on the Algerian secret service, and the disturbed status of Western Sahara, a mineral-producing desert region formerly called the

Spanish Sahara where Algerian-backed guerrillas of the Polisario Front have challenged Moroccan rule. With such geopolitical and historical baggage, building cooperation between Algeria and Morocco on the shared waters of the Bounaïm-Tafna groundwater basin has proved challenging, although both countries confront serious water scarcity.

Algeria's situation is worse because its per capita annual freshwater availability—324 cubic meters—is nearly three times lower than Morocco's (899 cubic meters).[23] The Bounaïm-Tafna Basin is spread over 2,650 square kilometers, traversing Algeria's northwest but more than two-thirds of it located on the Moroccan side. Instead of joint collaboration, the "politics of silence and noncooperation" has prevailed, with the result that the aquifers are threatened by overexploitation and degradation.[24] In the lower part of the Tafna River Basin in Algeria, many wells and some streams have dried up.[25]

In the historical dispute between Turkey and the now violence-torn Syria over the Arab-influenced enclave of Hatay, Syria's territorial revanchism has effectively precluded water sharing on the Asi-Orontes, the only notable river on which Turkey is not situated upstream. The river, which originates in the mountains of Lebanon, is indeed the only northern-flowing river in the subregion. On its way to the Mediterranean, the Asi-Orontes flows from Syria into the Turkish-held Hatay enclave—known as Alexandretta until the 1930s—where nationalist fervor and interethnic passions have traditionally run high. The Syrian civil war, paradoxically, forced tens of thousands of fleeing Syrians to take refuge in Hatay, home to a large and ethnically diverse Syrian minority.

Hatay, connected to the rest of Turkey by the mountain pass known since the time of Alexander the Great as the Syrian Gates, was ceded to Turkey in the late 1930s by Syria's then colonial ruler, France—an action that cut off northwest Syria's natural geographic access to the Levant coastline of the Mediterranean Sea. That transfer, along with the subsequent loss to Israel in war of the headwaters-controlling Golan Heights, contributed to an overweening sense of grievance in Syria, which has been independent since 1946.

Syria, with 69 percent of the Asi-Orontes basin, consumes the bulk of the river's waters, leaving only a meager flow of 12 million cubic meters per year for Turkey. Syria shied away from a water-sharing agreement with Ankara on the Asi-Orontes because such an accord would imply it recognizes Turkish sovereignty over Hatay. Syria, however, signed a 1994 accord with Lebanon, under which the latter is committed to limit its water withdrawal to 80 million cubic meters in a year if the Asi-Orontes' annual flow within Lebanese territory registered 400 million cubic meters or more, and to reduce its intake proportionately if the river's flow was actually less.

Syria's heavy utilization of the basin waters has long given Ankara a handle to draw an unflattering comparison between the Syrian and Turkish upstream utilization of the resources of the Asi-Orontes and the Euphrates respectively.[26] And the absence of water sharing on the Asi-Orontes has only emboldened Turkey—the dominant regional riparian—to continue unilaterally harnessing the larger Euphrates and Tigris rivers by building a series of dams and irrigation networks. However, Syria, as part of a broader effort before its civil war to repair damaged bilateral relations, signed an agreement with Turkey in 2009 for building a joint multipurpose "friendship dam" on the Asi-Orontes at a border site.

Today, water and oil resources are at the center of the cold war between Sudan and the newly minted, oil-rich Republic of South Sudan, from where the Nile flows to the north. Sudan, the largely desert north, populated mainly by Muslim Arabs, is an important consumer of the Nile Basin waters. By contrast, boreholes and unprotected wells serve as the main drinking-water sources for the black, largely Christian African tribes in the lush, green, but underdeveloped South Sudan, which is widely covered by grasslands, swamps, and tropical forests. In fact, three-fourths of the oil in the Sudanese region also lies in landlocked South Sudan, but the only way to get it to the international market is by pipelines passing through the north to a terminal on the Red Sea.

South Sudan's secession helped bisect the Nile Basin's black African "suppliers" from its Arab "recipients"—Sudan and Egypt—who use more than nine-tenths of the yearly water flows. Water is engendering growing mistrust between Africans and Arabs in the basin over the continuing colonial-era arrangements, under which the British gave Egypt the bulk of the Nile waters while ignoring the interests of the upstream communities, including Tanzanians and Ugandans.

After South Sudan became independent in July 2011 as part of a 2005 peace accord to end decades of bloody civil war, disputes between the north and the south over border demarcation, oil-sharing revenues, and water sharing have escalated along the deadly fault line of race and religion. Water is also a factor in the territorial dispute between South Sudan and Sudan over the strategic region of Abyei, whose status was left unresolved in the 2005 peace deal, with a proposed referendum to decide its future being postponed indefinitely. Abyei was effectively seized by the Sudanese army, in an action that displaced 109,000 people, just before South Sudan became independent.[27]

Abyei, straddling the middle section of the border between the north and the south, boasts a major oilfield and serves as a critical source of water for nomadic tribespeople from the north during the dry season. The United

Nations Security Council has called for demilitarizing Abyei. Despite the 2012 withdrawal of all Sudanese forces and intruding South Sudanese troops, Abyei remains a potential flash point for military conflict.

An example of a festering border dispute overlapping with a water dispute is the long-standing Iraq-Iran feud over the Shatt al-Arab waterway, formed by the confluence of the Euphrates and Tigris rivers. Disputes over the control of the 190-kilometer waterway and its use have long bedeviled relations between the two neighbors and served as one of the causes of their protracted war during the 1980s. So bitter are the disputes—which actually date back to the period when Iraq was under the Ottoman Empire—that the two countries call the waterway by different names: in Persian, it is the Arvand Rud ("Swift River"), but to Iraqis it is the Shatt al-Arab ("River of Arabs").

There are several other interstate water disputes linked to border issues. For example, the disputes in the Amu Darya and Syr Darya river basins in Central Asia are as much about water sharing as about the artificial borders from the Stalinist era, although the territorial aspect remains subtextual. During the reign of Soviet strongman Joseph Stalin, some ethnic groups in Central Asia and the Caucasus were intentionally split and territorially paired with traditionally rival nationalities as a way of assuring centralized control. Today, with the lack of a clear demarcation of frontiers in several areas in Central Asia helping to whet resource ambitions, violent border incidents have been recurring, often involving armed bands along the Tajikistan-Kyrgyzstan and Uzbekistan-Kyrgyzstan borders.

In the Horn of Africa, a severe drought and famine in 2011 not only forced tens of thousands to flee to Kenya and Ethiopia from the worst-affected areas of Somalia—battered by two decades of civil war and consecutive seasons of failed rains—but also helped highlight the violent water dispute near where Lake Turkana, the world's largest perennial desert lake, forms the border between Ethiopia and Kenya. This interethnic water feud between the Ethiopian Dassanech and the Kenyan Turkana groups is in the Ilemi Triangle, an area in the northwestern corner of Lake Turkana claimed by Kenya and South Sudan and also bordering Ethiopia.[28]

Drought and dwindling river inflows in recent decades have resulted in Lake Turkana's gradual retreat from Ethiopian territory, with the lake now mainly in Kenya. Because of this development and the receding waters of the Omo River Delta—which empties into Lake Turkana and constitutes the traditional border between the Dassanech and Turkana tribes—the Dassanech and other Ethiopian pastoralists have started cultivating land and fishing waters further south, triggering recurrent intertribal clashes, which left dozens dead in 2011–2012 and prompted Kenya to rush additional paramilitary troops to the region.

With Kenya charging that Ethiopian civilians backed by armed militias have illegally settled on Kenyan-controlled land, the water and territorial disputes in the region have become fused. Ethiopia's new $1.8 billion, Chinese-financed Gibe III Dam—the largest of three cascading hydropower projects on the upper Omo River—threatens to further reduce inflows into the dying Lake Turkana. The Kenyan-Ethiopian disputes, by underscoring the relationship between water scarcity and violent conflict, serve as a reminder of the dangers of a parched future.

The Horn of Africa example also illustrates how the linkage between water and territorial disputes can extend from the intrastate context to the interstate level, and vice versa. In fact, intracountry water issues can magnify intercountry water disputes, particularly when they center on water-rich regions racked by separatist unrest. Such examples include the Kurdish region in eastern and southeastern Turkey, Kyrgyzstan's Uzbek-influenced southern Fergana Valley, the Chinese-controlled Tibetan Plateau, and the Indian and Pakistani parts of the formerly princely state of Jammu and Kashmir, one-fifth of which is also held by China.

The intersection of intrastate and interstate water and territorial issues is most striking in Central Asia. The water-rich Fergana Valley—which Stalin divided among Kyrgyzstan (which holds almost two-thirds), Uzbekistan, and Tajikistan—is rife with resource-driven border disputes, with ethnic fault lines the source of periodic clashes among the Kyrgyz, the Tajiks, and the Uzbeks. The most conflict-torn regions are where the borders of the three countries converge, including Kyrgyzstan's Batken Oblast, Uzbekistan's Fergana Oblast, and Tajikistan's Sughd Oblast areas. Water and land issues—the main triggers of conflict—are inextricably linked in this subregion, with the rise of Islamist and other extremist elements only making the Fergana Valley a powder keg.

Indeed, of all the political problems facing Central Asia, none perhaps seems more intractable than interstate water sharing—an issue made explosive by the spillover effects of the intrastate ethnic quarrels and resource competition. The region's unsettled ethnic, territorial, and resource disputes actually threaten to destabilize large swaths of Central Asia, or even to set off local civil wars.

The bloody attacks by ethnic Kyrgyz in mid-2010 that killed several hundred Uzbeks in the Kyrgyzstan portion of the Fergana Valley were a grisly reminder of those dangers. The attackers accused not only Uzbeks of appropriating local land and water resources but also Uzbekistan of plotting to wrest control of the entire Fergana Valley.[29] The regional river waters originate in the mountain ranges of energy-poor Tajikistan and Kyrgyzstan, while the oil, gas, and mineral-ore wealth and much of the arable land are concentrated in downstream Kazakhstan, Uzbekistan, and Turkmenistan.

The intersection between territorial and water disputes is conspicuous across the Great Himalayan Watershed, which boasts the world's highest mountains, thousands of glaciers and underground springs, and some of the greatest river systems. This is also the belt where multiple international borders converge, including the heavily militarized frontiers between China and India, Pakistan and India, and Pakistan and Afghanistan. Because of the incredibly rich freshwater resources in the wider Himalayan region, most of the new dams and other water projects in Tibet, Nepal, Tajikistan, Pakistan, India, Bhutan, Burma, and Laos are also concentrated here, spurring political and environmental concerns. Little attention, however, has been paid to the specific transboundary lakes, wetlands, and glaciers that are under threat.

Some, such as the 134-kilometer-long Pangong Tso (or Lukung Lake) and the 6,300-meter-high Siachin Glacier, have even been battlefields between opposing armies. The Indian-controlled but Pakistani-claimed Siachin Glacier—strategically wedged between the Pakistani- and Chinese-held parts of Jammu and Kashmir—served as the world's highest and coldest battleground (and one of the bloodiest) from the mid-1980s until a cease-fire took effect in 2003. Since then, avalanches, blizzards, crevasses, and cold temperatures have claimed the lives of several hundred soldiers deployed around the glacier, on the northern tip of the Kashmir frontier.

The 4,350-meter-high Pangong Tso—the scene of bloody battles during the 1962 Chinese invasion of India—remains a hotly contested zone. China, in control of two-thirds of this endorheic lake, seeks to test Indian defense preparedness in the area by intermittently dispatching armed patrols across the line of control, including by motorboat. In 1999, while Indian forces were repulsing a major Pakistani military encroachment into the Kargil region of the Buddhist Ladakh region of Jammu and Kashmir, Chinese troops built a five-kilometer permanent track into Indian-administered territory along the lake, located on the opposite flank of the Ladakh border.

The India-Pakistan division of the waters that straddle their volatile border hasn't brought peace because water and territorial issues, at least from a Pakistani perspective, remain intertwined. The 1960 Indus Waters Treaty divided the six Indus-system rivers between the two countries, with Pakistan getting the largest three (the Chenab, the Jhelum, and the main Indus stream) that make up more than four-fifths of the total basin waters.

In the first and only such case in modern world history, the treaty effectively partitioned the rivers, reserving for Pakistan's use the main rivers of the Indian part of Jammu and Kashmir. Because these rivers flow to Pakistan from Indian-administered territory, the treaty severely limits Indian sovereignty over them by laying down precise restrictions on the use of their waters so that cross-border flows are not materially affected. The treaty drew a virtual line on the Indian map to split the Indus Basin into upper and lower parts and confine India's full sovereignty rights to the lower section (map 4.1). In other words, thirteen years after the subcontinent's territorial partition at the end of British colonial rule, a river partition was uniquely effected through a treaty brokered by the United States and the World Bank.

Despite the treaty, the two countries' festering territorial feud over Jammu and Kashmir and their complex internal water and political dynamics have hampered cooperative efforts and sharpened their water rivalry. Water is the only natural resource amply found in the otherwise resource-poor Indian part of Jammu and Kashmir. So, the gifting of its river waters to Pakistan by treaty has fostered popular grievance there over the denial of a resource essential for development—a situation compounded by a Pakistan-abetted Islamist insurrection in the Muslim-dominated Kashmir Valley against Indian rule. The water issue has triggered a long-simmering backlash in the Indian part, where many at the grassroots feel alienated from mainstream India, and prompted its elected legislature to call for revision or abrogation of the treaty.

Estimates of the economic losses incurred by the Indian part due to the treaty-imposed fetters on water utilization run into hundreds of millions of dollars per year.[30] India's belated moves to address the problems of electricity shortages and underdevelopment in its restive part by building modest-size, run-of-river hydropower plants on the three main Jammu and Kashmir rivers, however, have whipped up political passions in downstream Pakistan, although such treaty-sanctioned projects involving no reservoir storage are unlikely to materially alter transboundary flows. Run-of-river plants, as explained in chapter 2, use a river's natural flow energy and elevation drop to produce electricity, with their water inflows and outflows virtually the same.

The India-Pakistan water equation has been made murkier by China's construction of large dams and other strategic projects in an internationally disputed area—Pakistan-administered Kashmir, including close to Pakistan's cease-fire line with India. This has drawn protests not just from New Delhi but also from the communities in the dam-building areas, located mainly in the northernmost Gilgit-Baltistan region. China, in addition, has built dams in Tibet on the two Indus-system rivers that originate there—the Sutlej and the main Indus stream—and further damming activities by China could alter the region's cross-border flow patterns in the dry season. Both in the Pakistani and Indian parts of Jammu and Kashmir, local communities want to have greater control over their water resources.

The resistance to new dams in the Shiite-dominated Gilgit-Baltistan—one of several factors that have brought units of the People's Liberation Army into this thinly populated but rebellious Pakistani Kashmir region—is because these projects uproot local residents and benefit only the downriver province

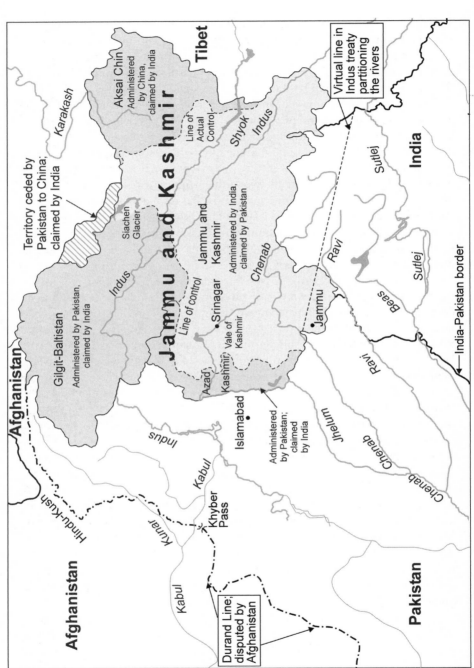

Map 4.1. Rivers of the Indus System, and the Treaty Line Partitioning Rivers

of Punjab, which dominates the economy and military of Pakistan, a mainly Sunni state. Pakistan has awarded to Chinese companies new megaprojects in Gilgit-Baltistan like the mammoth 7,100-megawatt Bunji Dam. The fact that China rules Gilgit-Baltistan's Shaksgam, Raskam, Shimshal, and Aghil valleys—ceded by Pakistan in 1963 to cement its strategic alliance with Beijing—has only added to the grassroots resistance against Chinese projects, which extend to mineral-resource extraction and new transportation links.

The Pakistani government has responded harshly to such opposition in Gilgit-Baltistan, tightening its direct military rule over the area. It even banned a book by local intellectual Muhammad Saeed Asad that referred to the Mangla Dam—built with World Bank aid on the Jhelum River in the 1960s—and other upstream dams as neocolonial instruments to expropriate water resources for Punjab Province at the expense of upriver communities.

The upstream Pakistani dams, through an extensive network of canals, irrigate an area larger than the size of Uruguay, England, Bangladesh, or Greece. The Pakistani government calls it "the largest contiguous irrigation system in the world."[31] This feat flows from a treaty that has no parallel in the world in terms of both the total volume and the share of waters reserved for the downriver party. Despite setting new standards in international water law and intercountry water relations, the Indus treaty did not resolve all Pakistan-India issues, reflected in the fact that some disputes have continued to linger even as new ones have arisen.

Worse still, the treaty's reservation of the main rivers of India's Jammu and Kashmir region—the so-called "western rivers"—for Pakistan has only whetted the territorial ambitions of hard-line Pakistani forces to wrest control of the Indian part, from where almost 94 percent of the Indus-system water flows to Pakistan, although only one river originates in that Indian region. The Pakistani military, lacking the capability to change the territorial status quo by force, has sought to bleed India by waging a proxy war by terror.[32] Even Pakistan's moderate civilian president, Asif Ali Zardari, while candidly admitting in 2009 that his country deliberately "created and nurtured" terrorist groups to help achieve foreign-policy objectives, drew a link between Kashmir and water.[33]

In an op-ed published in an American newspaper, Zardari wrote, "Much as the Palestinian issue remains the core obstacle to peace in the Middle East, the question of Kashmir must be addressed in some meaningful way to bring stability to this region . . . but also to address critical economic and environmental concerns. The water crisis in Pakistan is directly linked to relations with India. Resolution could prevent an environmental catastrophe in South Asia, but failure to do so could fuel the fires of discontent that lead to extremism and terrorism."[34] This was, however, not the first occasion when a Pakistani head of state drew a direct linkage between water and the Kashmir territorial dispute.

Shortly after the Indus treaty was signed, Field Marshal Mohammad Ayub Khan, Pakistan's president and first military dictator, publicly thundered that Pakistan could neither "trust India until the Kashmir question was settled" nor "afford to leave the Kashmir issue unresolved for an indefinite period."[35] Indeed, on the eve of Indian prime minister Jawaharlal Nehru's visit for the treaty-signing event in the Pakistani city of Karachi, Ayub Khan made plain Pakistan's intent: "The very fact that we will have to be content with the waters of the three western rivers will underline the *importance for us of having physical control on the upper reaches of these rivers* to secure their maximum utilization for the ever-growing needs of West Pakistan. The solution of Kashmir, therefore, acquires a sense of urgency" (emphasis added).[36] In other words, the Pakistani ruler viewed the treaty as justifying Pakistan's strategy for a change in the territorial status quo.

The treaty, however, was founded on the opposite logic, that a guaranteed share of the Indus-system waters for Pakistan would ease the territorial dispute and pave the way for subcontinental peace—an assumption that helped sway India to grant the lion's share of the waters to the lower-riparian party. David Lilienthal, the former chairman of the Tennessee Valley Authority and U.S. Atomic Energy Commission who helped sow the seeds of the Indus treaty with U.S. government backing, had emphasized the imperative of such a sharing pact to help dissipate Pakistan's desire to gain physical control of the rivers critical to its well-being.[37] "The starting point should be . . . to set to rest Pakistan's fears of deprivation and a return to desert," he wrote.[38]

The World Bank, the midwife in the partitioning of the six rivers, actually signed the treaty as a third party to lend its weight in support of several key provisions—the only instance of the Bank playing the role of a signatory in a binational treaty.[39] Yet the treaty, despite being tilted in Pakistan's favor and aimed at assuaging its concerns, ended up not only keeping the Kashmir territorial dispute alive, but also adding fuel to the quest of Pakistani hardliners to further redraw political frontiers, even if in blood. Today, while still hosting militant groups and terrorist-training camps close to its borders with India, Pakistan draws the full benefits of the Indus treaty.

Significantly, Pakistan is also beset by unresolved territorial and water issues with Afghanistan, which has long been a hostile neighbor, refusing to recognize the land frontier between the countries. Known as the Durand Line, the political frontier was an artificial, British-colonial invention that split the large Pashtun nationality.[40]

Because the Indus treaty only relates to the rivers that run from India to Pakistan, it does not cover the Kabul River, which flows from Afghanistan to join the main Indus stream in Pakistan and is thus part of the Indus system. If the Kabul River's inflow is factored in, Pakistan's share of the total Indus-system waters becomes larger than the 80.52 percent allocation made under the Indus treaty. The Kabul River, as its name suggests, rises near Kabul in the Sanglakh Range and flows north of the famed Khyber Pass to join the Indus northwest of Islamabad, the Pakistani capital.

The Kabul—the principal river in Afghanistan's mountainous east—has at least three important tributaries of its own, including the Kunar, whose flow contributes significantly to the main river's discharge. The Kunar, drawing a sizable portion of its waters from the melting snows and glaciers of the Hindu Kush mountains, actually flows into Afghanistan from Pakistan's northwestern corner and then runs parallel to the Durand Line before emptying into the Kabul River just to the east of the Afghan city of Jalalabad. The collective river then flows eastward into Pakistan.

For several Afghan cities, including national capital Kabul, the river is a critical lifeline. But Pakistan wants an agreement with Afghanistan governing the waters of the Kabul River so that its interests as the downriver state are protected, as in the Indus treaty. The absence of any bilateral accord indeed constitutes a source of potential water conflict, given landlocked Afghanistan's long-standing refusal to recognize the Durand Line, the sharpening Afghan-Pakistani competition over border-related water resources, and Pakistan's role in rearing and sponsoring Afghan militant groups, such as the Taliban and its affiliate, the Haqqani network.

Afghanistan, which is also locked in a long-running dispute with downstream Iran over sharing the waters of its longest river, the Helmand, has an abysmally small water-storage capacity—one of the lowest in the world. Its failure to harness the waters of the swift-flowing Kabul River for energy, development, and dry-season storage has exposed its eastern region to flash floods during the monsoons and drought in the lean periods. Once peace returns, Afghanistan will likely seek to build multiple dams on this river system, especially to create storage for the dry season—a development that may materially alter cross-border flows.

Another interesting case is Turkey's attempt to ease separatist discontent in its water-rich but underdeveloped Kurdish region through the multiphase Southeastern Anatolia Project, commonly known by its Turkish initials as GAP. The traditional Kurdish homeland, with its vast snowfields and powerful mountain springs, is the source of important rivers that are the lifeblood of lowland communities in a largely water-scarce region. The Karakaya Dam on the Euphrates, completed in 1988, was the first dam under GAP, which is concentrated in nine predominantly Kurdish provinces and billed as one of the largest water-resource development projects currently under way in the world (map 4.2).

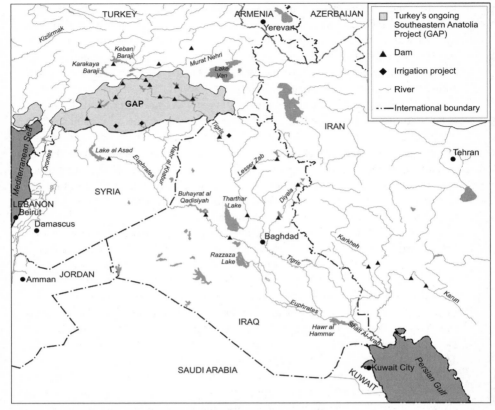

Map 4.2.   The Euphrates-Tigris Basin and Turkey's GAP Program

GAP was set up with the goal to irrigate 1.82 million hectares of land and generate 7,500 megawatts of electricity through a network of twenty-two dams and nineteen hydropower plants. But this ambitious, $32-billion program, designed as an engine of economic development in a region troubled by high unemployment and a farm crisis, carries significant implications for the transboundary flows of the Euphrates and Tigris. It also accentuates issues related to water and conflict in insurgency-torn southeastern Turkey.[41] The project, in fact, has created environmental and public-health challenges, including a growing threat from malaria and schistosomiasis, a disease caused by parasitic worms.[42]

More important, GAP, with its downstream impacts, has strained Turkey's riparian relations with internally torn Syria and Iraq, which no longer have control over their main sources of water. A broader and concerted Arab campaign, after GAP's launch in the 1980s, helped dissuade multilateral lending institutions from funding the Turkish program. International lenders, by insisting on a tripartite agreement among Turkey, Syria, and Iraq over uses of the twin rivers, forced Turkey to rely on its own budgetary resources.

The prolonged Arab campaign—backed by the six Gulf Cooperation Council member nations and Egypt—was fueled by concerns about the wider strategic ramifications of Turkey's water-appropriation plans and by suspicion that Ankara intended to use its growing ties with Israel, as symbolized by the 1996 Military Training Cooperation Agreement, to balance the Arab states. Turkey's completion of the southeastern Atatürk Dam—one of the world's ten largest dams whose reservoir extends across 817 square kilometers—was widely portrayed in the Arab media as a hostile riparian act, especially because the 1992 filling of its reservoir cut off the Euphrates' international flow for a month and triggered an Iraqi threat to bomb the structure, besides prompting Syria to step up support to the separatist Kurdistan Workers' Party (known as PKK).[43] The Turkish-Arab divide over water, compounded by the Kurdish factor, inflamed regional passions for years.

Ankara had long accused Syria of seeking to build leverage on the water disputes by actively aiding the Kurdish insurgency in Turkey, despite Syria's own restive Kurdish population—considered the country's most downtrodden ethnic group. In response to the Turkish interruption of the Euphrates' flow during the Atatürk Dam's initial filling, Syria itself drew a link between the water disputes and its Kurdish card against Turkey when its president, Hafez al-Assad, appeared at a PKK ceremony in the Syrian-controlled Bekaa Valley in Lebanon.[44] Syria even sought to link a proposed bilateral security protocol with Turkey to its demand for formal water-sharing arrangements on the Euphrates.[45]

In 1998, by massing 10,000 troops along the 625-kilometer border with Syria and threatening military reprisals against that country's "proxy war," Ankara precipitated a three-week showdown with Damascus that ended with Syria—fearful of facing a two-front war with Turkey and Israel—agreeing not to allow the PKK to "receive military, logistic, or financial support or to carry out propaganda" on Syrian soil.[46] Kurdish separatist leader Abdullah Ocalan, within months of his expulsion from Syria, was captured by Turkey with U.S. assistance. Slowly recovering from Ocalan's capture, the PKK—now known as the Kurdistan People's Congress, or Kongra-Gel (KGK)—has since 2004 revived the insurgency inside Turkey.

By 2011, however, the regional pendulum had swung in the opposite direction: it was the turn of an increasingly assertive Turkey not only to turn against its regional ally, Israel, but also to start a surrogate war of its own against Syria. It openly began providing sanctuary and military aid to a

Sunni Arab group, the Free Syrian Army, spearheading an armed insurrection against the government of Hafez al-Assad's son, Bashar.

A diplomatically weakened Syria, even before its civil war broke out, was compelled to place its territorial and water disputes with Turkey on the back burner. For its part, Turkey in recent years has sought to mend fences with a number of Arab states, even making common cause with Saudi Arabia, Qatar, and others in the armed campaign against Bashar al-Assad and serving as the main conduit for the supply of petrodollar-funded arms to Syrian rebels.

Successive Turkish governments—despite the heavy financial burden cast by GAP—have ensured bipartisan national support for the program's water-development projects by presenting them as integral to building a resurgent Turkey. The fact that the GAP's launch coincided with a rising tide of Kurdish separatism within Turkey only lent greater official support to the program's implementation so as to help end the economic marginalization and alienation of the Kurds. Indeed, GAP has become a symbol of national pride. The GAP case illustrates how water disputes may come in "handy to politicians in personifying real or perceived outside threats in the domestic context and in this way serve to unite the society against 'foreign enemies' and mobilize support for the government."[47]

Kurds—inhabiting the mountains and plateau regions where the borders of several nations converge—have for long been denied their identity and culture by the modern Turkish state, which emerged in 1923 from the wreckage of the Ottoman Empire.[48] Despite GAP's supposed benefits and Turkey's grudging offer to grant the Kurds greater linguistic and cultural freedom, the issue of Kurdish rights is proving as intractable as the larger Turkish political tussle, ongoing since 2003, that has pitted the ruling, Islamist-leaning Justice and Development Party against the military, which has carried out three coups since 1960 but now finds itself edged out of its long-standing role as guardian of the country's secular governing tradition. This tussle represents a broader and bitter battle for Turkey's soul between the politically besieged Westernized Turks who wish to preserve the secular traditions established by the republic's founder, Kemal Atatürk, and the grassroots-empowered Muslim conservatives now calling the shots.

The Justice and Development Party government led by Recep Tayyip Erdogan, in addition to jailing more journalists than China and empowering special courts to indefinitely detain citizens on suspicion of terrorism without charge or trial, has imprisoned hundreds of military officers on trumped-up charges that they planned a coup, and forcibly retired several dozen admirals and generals. Harsh judicial tactics and the curtailment of civil liberties in the name of establishing an Islamist-infused democracy have made it more difficult to co-opt Kurds and end separatism. Moreover, the benefits from the still-expanding irrigation under GAP have actually exacerbated intraregional disparities in the southeast between the disadvantaged Kurdish majority and other ethnic groups, including Arabs.

The emergence of a largely autonomous Kurdistan region in Iraq after the 2003 U.S. invasion of that country, plus the opportunity opened up by the Syrian civil war for Kurds in Syria's oil-rich northeast to exercise a degree of autonomy unheard of before, has helped to accentuate the Kurdish challenge for Turkey—home to about half of all Kurds—while presenting the Kurds living in the wider region with their best opportunity in almost a century to advance their cause. Oil is at the heart of the Iraqi Kurdistan region's wealth, although the issue of how to split oil revenues has dogged Iraq and exacerbated long-standing enmity between Kurds and Iraqi Arabs. But Turkey, having aided the Syrian insurrection, now faces blowback from the empowerment of Syrian Kurds.

The struggle of Kurds across several countries to become one nation has long unnerved governments in a broad regional belt that includes the Kurds' water-rich traditional mountain homeland (map 4.3). Many Kurds still dream of a greater Kurdistan, sprawling across the borders of Turkey, Syria, Iraq, and Iran, each of which has fought bloody battles with Kurdish separatists.

But no country faces a bigger Kurdish challenge than Turkey, whose regional riparian dominance is rooted in its control of the predominantly Kurdish areas and which today is seeking to fashion a new brand of neo-Ottoman clout in the Middle East. Turkey, reverting to hard-line anti-Kurdish policies on the heels of the withdrawal of U.S. forces from Iraq, renewed military incursions into northern Iraq in pursuit of alleged guerrillas but with the apparent aim of creating a cordon sanitaire on the Iraqi side of the common border. It might also seek such a protective corridor inside Syria to prevent Syrian Kurds from aiding Kurdish fighters in Turkey. Kurds, accounting for almost 20 percent of Turkey's total population, constitute the most important challenge to the unitary national identity that the Turkish state has long sought to fashion, as if it were a monoethnic society.

Turkey, in fact, is rich in water resources only in its eastern half, with water shortages becoming a regular feature in some of its western areas. It is like in China where the water wealth of the Tibetan Plateau contrasts starkly with the water poverty of the country's northern plains. And like China, Turkey is a major builder of dams, including on rivers flowing to other countries. However, the number of dams in Turkey pales in comparison with China's feat in having erected more dams than the rest of the world put together.[49]

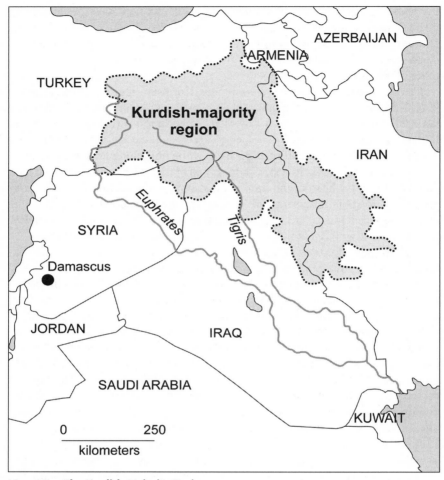

Map 4.3. The Kurdish-Majority Region

LESSONS FROM PAST WATER COOPERATION

Access to freshwater resources has historically been an instigating factor for conflict or cooperation. Water has even influenced the rise and decline of civilizations, their external relations, and their political and economic systems. Through the centuries, societies have struggled to effectively utilize and manage water—to build "cities around it, to transport goods upon it, to harness its latent energy in various forms, to utilize it as a vital input of agriculture and industry, and to extract political advantage from it."[50]

Control over water resources has usually symbolized power and influence. In fact, the term "rival" has its roots in the Latin word *rīvālis*, which means "one who uses the same stream" or "one on the opposite side of the river," implying the competitiveness of neighbors—people who share their lives or confront each other by the riverside.

It is no accident that the basins of major rivers, such as the Nile, the Tigris-Euphrates, the Yellow, the Indus, the Ganges, the Brahmaputra, and the Yangtze, were the seats of the world's oldest civilizations. More than 5,000 years ago, the earliest river-basin societies along the Nile, the Tigris-Euphrates, the Indus, and the Yellow developed agrarian practices by mobilizing water resources from those rivers. The "Fertile Crescent" region in Iraq, served by the Tigris and the Euphrates, is where Western civilization began.[51] The biblical Garden of Eden is said to be located there. India, or the land of the Indus, draws its name from that river, the largest portion of whose basin now falls in Pakistan, while the Yellow River served as the lifeline of the Han civilization since Neolithic times.

In addition to river basins serving as the cradle of human civilization, the control of water resources helped underpin the power of some empires, prompting one historian to coin the term "hydraulic civilizations."[52] Unlocking the hydraulic secrets of harnessing river waters for large-scale irrigation opened the path to the agricultural revolution and the advent of advanced civilization. The control and management of water resources indeed has been a central factor in human development and political power since ancient times. The ancient Egyptian civilization, for example, depended heavily on irrigated agriculture in the floodplains of the Nile, whose annual flooding brought nutrient-rich silt to revitalize overworked soils and helped to flush accumulated salts from the irrigated fields.

Two early hydraulic civilizations flourished in the Indus Basin.[53] The excavated city of Mohenjo-Daro ("Mound of the Dead" in the Sindhi language) was one of the world's oldest, with a remarkable water-management system. Built on a ridge in the middle of the floodplains of the Indus in the downstream Sind region, it came up about 2600 BCE, around the same time as the other river-basin societies in Egypt, Mesopotamia, and Crete. Another Indus city of that era was Harappa, on the Ravi River, in the upper Indus Basin. Mohenjo-Daro and Harappa were both built on a similar plan and had a sanitation system that included bathrooms linked to sewers. Whereas Mohenjo-Daro was devastated by severe flooding nine times, only to be rebuilt each time, recurrent destructive flooding may have contributed to the decline of the Harappan civilization after 2000 BCE.

In modern times, the location or rise of major cities has been greatly influenced by the availability of water resources. For example, the story of Los

Angeles is the story of water. The city's rise resulted from the building of an aqueduct in the early twentieth century that brought water from the Sierra Nevada Mountains in the Owens Valley—a construction that triggered the so-called California Water Wars, pitting Los Angles against the Owens Valley. Completed in 1913 after an underhanded, graft-ridden acquisition of water rights by Los Angeles left Owens Valley farmers in the lurch, the 375-kilometer aqueduct channeled the runoff from the Sierra Nevada range, drawing waters heavily from the Owens River.[54]

The diversion, which caused Owens Lake to virtually dry up by 1924, sparked an armed rebellion by farmers and ranchers, some of whom used dynamite to repeatedly blow up a part of the aqueduct.[55] Yet, for Los Angeles, the Owens Valley water diversion proved inadequate in less than three decades. To meet its fast-growing needs, Los Angeles in 1941 extended the aqueduct farther upriver to the hauntingly beautiful Mono Lake Basin, a vital habitat for millions of migratory and nesting birds. By siphoning off the tributary waters that used to flow into Mono Lake, the extended aqueduct caused the pristine lake to shrivel and ecologically retard.[56] A court battle and state regulatory action in more recent decades, however, has helped Mono Lake to escape the fate of Owens Lake.

Water also played a key role in the rise of another major American city, New York. In the early phase after the United States secured independence, New York and Philadelphia competed to be the new nation's leading city.

New York—by buying the rights to much of the water between it and Canada and bringing that water into town via a gravity-fed web of aqueducts—surpassed Philadelphia in population and importance. Its water entitlements and quality of supply helped present New York as a better and more hygienic place than its competitor, which lost roughly one-tenth of its population in a yellow-fever epidemic while serving as the temporary federal capital in the last decade of the eighteenth century.

Understanding the key role water played in shaping developments in history is useful in identifying the fundamental factors that can contribute to conflict or cooperation. Water conflicts tend to occur in situations of scarcity, when competition is fierce and tensions build up along the fault lines. Water conflicts also masquerade as battles for political and social justice or equitable development. Although water stress has begun to haunt many countries only in more recent decades, past history offers valuable lessons on the process of fashioning institutionalized cooperation. Some examples from modern history are particularly illuminating.

The Danube and Rhine Commissions

In modern history, intercountry collaboration on transnational rivers began in nineteenth-century Europe with the development of freedom of navigation on international rivers for vessels carrying economic goods. Until then, fluvial navigation was constrained by manifold burdens, extortions, and restrictions, and one riparian would arbitrarily deny shipping rights to another riparian.[57]

Entering an era of rapid industrialization and cross-border trade, Europe had little choice but to try to reconcile the conflicting interests of "independent nationalities and principalities which, up to that point, had jealously guarded their own stretches" of the rivers, especially the Rhine and the Danube.[58] The continued growth of intra-European trade to help promote European prosperity depended on the development and security of fluvial transportation, as well as freedom of navigation. Establishing open navigation on rivers to facilitate transboundary trade along fluvial corridors thus became a priority.

The Rhine and the Danube, as the longest rivers on the continent, serve as Europe's main fluvial arteries. Whereas the Danube flows eastward through many different lands on its way to the Black Sea, the Rhine flows northward to empty into the North Sea. The Rhine traditionally has boasted a larger volume of cargo than any other river in the world. It is the most important inland waterway in West Europe, and flows through heavily populated and industrialized areas. The Danube—Europe's largest river—has the distinction today of being the most international river basin in the world, encompassing territories of nineteen countries, including Austria, Bulgaria, Croatia, the Czech Republic, Germany, Hungary, Moldova, Romania, Slovakia, Slovenia, and Ukraine.

In the nineteenth century, expanding commerce made it imperative to develop and integrate large fluvial waterway systems over which significant trade could take place safely. That required, first and foremost, improving navigational conditions on the Danube and the Rhine by dismantling political barriers, halting extortions, and countering piracy and other threats to navigation in different stretches of these rivers. For example, the control of the Danube, the lifeline of Southeast Europe, was a source of constant struggle among the Ottomans, Hapsburgs, Hohenzollerns, and Romanoffs.

In 1878, through the Treaty of Berlin, the European Commission of the Danube (ECD) was established in what was one of the first efforts in modern times to promote international development of the resources of a common river. The Danube Commission was primarily focused on protecting safe, unimpeded navigation on the river, before it became a multipurpose authority in the post–World War II period.

The Danube Commission's counterpart, the less powerful Central Commission for the Navigation of the Rhine (CCNR), was set up even earlier. The decision to create the Rhine Commission was taken at the Congress of Vienna, a conference convened by the four European powers that defeated Napoleon so as to settle future political boundaries and establish a new balance of power in Europe. It was at this congress in 1815 that the doctrine of noninterference with shipping on transnational rivers was sought to be established. The treaty concluded by these great powers in 1815 and named the "Act of the Congress of Vienna" founded the principle of freedom and priority of fluvial navigation, applicable on a reciprocal basis to all states sharing a river.

Through a Rhine-centered Navigation Act that formed part of its annex, this treaty set in motion the process of evolution and codification of the principles of international law with regard to shipping on transnational rivers. In the face of opposition from some principalities and powerful merchant towns like Mainz and Cologne located along the Rhine, the Navigation Act did not abolish all the political barriers to free, unimpeded fluvial traffic. Their removal became a slow process.

The barriers were largely eliminated by the Convention of Mainz, signed in 1831. The Convention of Mainz was replaced in 1868 by the Revised Convention for the Rhine Navigation—also known as the Act of Mannheim—which was amended almost a century later by the 1963 Convention of Strasbourg and, subsequently, by several additional protocols.[59] The 1868 Act of Mannheim, however, served as an impetus to the establishment of the Danube Commission a decade later, leading to a major expansion of intra-European trade.

In fact, just a few years after the Danube Commission's formation, the movement toward freedom of fluvial navigation was extended by European colonial powers to Africa to help facilitate the consolidation and expansion of their colonies and to underpin their commercial interests there.[60] Through an 1885 treaty—the General Act of the Congress of Berlin—these powers opened two of Africa's key rivers, the Congo and the Niger, to free navigation among themselves. That treaty bestowed the principle of freedom of navigation on nonriparian states.

At the end of World War I, the 1919 Peace Treaty of Versailles opened all the navigable rivers in Europe to European states. In fact, the predecessor of the International Court of Justice, the Permanent Court of International Justice—established as part of the covenant of the League of Nations—held in a 1934 ruling on a case involving Britain and Belgium that freedom of fluvial shipping included freedom of commerce.

Freedom of fluvial navigation, however, was not embraced by the states in the Americas, where different interests were at play. In fact, the riparian states in the Americas remain entitled to forbid foreign navigation on the parts of transboundary rivers or lakes located within their frontiers. Interestingly, the United States became a temporary member of the Rhine Commission after World War II while West Germany was under Allied occupation.

The Danube and Rhine commissions, significantly, survived the unprecedented bloodshed of the two world wars. When Europe was racked by intense rivalries among its important powers, the Danube and Rhine commissions showed in different ways—despite the withdrawal of Germany and Italy from the Rhine Commission in 1936—that interstate cooperation on common waterways was sustainable even in the face of bitter political feuds and hostilities.

Yet the Cold War–era Iron Curtain that descended across Europe, creating two adversarial blocs, robbed the two commissions of their raison d'être—Europe-wide freedom of navigation. New modes of transportation by then, however, had lessened the importance of river navigation.

The U.S.-Canada and U.S.-Mexico Treaties

In North America, cooperative mechanisms between riparian neighbors have a long history, although not as long as those in Europe. North America, a region distinguished by the untrammeled dominance of the United States, bears little resemblance to Europe, a continent with several major competing powers where political borders have repeatedly been redrawn since the nineteenth century. Yet, significantly, the present International Joint Commission (IJC) of Canada and the United States dates back to a 1909 treaty, while the International Boundary and Water Commission (IBWC) of the United States and Mexico emerged from a water treaty concluded in 1944.[61]

Powered by its early economic advances, the United States exploited transboundary water resources at will to meet its needs. But when Canada and later Mexico began making demands on the same resources for their economic development, institutional mechanisms became necessary to regulate competition and protect the watercourses. It suited U.S. economic and diplomatic interests to build water cooperation with Canada and Mexico. The U.S. dependency ratio on transboundary water inflows is modest—just 8.2 percent—while Canada's is even smaller: 1.8 percent.[62] Canada and the United States, however, share the giant Great Lakes-St. Lawrence system, making demarcation and joint management necessary in order to avert injurious competition or conflict.

The unchallenged preeminence of the United States, with neither Canada nor Mexico in a position to take on the United States, paradoxically helped to fashion the two authorities—the IJC and the IBWC—and make them the world's most successful river-management institutions. Both institutions were largely molded by the United States, but it did make important concessions to its co-riparians. The two treaties that created these institutions were paradoxically shaped in an era in which the controversial "Harmon Doctrine" loomed large in the U.S. policy debate.

This doctrine, based upon an 1895 opinion of U.S. Attorney General Judson Harmon that the United States owed no obligations under international law to Mexico on shared resources, emerged during a dispute with Mexico over the American diversion of the Rio Grande (Spanish for the "Big River"), an important waterway in the arid areas of the southwestern United States and northeastern Mexico.[63] America also shares with Mexico the Colorado and Tijuana rivers. The Harmon Doctrine held that the United States enjoyed absolute sovereign rights over the portion of any international watercourse that fell within its borders, and therefore it was effectively free to divert as much of the shared waters as it pleased for its developmental needs.

Yet, under the 1906 Rio Grande Treaty, the United States agreed to supply 74 million cubic meters of water each year to Mexico from a dam to be built on the U.S. side of the border.[64] The pact, however, expressly stated that the United States neither recognized Mexico's prior legal claims to those waters nor did it "in any way concede the establishment of any general principle or precedent by the concluding of this Treaty."[65] In other words, the treaty, according to the United States, merely enshrined a bilateral political compromise.

Despite its implicit commitment to the principle of absolute territorial sovereignty as enunciated by the Harmon Doctrine, the United States also entered into a larger water-allocation treaty with Mexico in 1944. That treaty apportioned Colorado River waters to Mexico in what its text called an equitable manner.

With effect from 1950, the treaty guaranteed Mexico 1.85 billion cubic meters yearly of the Colorado River waters. However, if an "extraordinary drought" or serious accident were to make it "difficult" for the United States to meet its 1.85 billion cubic meter commitment, then "water allocated to Mexico will be reduced in the same proportion as consumptive uses in the United States are reduced."[66]

Canada, in the 1909 Boundary Waters Treaty, had to basically accept the U.S. position that absolute territorial sovereignty be retained by each country for the waters within its national territory, and tributaries thus should not come within the joint commission's purview.[67] Article 2 of the treaty effectively gives the upstream nation exclusive control over the use of waters on its own side of the border, but grants the downstream party—if materially injured—the right to legal remedies equivalent to those in force "in the country where such diversion or interference occurs."[68] Owing to the sovereignty principle, Lake Michigan is the only one of the five Great Lakes—which constitute a unique chain of interconnected glacial lakes—

not defined as boundary water, although Canada has been granted freedom of navigation on that lake.

In return, the United States agreed, even if reluctantly, to grant an arbitration role and special powers to the IJC, which has six commissioners, three appointed by each state party. Indeed, the IJC has been vested with quasi-judicial, investigative, administrative, and rarely used arbitral powers. Any national project that would materially affect the "natural" flow of boundary waters must be approved by the other government as well.[69]

The downstream party's right to seek redress for any harm caused, coupled with the IJC's powers, implies that, despite the principle of absolute territorial sovereignty tacitly embodied in the treaty, neither country is at liberty to use transboundary waters in a flagrant way to injuriously affect the other. The treaty indeed recognizes that each party is potentially affected by the other's actions on transboundary water resources, making joint cooperation indispensable.

The process leading to the 1944 treaty with Mexico was actually set in motion years earlier by the U.S. Congress when it passed an act in 1924 authorizing a study, "in cooperation with" Mexico, on the equitable use of the waters of the Rio Grande, two-thirds of whose approximately 3,000-kilometer course constitutes the U.S.-Mexico border. At Mexico's insistence, the scope of the study commission was broadened by a joint resolution of Congress in 1927 to include the Colorado River as well.[70]

The negotiations over the Colorado proved particularly contentious and protracted, especially with the upstream construction of the Hoover Dam—the world's biggest when completed in 1936 and named after Herbert Hoover, a trained engineer who initiated the project as U.S. president from 1929 to 1933.

The treaty that eventually emerged, besides allocating a portion of the Colorado River waters to Mexico, expanded the then-existing International Boundary Commission between the two countries to a new International Boundary and Water Commission. The IBWC, with its jurisdiction over transboundary water resources and joint water projects, was given a host of responsibilities: preservation of the lower Rio Grande as the international boundary; delivery to Mexico of the allocated share of the Colorado River waters; regulation of the Rio Grande flow through the joint construction, operation, and maintenance of storage dams, reservoirs, and hydropower plants; control of upstream activities that could affect downstream water flows into Mexico; and protection from river flooding by building levees and floodways. The IBWC, in essence, was designed to provide bilateral solutions to issues arising out of the common rivers and boundary demarcation.

Today, the two North American authorities, the IJC and the IBWC, serve as models of intercountry cooperation on transnational basins. This is despite noticeable deficiencies in their mandate and powers, especially of the IBWC.

One key reason they stand out as models is that these institutions were created as unique authorities: they cover not just one specific watercourse, but the full range of transboundary water resources. The omnibus authority of the two commissions, while increasing their operational burden, has given them "greater scope for the adjustment and compromise of water disputes."[71] In fact, thanks to the two commissions, intercountry water issues have been better handled in North America than intracountry water disputes within the United States, Canada, and Mexico.

Canadian-U.S. water resources are extraordinarily rich along what is one of the world's longest borders between any two countries. The Great Lakes-St. Lawrence system, straddling nearly half the total length of the U.S.-Canada border (excluding Alaska), contains almost one-fifth of the world's total unfrozen surface freshwater reserves and serves as the lifeblood for residents in eight U.S. states—Minnesota, Wisconsin, Illinois, Indiana, Michigan, Ohio, Pennsylvania, and New York—and for the large Canadian provinces of Ontario and Quebec. Only Baikal, the world's oldest and deepest lake situated in southeast Siberia, holds more freshwater than the Great Lakes.

The St. Lawrence River, navigable by deep-draft ocean ships, forms the border between eastern Canada and the United States and is the outflow of the Great Lakes to the Atlantic Ocean. In the 1950s, Canada and America built the St. Lawrence Seaway to facilitate vessels traveling between the Atlantic Ocean and the Great Lakes. Another great river the two countries share is the Columbia, located in the Pacific Northwest and famed for its salmon fisheries but now haunted by salmon depletion.[72]

The IJC assists in the management of these and other cross-border water resources, including setting the rules for the sharing of the Souris River in the Midwest by Saskatchewan, Manitoba, and North Dakota and establishing conditions for dam building on the Columbia, Osoyoos, and Kootenay rivers, which crisscross the northwestern U.S. states of Washington, Idaho, and Montana and the Canadian province of British Columbia. The IJC has helped determine how the waters of the St. Mary and Milk rivers are to be shared by Alberta, Saskatchewan, and Montana.

Protecting rivers and lakes from pollution has become a central mission of the IJC. In fact, with U.S. and Canadian manufacturing concentrated around the Great Lakes-St. Lawrence system, water contamination has been a major problem. The commission's antipollution mandate derives authority from the 1972 and 1978 U.S.-Canada accords to clean up the Great Lakes of toxic substances and ensure that wastewaters from industries and resident communities are properly treated before disposal. A follow-up bilateral protocol in 1987 called on the commission to review remedial action plans prepared by governments and communities as well as other proposed actions to improve the quality of water in all the Great Lakes—Erie, Huron, Michigan, Ontario, and Superior.[73] The commission, in addition, has worked to protect water quality in rivers ranging from the St. Croix to the Rainy.

Disputes over the division of waters, especially between Mexico and the United States, however, have remained recurrent. Water and economic rivalries among the U.S. states in the Colorado Basin, along with their hydroengineering projects, have contributed to exacerbating disputes with Mexico over the distribution of waters.[74]

Mexico has complained over the years that it has been denied its due share of the Colorado River waters, both in terms of volume and quality. The treaty did not address the water-quality issue while providing for an annual delivery, in accordance with schedules formulated in advance, of 1.85 billion cubic meters, plus any other waters arriving at the Mexican points of diversion.[75]

With the once-mighty river's waters siphoned by seven American states before it crosses into Mexico, the treaty's delivery commitment can scarcely be met in years when rainfall is deficient. In fact, the treaty was amended in late 2012 through a "Minute 319" accord that significantly rewrites rules on how the two countries share the Colorado River waters.[76] The five-year accord allows the United States to reduce its guaranteed water deliveries to Mexico when the river's level is low, with the elevation of Lake Mead serving as the trigger for such cuts. The cross-border annual delivery reductions would indeed be up to 154 million cubic meters if the projected surface level of Lake Mead—located along the Nevada-Arizona border—falls to less than 312.4 meters above mean sea level, or about 45 meters below average. Under the deal, Mexico, which has built little storage capacity of its own, will actually store some of its water in Lake Mead to help this drought-stricken reservoir stem its falling water level.

No less significant is the fact that the deal opens the way for the United States to effectively purchase a total of 153 million cubic meters of water by 2017 from Mexico's share of the Colorado River.[77] Mexico—in exchange for forfeiting this quantity of water, equivalent to 8.3 percent of its annual share—will receive $21 million from the United States through the IBWC to repair irrigation infrastructure damage in its Mexicali Valley caused by an Easter 2010 earthquake and to improve the lowermost basin's riparian and estuarine ecology.[78] U.S. interior secretary Ken Salazar expressed hope that the accord would end the two countries' "water wars" and lead to a longer-term agreement.

The Colorado—serving 1.6 million hectares of U.S. farmland and 30 million Americans, including Las Vegas, Phoenix, and Los Angeles residents—ranks as one of the most stressed rivers in the world.[79] It is also America's most diverted river system: the dams in its basin can store about four times the river's average yearly flow.[80] Because such large-scale upstream diversion has reduced the river's farthest downstream flow to a trickle, the Colorado Delta in Mexico has turned into a saline marsh. Under the 2012 accord, a joint pilot program is to try to generate minimum flows that could help restore the ecological health of the river's delta, with a binational coalition of environmental organizations also agreeing to raise money to buy water rights from Mexican farmers for the same cause.

The 1944 treaty dictates that, in return for the U.S. delivery of Colorado River waters, Mexico will allow a certain amount of water from the Rio Grande and its Mexican tributaries to reach southern Texas. But recurrent drought and overutilization of the Rio Grande system's resources resulted in Mexico accumulating a sizable water debt to the United States from 1992 to 2002. Following an IBWC decision, Mexico repaid that debt in 2005 by transferring 264 million cubic meters of water and making additional water available to southern Texas, besides carrying out a "paper transfer" of water already held in jointly controlled reservoirs.

The Great Depression–era Hoover Dam—which went on to inspire Soviet dictator Joseph Stalin and later Chinese strongman Mao Zedong to launch their own megaprojects to divert river waters to parched areas—has symbolized the power of an upstream project to transform an arid region. The Hoover Dam's storage reservoir—Lake Mead, the largest human-made lake in the Western Hemisphere—is supposed to hold almost a two-year supply of the Colorado River. Located at the tri-junction of the Mojave, Great Basin, and Sonoran deserts, Lake Mead, however, has begun to shrink in recent years because of overexploitation and drought. The Colorado River waters are also drawn off by other major U.S. projects, including the Glen Canyon Dam and the so-called All-American Canal, which diverts 3.2 billion cubic meters (which is greater than Mexico's share) across the desert of southern California to feed farms and cities in California's Imperial Valley.

The Hoover Dam, which helped bring water and electricity to America's arid Southwest, was once touted as an engineering wonder, rivaling, in the eyes of some, the pyramids. This is no different from the way China trumpets its Three Gorges Dam—now the world's largest—as the greatest architectural feat in history since the building of the Great Wall of China during the Ming Dynasty (1369–1644), despite the environmental devastation wrought by this dam project, which has contributed, for example, to a rapid shrinking of the country's largest freshwater lake, Poyang. And just as China now pursues megaprojects on international rivers without holding prior consultations with the riparian neighbors likely to be affected, the United States in more recent years presented Mexico a fait accompli by deciding to encase a portion of the unlined All-American Canal in concrete to prevent seepage—an action that will diminish the groundwater on which farmers on the Mexican side of the border are dependent, and increase the concentration of soluble salts in their aquifer. The 132-kilometer gravity-flow canal runs just north of the U.S. border with Mexico.

Indeed, as if it were harking back to the Harmon Doctrine, the U.S. State Department, while rebuffing the official Mexican protest that the All-American Canal's lining project violated international law, insisted that the water in the canal is American and the United States could do whatever it wished to conserve or utilize the water. The 1944 treaty requires its two parties to consult with each other if either side planned any action with transboundary implications.

The pact, however, was established—like the U.S.-Canada treaty—to deal with surface waters, not groundwater. New developments, including the now-abandoned U.S. proposal to build the Sierra Blanca nuclear-waste repository barely twenty-five kilometers from the Mexican border, indicate that the principles of customary international water law, including the norm of equitable utilization, must apply to groundwater that is linked with transboundary surface waters.[81] After all, with northern Mexican areas reeling under water scarcity and even U.S. parts of the Colorado Basin facing water stress, a lining project has turned into a contentious bilateral issue.

Building Transnational Water Institutions: An Evolutionary Process

The European and North American historical record shows that dynamic transnational institutions on water resources normally emerge from a lengthy evolutionary process that can stretch over many decades and even over more than a century. The record also indicates that significant transnational water institutions are generally born only after a convergence of interstate interests. The interests can converge on carrying out a strategic division of shared water resources, as happened in North America, or on a narrow issue, such as establishing freedom of fluvial navigation or combating river pollution, as in Europe. Commercial and strategic imperatives usually help foster such convergence of interests.

The two still-existing North American commissions, for example, were products of long-drawn-out efforts. Each was established after a century or more of treaty making on boundary and water issues. Before the 1909 treaty created the Canada-U.S. commission, the two neighbors had entered into a series of pacts from the late eighteenth century onward, including the 1794 Jay Treaty, the 1797 Agreement Relating to Naval Forces on American Lakes (also known as the Rush-Bagot Treaty), the 1842 Webster-Ashburton Treaty, the 1846 Northwest Boundary Treaty, the 1854 St. Lawrence River Treaty, and the 1871 Washington Treaty Relating to the Navigation of St. Lawrence River.

Negotiations that led to the 1909 treaty were carried out between Ottawa, Washington, and London because Canada was still a British dominion. Indeed, the 1909 pact, although formally known as the Treaty Relating to Boundary Waters between the United States and Canada, was signed by "His Majesty the King of the United Kingdom of Great Britain and Ireland and of the British Dominions beyond the Seas, Emperor of India."

The 1944 Mexican-U.S. water pact, which established the IBWC, traced its roots to eight earlier treaties concluded between 1848 and 1933. These include the 1848 Peace Treaty of Guadalupe Hidalgo and the 1853 Gadsden Treaty, which established temporary joint commissions to map and demarcate the new U.S.-Mexico border. The Convention of 1882 created another temporary joint commission to resurvey the western land boundary between the Pacific Ocean and the Rio Grande. Two conventions, one in 1906 and another in 1933, related specifically to the Rio Grande, which marks the entire Texas-Mexico border before emptying into the Gulf of Mexico.

The historical record also shows that the birth of a water institution does not necessarily end all differences or disputes. Often follow-up agreements become necessary to address outstanding issues. For example, a 1974 U.S.-Mexican agreement sought to resolve the issue of water quality, given that the 1944 binational treaty dealt only with water quantity and incorporated no promise by the United States to deliver usable water. Water quality became a subject of increasing bilateral acrimony in response to new irrigation projects in upstream U.S. states.

To meet its water-quantity obligation under the treaty, the United States overlooked the practice of some American users, such as the Wellton-Mohawk project in Arizona, in pumping highly saline drainage from irrigation into the Colorado River.[82] This affected agricultural production downstream in Mexico and triggered strong Mexican protests. It was only after newly elected Mexican president Luis Echeverria threatened to drag the United States before the International Court of Justice that President Richard Nixon's administration agreed to a salinity-control pact.[83]

It was Nixon's predecessor, Lyndon Johnson, who signed a bill approving construction of the Central Arizona Project, a 540-kilometer-long system of aqueducts, tunnels, and pipelines that siphons off 1.85 billion cubic meters of Colorado River waters per year—the same quantity as Mexico's treaty-apportioned share. Despite the 1973 salinity-control pact, the quality of surface waters flowing to Mexico has steadily deteriorated due to increased upstream water withdrawals, according to Mexican studies conducted by the Comisión Nacional del Agua (National Water Commission), Universidad Autónoma de Baja California, and El Colegio de la Frontera Norte. The Yuma Desalting Plant in Arizona, completed in 1992 to help the United States meet its obligations to Mexico, has remained largely dormant, with a pilot run being conducted during 2010–2011.

In Europe, the Danube and Rhine commissions evolved into active authorities through a stretched-out process. The Rhine Commission did not become active until the 1868 Convention of Mannheim—that is, more than half a century after the Congress of Vienna decided to set it up in 1815.

The Danube Commission, provided for in the 1878 Treaty of Berlin, traced its origin to the 1856 Treaty of Paris, which settled the Crimean War. It went through several name changes, from the Commissions of the Danube River, to the European Commission of the Danube, and then to the International Danube Commission. The international legal instrument governing shipping on the Danube, however, did not emerge until 1948, when a convention setting up the fluvial navigation regime was signed in Belgrade.

New challenges beyond navigation have spawned new conventions and institutions in the Rhine and Danube basins. A separate International Commission for the Protection of the Rhine (ICPR) was set up in 1950 as a permanent intergovernmental body among the co-riparian states. But the ICPR began fighting pollution of the Rhine in earnest only after a 1986 accident at a Basel plant. For a long time, industrial and domestic wastewater flowed untreated into the Rhine, earning it the sobriquet, "the Sewer of Europe." The Basel accident spewed thirty tons of herbicides, fungicides, pesticides, and dyes into the river, turning a large stretch of it red and destroying some fish species.[84]

Before the Basel accident, the antipollution legal instruments included the 1963 Agreement Concerning the International Commission for the Protection of the Rhine against Pollution, an Additional Agreement of 1976, and the 1976 Convention for the Protection of the Rhine against Pollution by Chlorides. Repealing these pacts, a new Convention for the Protection of the Rhine came into force in 2000. The nineteenth-century Rhine Commission on navigation continues its mission to this day. A separate organization, the International Commission for the Hydrology of the Rhine Basin, seeks to involve institutes in member states in preparing joint hydrological measures for sustainable basin development.

The 1994 Danube River Protection Convention was designed to ensure that ground and surface waters within the basin are managed and used sustainably and equitably. The convention has spawned a new transnational authority: the International Commission for the Protection of the Danube River (ICPDR), aimed at promoting an integrated, basinwide framework for protecting the Danube's water quality.[85]

The nineteenth-century European convergence of interests in making navigational uses a priority for cooperation on shared rivers was driven by economic imperatives. At that point in time, nonnavigational uses of rivers like hydropower generation and large-scale irrigation were still in the early stages of development, and greater intra-European trade hinged on freedom of fluvial navigation. But with accelerated industrialization, population growth, and the emergence of new technologies and modes of transportation, nonnavigational uses gradually became important and even found mention in the Treaty of Versailles before other developments ended the primacy of navigation.

Water sharing, however, has traditionally not been—and is still not—a major concern in Europe. For example, the states through which the Rhine flows on its way from the Alps to the North Sea have, by and large, "sufficient resources to meet all their legitimate needs for water."[86] And the Rhine Basin community, as it has evolved, is now primarily geared to combat pollution and promote sustainable practices rather than to work out water-sharing or conflict-avoidance arrangements.

Europe has had little experience with serious water-sharing disputes, unlike several other regions in the world. The European Union's Water Framework Directive, issued in 2000 with the goal of promoting sustainable resource management, was the product of the long experience Europe has gained in handling intercountry river issues other than water sharing. The directive requires all member states to establish water laws with common basic principles, as water is "not a commercial product like any other but rather a heritage which must be protected, defended, and treated as such."[87] The priority in Europe now centers on environmentally sound management of shared water resources, maintaining or improving water quality, controlling river pollutants, flood protection, and shielding freshwater biodiversity.

The extended evolutionary time frame in which water institutions in North America and Europe became operational must be borne in mind while considering the current transboundary water challenges in the developing world, where most nations are still young, having gained independence in the post–World War II period. In addition, the historical record shows that the effectiveness of such institutions depends on complex factors, including common basin goals, respect for the rights of a downriver state, and regional cooperation on larger issues.

Institutionalized basin cooperation demands coextensive restraint among all parties so that no state utilizes shared waters in a way to injuriously affect a co-riparian. But when a dominant riparian through unilateral actions on its own territory impinges upon the rights of a downstream nation—as illustrated by America's manipulation of Colorado River flows to Mexico—an existing institution may offer little protection to the injured state. Because of the wide power asymmetry between the United States and Mexico, the latter was left with little recourse against upstream actions.

Mexico's overall dependency ratio on transboundary waters, fortunately, is moderate—10.6 percent—and the weight of the impact of upstream diversions has largely fallen on its border regions that rely on such inflows.[88] But had a similar state of affairs prevailed in another region between two well-armed neighbors, with one state highly dependent on cross-border flows from the other, an explosive situation with unpredictable consequences would likely have developed. Basin organizations seek to rein in unilateral actions by presenting rules-based cooperation as being in the interests of even the dominant riparian. Yet the realities of power can overwhelm institutional logic.

One important historical lesson is the necessity of settling political boundaries in order to advance basin collaboration. America's water treaties with Canada and Mexico came after land boundary issues had been settled. Indeed, the formation of land boundary commissions preceded the establishment of water commissions in North America.

In the troubled parts of the developing world, territorial disputes remain the single biggest impediment to water sharing and cooperation. Settlements on political boundaries and shared water resources in these regions are also likely to follow an evolutionary path. In fact, the process of negotiating water accords in several basins, including the Ganges, Jordan, and Indus, took years, while institution building or interstate cooperation in many basins has still to take off. In the Senegal River Basin, the uppermost riparian, Guinea, withdrew from the nascent basin organization in 1971 amid regional tensions, only to rejoin almost thirty-five years later. Given that the processes of decolonization and subsequent national division created new international boundaries that split a large number of basins, building genuine intercountry water cooperation in the developing world remains a major challenge.

TODAY'S GRATING HYDROPOLITICS

Water has been a source of contention throughout history, but the interstate and intrastate hydropolitics of contemporary times are different in that they carry the seeds of interriparian conflict and are accompanied by the alarming trend of river depletion and rapidly falling groundwater levels. Past water shortages were either small in scale or temporary, and geographically confined. Now water stress has extended to large parts of the world and continues to spread wider. As a result, water competition is rife not only in arid and semiarid regions but also in some humid areas, which are generally better endowed with freshwater resources.

As countries and communities press against natural limits to the availability of freshwater for human use, the hydropolitics are becoming more charged. Shifts in economic development, income levels, population growth, and consumption patterns will likely continue to shape the relationship between water supply and demand to a much greater degree than climatic changes.[89] Given the scale of human impacts, the water cooperation witnessed amid intermittent discord in past centuries is unlikely to be replicated in a significant way at a time when the competition over scarce resources is becoming more grating. Acute water stress can actually trigger destabilizing conflict.

In fact, when natural replenishments cannot keep pace with human demand and the ensuing water scarcity begins to cause severe social and economic distress, the sanctity of existing intercountry and intracountry water-sharing agreements may also come under challenge. After all, a number of these accords were concluded when water stress was absent or not serious. Now overexploitation imperils the future of many watercourses.

A water-shortage crisis, if unmitigated, will sooner or later impinge on economic modernization and basin-level cooperation among co-riparians. Present trends suggest that issues related to managing and sharing water resources will increasingly become politicized, both within and between countries. More often than not, hydropolitical agendas already tend to trump sound environmental policy making and sustainable practices. The current politics extend to the building of dams, irrigation canals, and other storage or diversion works, as well as to issues that have a bearing on water consumption, like the type of crops grown or energy plants set up.

The Intersection of Water Scarcity, Overpopulation, and Terrorism

One country's access to a transnational watercourse can be affected by another nation's actions. The water competition is sharpening to such an extent that rival national plans to exploit resources of shared watercourses raise serious concerns about future peace and stability in some regions. The level and intensity of competition, of course, varies significantly from region to region.

In South America, the world's richest continent in per capita water availability (table 4.1), water disputes are less common or intense than in other developing regions. For a region home to barely 5.7 percent of the global population, South America (excluding Central America and the Caribbean) boasts 28.9 percent of the world's freshwater resources.

Table 4.1.   Internal Freshwater Resources in Different Continents and Subregions

	Volume per Year (billion m³)	Annual per Capita Availability (m³)
World	42,370	6,079
Africa	3,931	3,764
Northern Africa	47	279
Sub-Saharan Africa	3,884	4,431
Asia	11,865	2,816
South, Southeast, and East Asia	11,139	2,924
Central Asia	242	2,576
Middle East	484	1,559
Europe	6,578	8,884
Western and Central Europe	2,128	3,999
Eastern Europe	4,449	21,389
North America	6,077	13,147
Northern America	5,668	16,314
Mexico	409	3,563
Central America and Caribbean	781	9,328
South America	12,246	30,890
Oceania	892	30,447
Australia and New Zealand	819	30,310
Other Pacific Islands	73	32,065

Source: Aquastat, December 2012.

Note: Middle East encompasses Arabian Peninsula, Caucasus, Iran, Iraq, Israel, Jordan, Palestinian territories, Syria, and Turkey, while Eastern Europe includes the entire Russian Federation.

Not all parts of this continent, however, are endowed with bounteous water resources. There are major spatial, interseasonal, and interannual differences in water availability, with water stress a source of intersectoral conflict in the region's arid and semiarid zones. Given the high percentage of internationally shared water resources in South America—more than one-third—managed or low-level competition and conflict in some basins does rage. Yet the fact is that South America collectively, along with Canada, Australia, New Zealand, the Pacific Islands, Gabon, and some of the Nordic countries, boasts the world's highest per capita water availability.

The arc of Islam, by contrast, is on the opposite end of the spectrum—a region torn by the world's most serious water challenges, competition, and conflict, engendering greater volatility and violence. This arc of water crisis and political instability stretches from the Maghreb and the Sahel to the Arabian Sea region, through the Horn of Africa, the Arabian Peninsula, and the Iraq-Iran and Afghanistan-Pakistan ("AfPak") belts.

This is easily the most water-scarce region in the world, with water more precious than oil in some areas. Parts of the arc of Islam are endowed with some of the world's greatest hydrocarbon reserves but are woefully short of the most essential resource for survival. The Middle East has 1.1 percent, North Africa 0.1 percent, and Central Asia 0.6 percent of the world's freshwater resources.

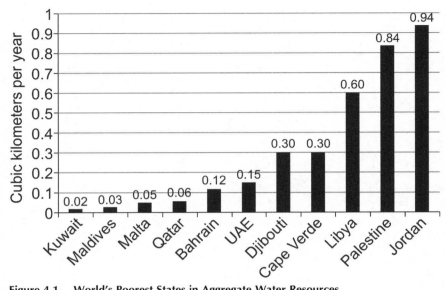

Figure 4.1.   World's Poorest States in Aggregate Water Resources

Source: Based on Aquastat online data, 2013.

Note: Palestine is not a United Nations member state but a "nonmember observer state," and the term here refers to the Palestinian territories of the West Bank and Gaza Strip.

In fact, most of the world's poorest states in water resources, as figure 4.1 shows, are Islamic, including Kuwait, the Maldives, Qatar, Bahrain, the United Arab Emirates, Djibouti, Libya, and Jordan. So, it is scarcely a surprise that many domestic conflicts in the largely hyper-arid arc of Islam revolve around water-related issues, including freshwater access and food production, with the battle lines of internal war tending to follow the lines of watercourses. In some parts, water reserves—like oil—are the target of political and even military intrigue.

To make matters grimmer, this contiguous arc of water crisis is also the arc of international terrorism (September 11, 2001, was a wake-up call) and the arc with the world's highest collective population growth rate for any region. A study by the Washington-based Pew Research Center predicted that the world's Muslim population would grow "at about twice the rate of the non-Muslim population" between 2010 and 2030, despite falling fertility rates in many Muslim-majority countries.[90]

It said that the aggregate Muslim population by 2030 will have doubled since 1990, when the world had about 1.1 billion Muslims. Sunnis, according to the report, would continue to make up the bulk of the Islamic population (about 87 to 90 percent), especially as Shiite numbers decline because of relatively low birth rates in Iran, where couples are required to attend a family-planning class before applying for a marriage license. However, most other countries in the arc of Islam, according to United Nations figures, still maintain relatively high total fertility rates.[91] The ticking population bomb is one of the factors imperiling the Islamic world's water future.

The arc of Islam indeed has become the arc of unending conflict. It is caught in a vicious circle in which water scarcity aids the very factors behind the growing water distress (figure 4.2). This arc of water scarcity, growing food insecurity, burgeoning populations, greater fundamentalism and militancy, and rising unemployment serves as a reminder that violence is often rooted in issues that extend beyond culture and politics. The international-security implications of this situation are particularly ominous: as British prime minister Tony Blair warned in 2006, an "arc of extremism" stretching across the Middle East poses a threat to countries far outside that region.

It is not an accident that the main springboards of international terrorism or ocean piracy—the AfPak belt, Yemen, and Somalia—are troubled by growing water scarcity, exploding populations, a pervasive lack of jobs, high illiteracy, and fast-spreading extremism. The intersection of water stress, food insecurity, political instability, popular discontent, population and resource

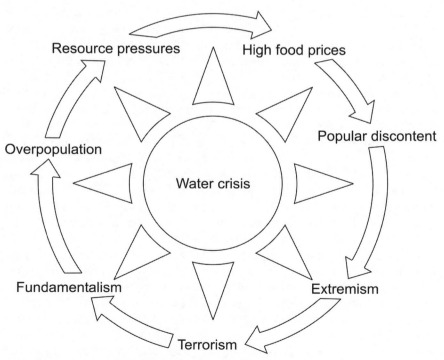

Figure 4.2.   The Arc of Islam Faces Vicious Circle

pressures, extremism, and terrorism forms a deadly cocktail of internal disarray, spurring a cycle of unrest and violence and fostering a pervasive jihad culture. Further compounding the situation is a sectarian explosion that has brought the fragile state structure under increasing pressure in this arc, where a number of states were created by European colonial powers in the past century with little regard for ethnic, sectarian, and cultural fault lines.

When popular uprisings spread like fire across the Middle East and North Africa from early 2011, several factors were cited as probable causes, ranging from political injustice and repression to economic stagnation and unemployment. One less-noticed trigger for the serial Arab Spring uprisings—rising food prices—directly relates to the worsening regional freshwater crisis.[92] Water and environmental stresses clearly are as important as political and economic factors in spurring continued unrest in this vast region. If not mitigated, these stresses could act as threat multipliers, leading to everlasting instability and turmoil in a region where Islamism has now supplanted Arab nationalism.

With water tables falling alarmingly in the arc of Islam, grain production has declined in several countries despite soaring domestic food demand, forcing them to rely increasingly on imports. But record international food prices have heavily increased the subsidy burden of these countries, many of which also provide cheap or free water and subsidized gasoline to help buy peace at home. In instances where food and other subsidies have been cut, this has often fueled popular anger against autocratic rulers and stirred up renewed instability. Many Arab economies depend on oil and gas export prices for revenue, yet, paradoxically, higher international hydrocarbon prices—by increasing the costs of fertilizers and agricultural transportation—make their imported food dearer.

In the next two decades, the real wild card for greater unrest in this region, according to one study, "is not war, terrorism, or revolution—it is water. Conventional security threats dominate public debate and government thinking, but water is the true game-changer in Middle Eastern politics."[93] Abundant oil helped build the modern Arab world, but scarce water threatens its future. Almost half of all Arabs depend on water flows from non-Arab countries, such as Turkey and the upstream Nile Basin states.

The high population growth rates in the arc of Islam—with the striking exception of the birth-control-promoting Iran, whose total fertility rate is lower than even that of the United States—mean that the already-unsustainable patterns of water extraction will get worse, exacerbating water scarcity and domestic instability. A growing youth bulge, coupled with income disparities and slackening creation of new jobs, will also foster greater discontent and alienation, fueling radicalization. Nowhere in the world does the prospect of a dreadful collision between population growth and water supply look more real than in this arc.

The higher the population growth rate, the greater will be the fall in per capita freshwater availability in a region where most states are already either below the international water-poverty line of 1,000 cubic meters per head per year or close to slipping into that category. According to the United Nations Development Programme, the burgeoning regional populations by 2025 are likely to reduce the average freshwater availability in the Middle East and North Africa to 460 cubic meters per person per year, or less than half the water-poverty threshold.[94]

But in Palestinian-majority Jordan, without demand management and water-efficiency measures, per capita water supply is projected to fall from the current measly 150 cubic meters per person to under 100 cubic meters by 2025—a situation that already prevails in conflict-torn Yemen, which, with one of the world's fastest-growing populations, faces potential sociopolitical collapse. These states lack the financial resources of water-distressed Saudi Arabia, Qatar, Kuwait, and the United Arab Emirates to mitigate their crisis through technical and other means.

Demographic pressures and profligate resource use are already creating significant threats to human security in the arc of Islam. Communities in the extensive drylands within this arc rely on natural resources for subsistence, making them highly vulnerable to the growing impacts of environmental degradation. More than 35 percent of the Arab population is still engaged in agriculture, even as the overuse and mismanagement of farmland and freshwater resources undercuts the relatively low fertility of soils. The pressures on water and land resources indeed are increasing social violence in a region where the majority of states are already racked by serious internal conflicts. Furthermore, upstream diversion of the waters of the region's very few rivers is wreaking environmentally damaging impacts in downriver areas.

Government policies and projects are aiding the environmental damage. A 330-kilometer canal constructed by ousted Iraqi president Saddam Hussein's regime from near Baghdad to close to Basra has come to symbolize the environmental havoc and human suffering wrought by a recklessly designed water project. The project, by draining waters from the Mesopotamian Marshlands and rendering large parts of those wetlands dry, forced as many as two-thirds of the "Marsh Arabs" into exile in southern Iran or into refugee camps within Iraq.[95]

The Shiite Marsh Arabs—heirs to the ancient Babylonian and Sumerian civilizations who were persecuted under Saddam Hussein—have evolved a unique subsistence culture tied to their aquatic environment. In addition to the canal that was built, the upstream damming of the Tigris and Euphrates rivers contributed to the drying of the marshlands, which by 2002 had shrunk from 9,000 square kilometers to barely 760 square kilometers.[96] But after America's military invasion drove Hussein out of power in mid-2003, some Marsh Arabs and other local residents broke the drainage embankments and opened the floodgates to bring the water back into the wetlands—home to rare and unique species like the sacred ibis and African darter.

Some of the world's most important aquifer systems are located under the desert sands in the arc of Islam. But unbridled water extraction has endangered the deep fossil-aquifer systems. The greatest groundwater basin in the arc is spread over 1.5 million square kilometers in the Arabian Peninsula, covering much of Saudi Arabia other than its midwest and west and extending into the Gulf countries in the east and Jordan in the northwest. This multilayered sedimentary basin—ranging from Cambrian-age sandstone aquifers in the north to sand-and-limestone aquifers from the Neogene age along the peninsula's eastern margins—is estimated to hold water reserves totaling 2,000 billion cubic meters, with its actual replenishment capacity being only 1 billion cubic meters per year.[97] However, just between 1975 and 2000, about 380 billion cubic meters of its waters were extracted.[98] And there is little evidence to suggest the rate of abstraction has significantly slowed since 2000.

In no part of the world are transboundary water resources so critical for human survival as in the arc of Islam stretching up to Pakistan. In fact, the countries with the world's highest dependency on incoming flows from across their borders are all located in this arc, including Kuwait (100 percent), Turkmenistan (97.1 percent), Egypt (96.9 percent), Mauritania (96.5 percent), and Bahrain (96.6 percent).[99] Several countries in this arc also share groundwater resources with a neighboring Islamic state. As overexploitation of shared water resources continues unchecked, interriparian relations in the arc are likely to increasingly come under strain.

Saudi-Jordanian water tensions, for example, have escalated over Jordan's U.S.- and European-funded pipeline project to take 110 million cubic meters of water per year to its north from the transboundary aquifer al-Disi in the deep south.[100] The Turkish-constructed and French-operated pipeline exemplifies a silent pumping race over a shared aquifer, which Saudi Arabia has been heavily exploiting since the 1980s.

Called Saq by Saudi Arabia, al-Disi is one of the aquifers in the Cambrian-age subbasin. The 340-kilometer, $1.7-billion pipeline was designed largely to improve water supply in the Jordanian capital of Amman, where most residents get tap water only once a week and buy essential supplies from water trucks. The Jordanian-Saudi pumping race promises to completely deplete al-Disi within a few decades.

Water competition and conflict in the region extends to the main transboundary rivers—the Euphrates-Tigris, which empties into the Persian Gulf; the Nile and the Asi-Orontes, draining into the Mediterranean Sea; and the Jordan, flowing to the Dead Sea. Turkey, although a Muslim state in the region, lies outside the Islamic arc of political instability and water stress. Turkey indeed is the upstream water controller. Its chain of upstream dams on the Tigris and the Euphrates has fostered conflict-ridden relations with Syria and Iraq, whose own dams have only compounded the water challenges in this twin river basin. The bulk of the large dams in the Middle East are in these three countries, with Turkey, of course, in the lead.

Water is likely to remain a deeply divisive issue between Turkey and its downriver neighbors, just as the split between Arab and non-Arab states in the Nile Basin promises to roil interriparian relations. Indeed, the risk of water becoming a trigger for war or diplomatic strong-arming is high across the arc of Islam.

Fashioning Water Pipelines for Peace

Nothing can better alleviate water insecurity and build interriparian peace than institutionalized cooperation and joint projects on shared water resources. Joint water projects between competing or disputant nations (or provinces) indeed can greatly help to reduce suspicions and tensions. Between co-riparian countries, such peace projects can even contribute to building common interests on sustainable management of shared resources, besides helping to mute or ease territorial disputes.

If water-related challenges are to be managed, weighing real-world costs and benefits of interriparian cooperation will have to come ahead of the pursuit of narrow geopolitical objectives. If not, a new Great Game over water is likely to parallel, and probably surpass, the geopolitical competition over energy resources.

Just as energy pipelines of peace have been proposed, water pipelines or canals of peace are also conceivable to bring water from wet areas to dry areas. In fact, as part of the Israel-Jordan water arrangements embodied in Annex 2 of their 1994 Peace Treaty, a pipeline was completed on schedule to take water from the Jordan River—immediately south of its exit from Lake Tiberias in Israel—to the King Abdullah Canal (formerly the East Ghor Canal) in Jordan (map 1.1).

However, most proposals to establish transnational water pipelines have evoked suspicion or even opposition. Even purely commercial proposals to transfer surplus water by pipeline between two closely tied neighbors can stir up public passions in the country of intended origin, as illustrated by the strong Canadian opposition to bulk-water exports to the United States—a case examined in chapter 3.

A classic example of a "peace pipeline" proposal stoking suspicions and ill feelings in a region was the idea Turkish president Turgut Özal put forward in 1986 to take surplus river waters from his country by pipeline to eight Arab countries: Bahrain, Jordan, Kuwait, Oman, Qatar, Saudi Arabia, Syria, and the United Arab Emirates (UAE). His proposal was widely seen in the Arab world as an ingenious Turkish move to build greater water leverage against Arab states.

The pipeline project's intended funders—the oil-rich Arab Gulf nations—publicly dubbed the Turkish proposal as cost ineffective, contending that desalination would be a cheaper alternative for them than to invest billions of dollars in piping water all the way from Turkey. But more than the economic costs, it was the political risks of building water dependency on Turkey and potentially suffering supply disruptions—including by acts of sabotage—that weighed heavily on their thinking. Indeed, in private, Arab officials derisively portrayed the proposal as an attempt to extend Turkey's water-supply hegemony to states outside its natural sphere of cross-border river flows. The proposal, furthermore, was viewed as a ploy to blunt Arab criticism of the launch of Turkey's controversial GAP program.

The intended 2,750-kilometer pipeline to the Gulf states through Syria and Jordan was to convey 6 to 10 million cubic meters of water per day from the Ceyhan and Seyhan rivers, which drain into the Mediterranean. One pipeline was to be dedicated to supplying Bahrain, Kuwait, Oman, Qatar, Saudi Arabia, and the UAE, and another pipeline was to lead to Syrian and Jordanian cities and the Saudi areas along the Red Sea. Because the idea never took concrete shape, it remained unclear whether the flows of the Ceyhan and the Seyhan—both internal rivers of Turkey—would have sufficed for such large transfers or whether water would also have had to be drawn from the transnational Euphrates Basin.

In fact, the proposal perished amid persistent questions in the Arab world about Turkey's intentions in offering to sell water by pipeline while simultaneously impeding the natural cross-border flows of the Euphrates and the Tigris to Syria and Iraq through its hydroengineering projects. There is no water-sharing treaty in the Euphrates-Tigris Basin largely because Turkey long refused to accept even the international character of the twin rivers, which it has linked by building canals. It wished to discuss only the scientific and technical aspects relating to the rational utilization of transboundary waters.[101] Turkey even voted against the 1997 United Nations Convention on the Law of the Non-Navigational Uses of International Watercourses on the grounds that it unjustly grants lower riparians a right to "veto" upstream water-development plans.

To be sure, Turkey pledged in 1987 to leave a yearly average Euphrates flow of 500 cubic meters per second—aggregating to a flow of 15.75 billion cubic meters in a year—for Syria, which in turn agreed in 1989 to reserve for Iraq 58 percent of the Euphrates waters. The Turkish pledge, in exchange for antiterrorist assurances by Syria, was contained not in a water-sharing treaty but in a protocol of economic cooperation signed by Özal during a visit to Damascus months after he unveiled his peace pipeline proposal.[102] Not only was the pledge publicly flayed as injudicious by Özal's successor, Süleyman Demirel, but also the protocol's implementation ran aground as simmering Turkish-Syrian tensions erupted in bitter mutual recriminations.[103]

Critics in the Arab world charged that the pipeline project would arm Ankara with greater latitude to potentially turn water into a political weapon against Arab nations. In addition, they contended that the project could help underpin a latent Turkish-Israeli partnership against Arab states, especially if Israel was later made part of the water-supply arrangements.[104] The original Turkish proposal actually envisioned Israel's inclusion in the supply network, but the strong Arab reaction led to its modification, with Israel's participation made contingent on peace settlements with its Arab neighbors.[105]

Years after the pipeline proposal collapsed under the weight of unrelenting Arab opposition, Turkey, however, offered to export 50 million cubic meters of water per year directly to Israel. But Israel, as discussed in the preceding chapter, backed out of a 2004 agreement-in-principle to import Turkish water for the same reason that the Arab states didn't embrace the pipeline proposal—it would arm Turkey with undue hydro-leverage.

Building intercountry water pipelines typically has proven more problematic than constructing them within a nation. And even when intercountry pipelines have been built, friction between parties has often arisen subsequently over the price of water and other issues, as in the Malaysia-Singapore case. Within nations, projects to take water to water-scarce areas by pipelines or canals have a long history. Such projects have aided domestic peace, stability, economic development, and even discovery of mineral wealth.

For example, at the beginning of the twentieth century, Western Australia built a 530-kilometer-long water pipeline to remote communities to facilitate the search for gold in the arid and desolate Kalgoorlie and Coolgardie areas. The provision of water did help in the discovery of gold. That pipeline continues to operate today, supplying more than 100,000 residents. In more recent times, water-stressed Iran opened a 333-kilometer pipeline that transports water from the central Zayandeh River to the desert province of

Yazd. And in the United States, a new pipeline has been built to transport Missouri River water to some communities in the states of South Dakota, Minnesota, and Iowa.

Israel, in fact, boasts a national water-pipeline grid that allows supply to be measured and controlled. Completed in 1964, this cross-country web starts from Lake Tiberias and terminates in the Negev Desert, with large pipelines conveying water across the Jezreel Valley and the southern region along the coastal plain. Smaller pipelines branch out to take water to rural communities and farmlands.

Turkey, decades after its forces invaded Cyprus in a dispute with Greece over the island, is building an undersea pipeline to supply 75 million cubic meters of potable water annually to water-scarce Northern Cyprus, which much of the world shuns. A protracted ethnic conflict has split Cyprus between the mainly Turkish-speaking north, which is controlled by Turkey, and the internationally recognized, mainly Greek-speaking Republic of Cyprus in the south, with the recent discovery of natural gas wealth beneath the seabed of the Levant Basin near Cyprus adding to the Turkey-Greece tensions.

The world's largest underground water-pipeline network is Libya's Great Manmade River Project (GMRP), a Muammar Qaddafi legacy. This 3,500-kilometer system of pipelines conveys groundwater from the Libyan Sahara in the south to the coastal cities in the north. However, aquifers in Libya are recharged only in the north. So, with fossil reserves in the south serving as the country's dominant source of water supply, internally torn Libya risks running out of groundwater in the coming decades.[106]

As water stress in the world accentuates, pipelines are set to become more common to transport water within countries from areas with abundant resources to zones seared by scarcity. Water is also likely to be increasingly moved by canals, using the principle of gravity. But unlike river-water diversion by open canals—as exemplified by China's Great South–North Water Diversion Project and India's taking of water southwestward to Rajasthan from the confluence of the Beas and the Sutlej rivers—pipelines convey water across great distances without significant loss due to evaporation or seepage. Dilapidated pipes, however, lose water through leakage. Measuring up to six meters in diameter and made from different materials, water pipelines can be laid underground, underwater, or above ground.

In addition to conveying clean water, pipelines are also essential for recycling water—first by transporting wastewater for chemical treatment and then taking the cleansed water to users. In fact, with an increasing number of desalination plants coming up, the construction of extensive water pipelines is becoming normal in some countries. Saudi Arabia, the world's largest producer of desalinated water, is building a four-hundred-kilometer pipeline to transport 1 million cubic meters per day from the Ras Al Zour desalination plant to its capital, Riyadh.

Pipelines, however, are expensive to build and maintain, and they demand energy-intensive pumps to keep the water flowing. In areas that are mainly flat or gently sloping down, piped water can also be conveyed by gravity flow. The cost factor can deter the building of pipelines when alternative options are available. For example, the UK Environment Agency, in a 2006 report, advised against piping water from the northern Pennines to water-stressed areas in southeast England, including London, because it would cost between five and eight times more than developing extra infrastructure in the southeast.[107]

A WATER HEGEMON WITH NO MODERN HISTORICAL PARALLEL

International discussion about China's rise has focused on its increasing trade muscle, growing maritime ambitions, expanding capacity to project military power, assertive mercantilism centered on resource acquisition as a vital strategic interest, and aggressive push for regional hegemony. China has also aroused international alarm by using its virtual monopoly of rare earths as a trade and political instrument and by stalling multilateral efforts to resolve disputes in the South China Sea.

One critical issue, however, usually escapes international attention: China's rise as a hydro-hegemon with no modern historical parallel. No other country has ever managed to assume such unchallenged riparian preeminence on a continent by controlling the headwaters of multiple international rivers and manipulating their cross-border flows. In fact, it wasn't geography but guns that established China's chokehold on almost every major transnational river system in Asia, the world's largest continent in both area and population.

With the world's most resource-hungry economy, China has gone into overdrive to corner natural resources, yet bristles at criticism that it is following in the footsteps of European colonial powers by extracting resource riches in Africa and elsewhere. The fact is that capturing or controlling resources has become an important driver of its policies, which seem pivoted on the Malthusian logic that natural resources will become scarcer over time and thus the strategic advantage will lie with those who hold them.

China's pursuit of an aggressive strategy at home and abroad to secure (and even lock up) supplies of key resources like water, fuel, and minerals appears aimed at gaining a long-term advantage that its peer rivals would not be able to offset. China knows that its military capabilities, although vast and rapidly growing, can be counterbalanced by its peers, especially by forming countervailing geopolitical partnerships or alliances. Even its nuclear arsenal can be checkmated through deterrence. With nuclear-armed Russia and India located on its opposite flanks, China's nuclear armory gives it no decisive edge. But no rival can neutralize, for example, China's throttlehold over the headwaters of Asia's major transnational rivers and its capacity to serve as the upstream controller by reengineering transboundary flows through dams and other structures.

Just as Saudi Arabia sits over immense reserves of oil and is a critical international supplier, so China controls vast transnational water resources, giving it significant, even if latent, clout over the states to which rivers flow from its territory. At the hub of Asia, China shares land borders with fourteen states, thirteen of them its co-riparians. There is deep concern among its riparian neighbors that, by building extensive hydroengineering infrastructure on upstream international basins, it is seeking to turn water into a potential political weapon. After all, China is a rising great power whose muscular confidence is increasingly on display.

By controlling the spigot for much of Asia's water, China is acquiring major leverage over its neighbors' behavior in a continent already reeling under very low freshwater availability. Although Asia now serves as the locomotive of the world economy, its per capita annual water supply (2,816 cubic meters) is not even half of the global average (6,079 cubic meters). In fact, as table 4.1 shows, it is almost one-eleventh of that in South America, Australia, and New Zealand; not even one-fifth of Northern America's; barely one-third of Europe's; and a quarter less than Africa's. Yet the world's most rapidly growing demand for water for industry, food production, and municipal supply is in Asia, which geographically stretches from the Japanese archipelago all the way to the Bosporus and is home to three-fifths of the global population.

China's Establishment of Hydro-Hegemony

Asia's water map fundamentally changed after the 1949 Communist victory in China. China became the source of transboundary river flows to the largest number of countries in the world, extending from the Indochina Peninsula and South Asia to Kazakhstan and Russia.

This unique riparian status is rooted in China's forcible absorption of sprawling ethnic-minority homelands, which make up 60 percent of its landmass. In the same way Israel changed the subregional water map through the 1967 Six-Day War, almost all the important international rivers of mainland Asia originate in territories that the new People's Republic of China grabbed while extending its control to regions that were imperial spoils of the earlier foreign rule in China under the Manchu Qing dynasty (1644 to 1911) and the Mongol Yuan dynasty (1271 to 1368).[108]

The annexed Tibetan Plateau, for example, is the world's largest repository of freshwater after the Arctic and Antarctica and is the source of Asia's greatest rivers, including those that are the lifeblood for mainland China and South and Southeast Asia. Whereas the water in the polar icecaps is all locked up, much of the renewable freshwater on the Tibetan Plateau—often referred to as the "Third Pole"—is accessible and flows naturally into watercourses.

The world's highest and largest plateau thus boasts the world's greatest concentration of available freshwater resources. Asia's ten great river systems—fed by thousands of Himalayan-cum-Tibetan glaciers and mountain springs and flowing down from the Tibetan highlands—highlight the role played by this plateau as the continent's "water tower." The numerous lakes on the plateau alone store, according to the Chinese Ministry of Water Resources, 608 billion cubic meters of freshwater.

China's control over this giant incubator of Asia's main river systems, coupled with its damming programs, arms it with transboundary hydro-leverage that it could potentially take advantage of. Another large ethnic-minority homeland, Xinjiang, is the territory from where the Illy River flows to Kazakhstan and the Irtysh to Kazakhstan and Russia. In Northeast Asia, the Amur River divides Manchuria into Chinese- and Russian-held parts until it enters Russia, making China a transnational water supplier even from the Amur Basin (table 4.2). Smaller quantities of water also flow from China's ethnic-minority regions to Kyrgyzstan and Mongolia.

China is thus the source of cross-border water flows to more countries than any other upstream power in the world. Its regional hydro-hegemony is clearly on a much higher scale than that of the United States, Brazil, Turkey, Israel, Egypt, Uzbekistan, or any other regionally dominant riparian power, whether located upstream or downstream. China indeed controls the headwaters of more than a dozen major rivers, which are the lifeblood for nearly half of the global population, including the people living in the Yangtze and Yellow basins in the Han heartland.

Yet China rejects the very concept of water sharing or institutionalized cooperation with downstream countries. Consequently, China does not have a single water-sharing treaty with any co-riparian country. By contrast, riparian neighbors in South and Southeast Asia are bound by water treaties that they have negotiated between themselves, while Soviet-era water arrangements continue in Central Asia, even if falteringly.

Getting Asia's paramount riparian power to accept water-sharing arrangements or other cooperative institutional mechanisms has proved unsuccessful so far in any basin. China was one of only three countries that voted against

Table 4.2. Major Rivers Flowing Out of Chinese Territory to Other Countries

River System	Direct Destination	*Average Annual Runoff Volume Flowing Out (km³)*
Amur and Suifen	Russia	119.04
Brahmaputra (Yarlung Tsangpo) and tributaries	India	165.40
Indus, Sutlej, and other rivers from southwestern and western Tibet	India	181.62
Arun and other rivers from southern Tibet	Nepal	12.00
Salween (Gyalmo Ngulchu, or Nu)	Burma (Myanmar)	68.74
Mekong (Zachu or Lancang)	Laos and Burma	73.63
Red River (Yuan)	Vietnam	44.10
Tibetan rivers flowing out from Yunnan's west (southeastern Tibet Plateau)	Burma	31.29
Illy flowing out of Xinjiang	Kazakhstan	11.70
Black Irtysh (Ertix He), originating in Mongolia-China borderland	Kazakhstan	9.53
Rivers from the western side of the Barluke Range	Kyrgyzstan	0.56

Source: Aquastat online data, 2013.

Note: There are also several border rivers. They include the Yalu and the Tumen, which form the China–North Korea border and have a total average discharge of 20.3 cubic kilometers per year, and the Argun (Ergun He) and the Ussuri (Wusuli) along the China-Russia frontier in Manchuria. One cubic kilometer (km³) equals 1 billion cubic meters.

the new international water law—the 1997 United Nations convention, which lays down rules on the shared resources of international watercourses.

Yet water is fast becoming a cause for competition and discord between countries in Asia, where intranational water disputes are already widespread. Water increasingly is turning into a divisive issue. This strengthens the risks of water becoming a flash point in intercountry relations. The growing water stress threatens Asia's rapid economic growth and sociopolitical stability.

Despite its centrality in Asia's water map, China has declined to open negotiations on a water-sharing treaty with any co-riparian country. Worse still, while promoting multilateralism on the world stage, China has given the cold shoulder to multilateral cooperation among river-basin states. Indeed, having its cake and eating it, China is a dialogue partner but not a member of the Mekong River Commission, underscoring its intent to stay clued in on the discussions, without having to abide by the Mekong Basin community's rules or take on any legal obligations.

China publicly favors bilateral water initiatives over multilateral institutions, yet it has not shown any enthusiasm for meaningful bilateral action. Indeed, as epitomized by its construction of upstream dams on international rivers such as the Amur, Arun, Brahmaputra, Illy, Irtysh, Mekong, and Sal ween, China is increasingly bent on unilateral actions, impervious to the concerns of downriver nations. As a result, freshwater has become a new political divide in the country's relations with neighbors like India, Russia, Kazakhstan, Nepal, Burma, Vietnam, and others. The lower-Mekong countries view the Chinese strategy as an attempt to "divide and conquer."

While reluctant to enter into any treaty to share waters, China is willing to share flow statistics—but at a price. In fact, it deflects attention from its refusal to share waters, or to enter into institutionalized cooperation to manage common rivers, by flaunting the accords it has signed on sharing flow statistics with riparian neighbors.[109] These are not agreements to cooperate on shared resources but commercial accords to sell hydrological data, which several other upstream countries provide free to downriver states. The 1997 United Nations convention, in fact, calls for regular exchange of hydrological and other data between co-basin states, directing that only when any requested data or information is "not readily available" may a state charge its neighbor "reasonable" collection and processing fees.[110]

Another striking fact is that China is the biggest dam builder at home and abroad and boasts a greater number of large dams on its territory than the rest of the world combined. Before the Communists came to power in 1949, China had only twenty-two dams of any significant size. Now the country— as part of a program to impound river waters to generate electricity, irrigate crops, support mining and manufacturing, and slake the thirst of its rapidly growing cities—has slightly more than half of the world's roughly 50,000 *large* dams, defined as having a height of at least fifteen meters or a storage capacity of more than 3 million cubic meters.[111]

Thus, China has completed, on average, at least one large dam per day since 1949.[112] The United States, the world's second-most-dammed country with about 5,500 large dams, has been left far behind, although for much of the twentieth century it pursued building as many dams as possible. Indeed, it was the American megaprojects—the Great Depression–era Hoover Dam, Tennessee Valley Project, and Grand Coulee Dam, and the post–World War II Ford Randall and Garrison dams—that inspired Chinese strongman Mao Zedong to build his country's first big dams, including the Sanmenxia, Liujiaxia, and Danjiangkou dams.

If dams of all sizes but with a reservoir are counted, the number in China totals close to 86,000 today. The figure surpasses 90,000 with the inclusion of run-of-river dams generating power without the aid of an artificial reservoir. The majority of the dams have actually been built in the post-Mao period, during much of which engineers have dominated the top echelons of Chinese power. After being appointed premier in 1987, Li Peng—a hydroengineer—

revived the Three Gorges Dam project and single-mindedly promoted other water megaprojects. China's once-a-decade power transitions have brought to office as president a succession of engineers—Jiang Zemin, Hu Jintao, and Xi Jinping.[113]

Engineers are trained to provide supply-side solutions, which has exactly been China's policy focus. Even America's first mammoth dam—the Hoover Dam, which still attracts almost a million visitors every year and which set the stage for other water megaprojects—was fathered by an engineer president, Herbert Hoover, who early in his career worked in China for a private corporation as "China's leading engineer," according to a White House biography.[114] The Hoover Dam, despite its downstream environmental impacts, has thus always carried a special resonance in China.

Data compiled by the United Nations Food and Agriculture Organization shows that China's dam reservoirs by 2005 numbered 85,108, with a capacity to store 562 billion cubic meters of water, or slightly more than 20 percent of the country's total renewable water resources in a year.[115] Since then, China has completed scores of new dams, including the world's largest—the Three Gorges Dam, a project on the Yangtze River that officially uprooted 1.7 million Chinese, besides causing still-persistent environmental problems. In the fall of 2011, China unveiled a mammoth $635-billion fresh investment in water infrastructure over the next decade, more than a third of which is to be channeled for building dams, reservoirs, and other supply structures. China's plans, in fact, call for constructing more new large dams by 2025 than the United States managed to build in its entire history.

In addition to being the world's most dammed country, China is the global leader in exporting dams. Its state-run companies have built or are building more dams overseas than all other international dam builders put together. In 2013, thirty-seven Chinese financial and corporate entities were involved in more than one hundred dam projects in the developing world, extending from South America to Southeast Asia. Some of these entities are very large and have multiple subsidiaries. For instance, Sinohydro Corporation—which comes under the State-Owned Assets Supervision and Administration Commission of China's State Council—is made up of ten holding companies and eighteen wholly owned subsidiaries, and boasts fifty-nine overseas branches. The Export-Import Bank of China is a major dam financier.

Both a profit motive and a diplomatic strategy to advance larger economic interests by showcasing engineering prowess drive China's overseas dam building. Its declared policy of "noninterference in domestic affairs" serves as a virtual license to pursue dam projects that flood ethnic-minority lands and forcibly uproot people in other countries, just as it is doing at home by shifting its dam-building focus from dam-saturated internal rivers in the Han heartland to international rivers that originate in the Tibetan Plateau, Xinjiang, Inner Mongolia, and Manchuria as part of an ever more desperate quest to exploit water resources in remote, ecologically sensitive regions.

China contends that its role as the global leader in exporting dams has created a "win-win" situation for host countries and its own companies. But evidence from a number of project sites shows that the dams are exacting a serious toll on the environment in those countries. Grassroots protests have flared at multiple project sites in Asia, Africa, and Latin America. By often using a largely Chinese workforce to build dams and other projects abroad (a practice that runs counter to its own 2006 regulations calling for "localization"), Beijing has spawned a perception that it is engaged in exploitative practices, especially because dams create the necessary energy and water infrastructure for extraction of what China usually covets—mineral ores and fossil fuels.[116] Its overseas dam building indeed often accompanies other infrastructure and resource-extraction projects. Chinese convicts have also been used as laborers on projects in countries too poor and weak to protest.

At home, China has graduated from building large dams to building mega-dams, exemplified by its latest additions on the international Mekong River: the 4,200-megawatt Xiaowan, which dwarfs Paris's Eiffel Tower in height, and the 5,850-megawatt Nuozhadu, a greater water appropriator whose 190-square-kilometer reservoir has been designed to hold nearly 22 billion cubic meters of water. New dams planned for construction, according to the state-run hydropower industry, include a 38-gigawatt project at Metog (or "Motuo" in Chinese) on the world's highest-altitude major river, the Brahmaputra (or Yarlung Tsangpo, "the Purifier," in Tibetan).[117] The power-generating capacity of the proposed dam at Metog—close to the disputed, heavily militarized border with India—will be nearly double that of the Three Gorges Dam, the length of whose reservoir is longer than North America's Lake Superior.

In addition, China has identified another mega-dam site on the Brahmaputra at Daduqia, which, like Metog, is to harness the force of a nearly 3,000-meter drop in the river's height as it takes a sharp southerly turn from the Himalayan range into India, forming the world's longest and steepest canyon in the process. Twice as deep as the Grand Canyon in the United States, the Brahmaputra Canyon (labeled the "Yarlung Tsangpo Great Canyon" by China) holds Asia's greatest untapped water reserves, making it a powerful magnet for Chinese water planners.

The Brahmaputra was one of the world's last undammed rivers until China began constructing a series of midsized dams on sections upstream from the famous canyon. China's latest plans to build mega-dams in a seismically active region just before the Brahmaputra enters Arunachal Pradesh—the northeastern Indian state to which Beijing lays claims—poses important risks.

In fact, the massive 2008 earthquake that struck the Tibetan Plateau's eastern rim, killing more than 87,000 people, mostly in Sichuan Province, drew international attention to the phenomenon of reservoir-triggered seismicity, or RTS. Chinese experts identified the new, 156-meter-tall Zipingpu Dam—located next to a geological fault line and barely five kilometers from the quake's epicenter—as the likely cause, saying the dam triggered severe tectonic stresses by impounding several hundred million cubic meters of water in its reservoir and unleashing the combined pressures of gravitational load and pore water diffusion.[118]

The Wages of Dam Frenzy

The consequences of China's frenetic dam building on international rivers are ominous. First, China is now locked in water disputes with almost all its riparian neighbors, even a weak client state such as North Korea. Second, China's new focus on water megaprojects in the homelands of ethnic minorities has triggered tensions over displacement and submergence at a time when the Tibetan Plateau, Xinjiang, and Inner Mongolia have all been racked by protests against Chinese rule, under which a state-sponsored influx of Han Chinese settlers is altering the demographics of these vast regions. Third, the projects threaten to replicate in international rivers the degradation haunting China's internal rivers. The "export" of river degradation and fragmentation is bound to exacerbate China's water disputes with its neighbors.

Yet, as if to declare itself the world's unrivaled hydro-hegemon, China takes pride in being the largest dam builder at home and overseas. From Pakistan-held Kashmir to Burma's troubled Kachin and Shan states, China has widened its dam building to disputed or insurgency-torn areas abroad, despite local backlash. China's dam building in Burma contributed to ending a seventeen-year cease-fire between the Kachin Independence Army and government forces before the renewed fighting and other concerns compelled the Burmese government in late 2011 to halt work on the largest and most controversial Chinese project—the $3.6 billion Myitsone Dam, located at the headwaters of the Irrawaddy River, Burma's lifeline.

Designed by China to generate electricity for export to its own market even as much of Burma suffers from long daily power outages, this 3,600-megawatt dam had been hailed by the State-Owned Assets Supervision and Administration Commission in Beijing as a model overseas project serving Chinese interests.[119] But in Burma, it became a symbol of China's resource greed and a trigger for renewed ethnic conflict. The Myitsone Dam indeed was the biggest hydropower project in the entire Southeast Asian region.

The Burmese decision to abandon the project shocked Beijing, which had begun treating Burma as a reliable client state—one where it still has significant interests, reflected in its construction of six other dams and a multibillion-dollar pipeline to take energy supplies to southern China. The bold decision on Myitsone was followed by major political developments within Burma, setting in motion an easing of long-standing Western sanctions and ending the country's international isolation—best symbolized by Barack Obama's late 2012 visit to Burma, the first ever by a U.S. president.

China has the world's fifth-largest aggregate water resources, after Brazil, Russia, the United States, and Canada. Whereas many Asian countries receive a significant portion of their river and aquifer waters through cross-border inflows, China has one of the world's lowest dependency ratios on external inflows—less than 1 percent, lower even than Canada's.[120] China's population is not even 10 percent larger than India's, but its internally renewable water resources (2,813 billion cubic meters per year) are almost twice as large as India's. In aggregate water availability, including external inflows (which are sizeable in India's case), China boasts virtually 50 percent larger resources than India.

Despite this seemingly happy situation, China, in hydrological terms, is not one but two countries. The hydrological divide in the ethnic Han heartland actually runs along the historical north-south cultural and linguistic fault lines. In contrast to copious water resources in the humid south—which includes the Yangtze Basin and everything south of it—the arid north has just 19 percent of China's water. Because of this spatial imbalance in water availability and the overall size of its population, China supports 19 percent of the world's population on its territory with a 6.7 percent share of aggregate global water resources. By contrast, Canada, with 6.85 percent of the world's water resources, is widely considered a freshwater superpower because its population size is barely 2.6 percent that of China.[121]

China's parched north accounts for almost half its population and 64 percent of its cultivated land.[122] Just the Huai, Yellow, and Hai-Luan river basins have 35 percent of China's population but only 7 percent of its water.

Yet, as a result of the Mao-era policies that helped treble the country's total irrigated acreage between 1950 and 1980, the north has become the country's food bowl. The Gobi Desert's advance toward densely populated northern areas has been aided by government-sponsored intensive irrigation in arid plains, which has led to soil degradation and depletion of surface and subterranean water resources.

The 5,830-kilometer Yellow River—the cradle of the Han civilization and the main nurturer of the northern plains—was once known as China's sorrow because of its annual flooding. But with booming cities, industries, and farms siphoning off its waters, the Yellow—one of the world's most heavily dammed and muddiest rivers—is today the source of sorrow for exactly the opposite reason: it is running dry.

With the country's political power center located in the north, the government has sought to underpin an environmentally damaging paradox: the northern plains are to increasingly rely on huge water transfers from the south via the Great South–North Water Diversion Project while continuing to grow food for the fertile south. The South–North Water Diversion Project, changing the face of China's countryside through colossal infrastructure development, is the most ambitious interriver and interbasin transfer program ever conceived in the world. It has been designed to transfer water all the way to the megacities of Beijing and Tianjin.

China's overdamming of rivers and its interbasin water transfers have wreaked havoc on natural ecosystems. The social costs have been even higher, a fact reflected in Chinese premier Wen Jiabao's stunning admission in 2007 that, since 1949, China had relocated a total of 22.9 million Chinese to make way for water projects—a figure larger than Australia's population.[123] Since then, another 350,000 residents—mostly poor villagers—have been uprooted. So, by official count alone, 1,035 citizens on average have been forcibly evicted for water projects every day for more than six decades.

The wages of being the world champion in dam building are also reflected in the record number of dam collapses in China. In the world's worst dam-related disaster, at least 83,000 Chinese were killed and millions left homeless in August 1975 in serial dam collapses after a typhoon hit Henan province, with consequent health epidemics and famine raising the fatality toll to as high as 230,000.[124] A total of sixty-two dams caved in, including the large Banqiao and Shimantan dams.

Since then, scores of other dam failures have been reported in China, where the location of many reservoirs near large population centers means that dam breaches often leave a trail of death and destruction. State media has blamed dam failures on continuing shoddy construction and embezzlement of funds, saying that many dams remain at risk of being breached, with one official—Xu Yuanming, China's water reservoirs department director—admitting, "These reservoirs are a major risk and will ruin farmland, railways, buildings, and even cities when they collapse."[125]

China's frantic efforts to tap water resources also involve setting up projects in the remote mountains, forests, and valleys of its borderlands, even at the risk of damaging the last sanctuaries of biological and ethnic diversity and diminishing cross-border flows. Water nationalism at a time of growing water stress in China's north and northwest has only whetted the government's appetite for increased resource utilization of transnational rivers.

For downriver countries, a key concern is that China remains opaque on its dam projects and loathes even rudimentary forms of transparency. It usually begins work quietly, almost furtively, and then presents the project as unalterable and as holding flood-control benefits for the downstream basin. The fact is that the annual flooding cycle of many Asian rivers, despite the human displacement and suffering often wrought by it, performs key ecological functions, including refertilizing overworked soils by spreading nutrient-rich silt into the floodplains and opening giant fish nurseries. Heavy upstream damming can permanently disable the annual flooding cycle (which creates optimal, natural-pond conditions for cultivation of rice paddies) and damage the fluvial ecosystem. And by affecting sediment transport and downriver flows, it tends to impede silt from reaching the oceans, where marine life depends on the nutrients and minerals disgorged by rivers. When farmers can no longer rely on nature's annual gift of silt—as is the case in the lower Yangtze Basin after the building of multiple large dams, including the Three Gorges Dam—they are forced to use more chemical fertilizers every year, leading to degradation of land and water resources.

The countries likely to bear the brunt of China's diversion of transboundary waters are those located farthest downstream on rivers like the Mekong and the Brahmaputra—Vietnam, a rice bowl of Asia, and Bangladesh, whose very future is threatened by climate and environmental change. The Mekong is the main river of southern Vietnam, while the Red River, which originates on the southeastern edge of the Tibetan Plateau and derives its name from its reddish-brown, silt-laden upstream water, is the lifeblood for northern Vietnam. A day after China commissioned its largest dam on the Mekong at Nuozhadu in 2012, Vietnamese president Truong Tan Sang warned that "tensions over water resources are threatening economic growth in many countries and representing a source of conflict." Addressing business leaders on the sidelines of the Asia-Pacific Economic Cooperation summit in Vladivostok, Russia, Truong said—without naming China—"Dam construction and stream adjustments by some countries in upstream rivers constitute a growing concern for many countries and implicitly impinge on relations between relevant countries."[126]

China's water appropriations from the Illy River, meanwhile, threaten to turn Kazakhstan's Lake Balkhash into another Aral Sea, which has shrunk to almost a quarter of its original size.[127] China has also diverted waters from the Irtysh—which feeds Russia's Ob River—for its projects in Xinjiang, including a canal supplying the booming oil town of Karamai.

In addition, China has planned the "Great Western Route," the proposed third (and most ambitious) leg of the South–North Water Diversion Project (map 4.4). The $62-billion project's first two legs—designed to take a total of 29 billion cubic meters of water to the north every year by connecting the Yangtze with the Huai, the Hai, and the Yellow—involve internal rivers in the ethnic Han heartland. But the third leg is to divert up to 200 billion cubic

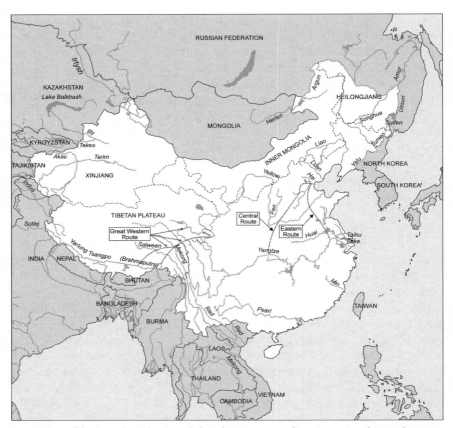

Map 4.4. China's Major Rivers and the Three Routes of Its Great South–North Water Diversion Project

meters of waters from international and internal rivers on the Tibetan Plateau to the Yellow, which also originates in Tibet.

The first leg—the Eastern Route—has already been completed. Work on the third leg to take river waters cascading from the Tibetan highlands to the Han heartland is slated to begin after completion of the second leg—the Central Route, whose longest section is to be ready by the revised deadline of 2014. Built alongside an ancient waterway for barges called the Grand Canal, the Eastern Route—although the shortest and least challenging of the three legs—symbolizes the problems of transporting water by canal across basins: 426 energy-hogging treatment plants along the route to tackle water degradation and pollution account for 44 percent of the $5-billion construction cost of this leg, which is also promoting deterioration in the water quality of the source river, the Yangtze.[128] Given the South–North Water Diversion Project's energy intensity, ballooning costs, and environmental impacts, a better alternative would have been desalination, wastewater treatment and recycling, and reduced irrigated farming in arid plains.

As the world's most dam-dotted country, with an unparalleled water-diversion infrastructure in place, China is already the largest global producer of hydropower, with an installed generating capacity of more than 245 giga-watts. Chinese plans call for increasing this capacity further to 450 gigawatts by 2030. Consequently, far from slowing its dam-building spree, China has stepped up its reengineering of river flows for irrigation, drinking water, flood control, navigation, and electric power.

In contrast, America's dam building, spearheaded by the U.S. Army Corps of Engineers since the early twentieth century, peaked in the 1960s, with the number of dams now totaling almost 84,000.[129] The rate of decommissioning of U.S. dams has overtaken the pace of building new ones, largely because of aging infrastructure, environmental impacts, and diminished availability of good damming sites. An environmental movement has also helped raise American consciousness about the dams' damaging legacy.[130]

In China, however, the fusion of autocratic politics and capitalist econom-ics under a largely technocratic civilian leadership has spawned a system that still seeks to bend nature to its will. The overexploitation of the Yangtze resources, for example, has made extinct the *baiji*, China's famed White River Dolphin that was revered as the river goddess. Yet China is ambi-tiously working to increase its present hydro-generating capacity by almost two-thirds within a decade by building new mega-dams at home and pumping electricity into its southern grid through dam projects in neighboring Laos, Cambodia, and Burma.

More fundamentally, China's dam frenzy has spawned two important de-velopments. First, Chinese companies now dominate the global hydropower-equipment export market, with power-generation equipment now ranking as the country's second-largest export earner after electrical machinery and equipment. Sinohydro, having eclipsed Western equipment suppliers like ABB, Alstom, General Electric, and Siemens, claims to alone control half the world market for hydropower projects. And second, the growing domestic clout of the state-run hydropower industry has led Beijing to increasingly eye resources of international rivers and to aggressively campaign for overseas dam projects by offering attractive, low-interest loans to other governments. It takes on foreign projects in places where multilateral lending institutions and Western firms hesitate to go because of environmental and other concerns.

China's portentous shift in its dam-building focus at home from internal to transnational rivers compounds the need for concerted outside pressure on Beijing to halt its appropriation of shared waters, accept some form of institutionalized cooperation, and embrace international environmental standards. The plain fact is that China's strategy to exploit its unmatched control over transnational water sources poses the biggest obstacle to building Asian cooperation to harness internationally shared rivers for mutual and sustainable benefit.

It also exacerbates regional water challenges and vulnerabilities. The threat that China may use its riparian preeminence as a diplomatic and political weapon was highlighted by its 2010 action of turning its dominant control over the global supply of rare earths into an instrument of economic warfare by slapping an unannounced export embargo on Japan. It could similarly employ its geographic chokehold on river systems to inflict economic damage on a downstream rival state, like Vietnam or India.

In fact, as in the case of its territorial and maritime disputes with neighbors, so also on international river flows, China is actively seeking to disturb the status quo. And just as it has preemptively encroached on disputed territories in the past to present a fait accompli—for example, in the case of the Paracel Islands (in 1974), Johnson Reef (1988), and Mischief Reef (1995) in the South China Sea and the Aksai Chin Plateau (1950s) of the original princely state of Jammu and Kashmir—so now China is seeking to materially alter cross-border river flows by keeping its dam projects secret until they can no longer be denied. Acceptance of the status quo—territorial and hydrological—is a prerequisite for peace and stability.

At a time when shared water is becoming an important geopolitical issue in water-stressed Asia, persuading China to halt its unilateral appropriation of transnational waters holds the key to rules-based riparian cooperation and peace on the continent. Otherwise, China would consolidate its power and control over international water resources to the detriment of legitimate users downstream, stoking bitter rancor and spurring destabilizing tensions. The extent of transboundary flows during the dry season in this situation could even depend on its political concession. In effect, China would emerge as the master of Asia's water taps.

The choice before China is to bolster its hydro-hegemonic approach by exploiting prevailing power asymmetries or to play a cooperative leadership role in Asia on the strength of its unique riparian position. Rather than make water an increasingly contentious issue in its relations with neighbors and thereby foster low-intensity water conflicts or overt tensions, China has the option to set an example by investing in building institutionalized water sharing and other cooperation—arrangements that would fill the diplomatic vacuum currently characterizing its water relations with neighboring states.

Chapter Five

Shaping Water for Peace and Profit

In the coming years, as communities adapt to water scarcity with new mechanisms, fleets of water tankers delivering water to households, hotels, and factories—a routine sight already in some Middle Eastern and North African states—are expected to become common in several other parts of the world. Environmentally questionable canal projects to transfer water across rivers and basins—as China has carried out—will likely be initiated in countries seared by water scarcity, overlooking ecological concerns about transferring water from one ecosystem to another. Ideas controversial today, such as interlinking domestic rivers, will gain wider acceptability, despite their environmental costs and potential transboundary implications. Massive new dams and other storage facilities are expected to be erected to stockpile water for the dry season, even as such projects designed to offer structural solutions are set to make intercountry and intracountry water issues more contentious.

Such measures notwithstanding, the market for water promises to be the next big investment opportunity of the twenty-first century, as countries seek to reconcile plans for increased economic output with decreasing water security and as the widening gap between freshwater demand and supply translates into higher prices. Water remains not only an undervalued commodity, but the policy choices for water-distressed countries are also limited. Unlike the fossil fuels, mineral ores, and timber that resource-hungry economies import even from distant lands, they must make do with their own water resources, a significant share of which is in transnational watercourses.

One option for water-constrained economies is to sustainably optimize the availability of water resources within their national frontiers, including through environmentally sound policies and better management to help achieve increased water efficiency and productivity. If a state has a significant dependency on cross-border inflows, it could use diplomatic tools and other

243

means to dissuade the upper riparian from materially altering transboundary flows. Another option—predicated on sufficient and maintainable hard-currency reserves—is to import rather than produce water-intensive products in which substantial virtual water is embedded. But as discussed earlier in the book, multiple economic, political, and social factors limit the attraction of—and gains from—virtual-water trade for a water-stressed nation.

Freshwater, as a vital resource for individual and national survival, is a magnet for competition. The way to manage or mitigate such competition is to build stable, institutionalized cooperation among contending parties for mutual benefit. Yet long-standing water tensions and recurrent disputes, coupled with soaring water demand, have already made a number of shared basins potential flash points for water conflict between nations or provinces.[1]

In a world in which transnational watercourses are virtually ubiquitous (148 nations have territory within an international basin), the troubling lack of mechanisms in most international basins to facilitate transboundary water sharing or cooperation poses significant geopolitical risks. There are, of course, a number of important basin-level agreements already in place. They cover two lake systems—the Great Lakes on the U.S.-Canada border and Lake Tanganyika in Africa—and quite a few river basins, including the Al-Asi/Orontes, Amazon, Araks, Atrek, Colorado, Danube, El-Kaber, Euphrates, Gandak, Ganges, Indus, Jordan, Komati, Kosi, La Plata, Mahakali, Mekong, Mesta/Nestos, Niger, Nile, Quaraí/Cuareim, Rhine, Senegal, and Senqu/Orange.

But not all of these accords incorporate water sharing. For example, the 1969 La Plata Basin Treaty involving Argentina, Bolivia, Brazil, Paraguay, and Uruguay and the 1978 eight-nation Amazon Cooperation Treaty in South America were designed to promote joint development of the respective basins, while the Danube and Rhine agreements in Europe relate to navigation and water quality. Several other important river-basin accords, stretching from the Senegal to Mekong basins, also specify no national water-sharing allocations. The 2003 Convention on Sustainable Management of Lake Tanganyika—signed by co-riparians Burundi, Congo, Tanzania, and Zambia—is part of a biodiversity-protection project backed by international agencies.[2]

Not even twenty of the existing agreements involve genuine water sharing. Some that include water sharing (such as on the Jordan, Euphrates, Araks, and Al-Asi/Orontes) are limited to two river-basin states, rather than covering all the important co-riparians. Groundwater-sharing pacts are rare, and only a few intercountry water treaties have any groundwater-related provisions.

According to United Nations data, more than three-fifths of the 276 international river and lake basins—and nearly all the transboundary aquifer systems outside Europe and North America—lack any type of cooperative framework, including several basins that are widely seen as potential flash points for conflict. In the majority of the cases where an existing accord permits some form of basin cooperation, one or more key riparian state has been left out or has refused to join, undercutting the value of the initiative. Of the 105 international surface-water basins identified by the United Nations that have an agreement of cooperation of some kind in place, only a few arrangements stand out for their comprehensive membership and mandate.

In fact, as water stress has escalated in many parts of the world in recent decades, newer water-sharing arrangements have become more tightfisted than the older ones. Conspicuous upper-riparian generosity toward a subjacent state is less likely to occur in an era of growing water shortages and sharpening intercountry resource competition.

The spreading water stress also makes it more difficult to replicate models of comprehensive cooperation extending not just to a specific basin but to the full range of transboundary water resources, as exemplified by the U.S.-Canada and U.S.-Mexico treaties. These treaties were originally designed to govern surface waters, but, as illustrated by the Abbotsford-Sumas Aquifer (U.S.-Canada) and the Santa Cruz Basin (U.S.-Mexico), existing institutions have been adapted for groundwater cooperation on an ad hoc basis. The All-American Canal, on the other hand, shows that bilateral disputes persist.

Whereas water security is a great economic-development asset and stabilizer, water insecurity is a destabilizing factor carrying potentially corrosive effects on peace and social cohesion at the regional, national, and substate levels. Sowing the seeds of potential conflict at the interstate level is the rush to dam transnational rivers, overexploit aquifers that straddle international borders, and create an upstream hydroengineering infrastructure to support the use of water as an asymmetric political tool. Transnational water resources remain the most important area not yet regulated by a binding international convention in force, although transboundary disputes are "on the increase, and now cover a wide array of issues that go beyond quantity and quality of the shared waters."[3] That this trend is most visible in the world's water-stressed regions only underscores the risks of overt conflict.

In this light, the purported absence of water-centered wars in recorded history is no assuasive sign of a peaceful water future. The post–Cold War upsurge of international terrorism, including the use of novel targets and methods, is a reminder that history usually is no reliable guide to the future. Indeed, counterterrorism measures today seek to preempt attacks even of the type that have never occurred thus far. To focus on the past and to overlook the underlying trends and threats of today, therefore, is to invite nasty surprises.

A majority of nations surveyed by the United Nations actually believe that water competition and related risks have increased since the early 1990s.[4] Unless steps are taken to mitigate and manage the increasing water competition and tensions, the threat of water disputes escalating to open conflict will impose larger costs by impeding regional cooperation and integration.

Spurred by worsening demand-supply deficits, the battle for control of freshwater resources has already begun in earnest, with regional heavy-weights flexing their muscles to gain a larger share through upstream hy-droengineering projects or, if they are located downriver, to hold on to their lion's share of the common waters. Such moves come atop mercantilist efforts to lock up long-term supplies of other strategic resources—energy, metals, and mineral ores. Indeed, gaining access to—and being in command of—supplies of natural resources continues to drive great-power policies, which regard control of resources as a key strategic advantage. The powerful are likely to try to underpin their might by controlling the most important of all strategic resources: water.

BUILDING INTERNATIONAL WATER NORMS AND RULES

The prevention and management of conflict on transnational waters demands at least three basic elements. The first is a clear set of international or basin-level norms and rules. Another necessity is the establishment of cooperative institutional mechanisms among states sharing a watercourse. And a third requirement is prudent, environmentally sustainable water management, cen-tered on prevention of degradation of shared resources, water conservation and recycling, and productivity gains through greater efficiency of water use by farms, industries, and households.

The focus of international rule-making efforts in history was on naviga-tional uses of rivers and lakes because water shortages were relatively un-known in most parts of the world. Europe's rise during the industrial revolu-tion helped promote the growth of European river trade, prompting separate intra-European pacts in the nineteenth century on freedom of navigation on the continent's main fluvial arteries, the Rhine and the Danube. Also in the nineteenth century, colonial powers reached accord to open up the naviga-tional use of the Congo and Niger rivers in Africa among themselves.

International efforts to frame rules on the nonnavigational uses of trans-national watercourses actually did not begin until the 1950s. Indeed, they picked up traction only in the 1990s, even as the global population doubled and water withdrawals tripled just between the 1950s and 1990s. Yet today, no binding global law is in force to govern the use and management of inter-national waters.

Very few international boundaries follow natural physical features, espe-cially in the developing world. So it is common for water resources to sit astride political frontiers or for the main catchment region of a watercourse to extend beyond the borders of a single nation. To help determine a reasonable and equitable share of any transnational waters and avert conflict, norms and rules are more important than the measurement of hydraulic parameters, such as the correct quantification of transboundary flow. The absence of a binding agreement between co-basin states, coupled with the weakness of international law, can promote a dam race or a pumping race that depletes or degrades shared water resources.

With the resources of the vast majority of transnational watercourses coming under growing pressure, it is becoming increasingly important to develop international legal standards, monitoring provisions, compliance mechanisms, and water-sharing rules to help moderate competition and conflicting uses, address hydrologic variability and changing basin dynamics, and shield environmental sustainability. Much of the global freshwater is shared water: transnational river and lake basins extend to nearly half of the world's land surface, while underground freshwater basins straddle many international borders. This fact alone underscores the necessity for rules-based, institutional arrangements among nations to govern shared waters.

Yet just 18 of the 276 transnational river and lake basins are currently covered by a genuine, mutually binding water-sharing agreement (see appendix B). There is also one interim groundwater accord between Israel and the Palestinian Authority. A number of these water-sharing accords do not provide for institutionalized collaboration or dispute settlement. Worse still, not one basin involving the world's highly disputed transnational resources is covered by an institutional arrangement encompassing all the nations with a stake.[5]

In fact, there is no reliable figure on the total number of aquifers in the world whose natural subsurface path of water flow is intersected by an international boundary. Unlike surface water, groundwater-resource boundaries are often poorly known, and "many transboundary aquifers remain only partly recognized."[6] However, UNESCO's International Hydrological Program has inventoried at least 274 aquifers that span across international borders, with the number set to grow in response to new initiatives. Of these, 155 have been identified just in Europe, where groundwater accounts for at least two-thirds of all the water used.[7] However, there are also many transboundary aquifers in Africa, the Middle East, and South America.[8]

Hidden from the surface but straddling political frontiers, groundwater sources are the lifeblood for about 2 billion people. In several countries, such as Libya, Malta, Saudi Arabia, and Kuwait, aquifers are the only natural source of freshwater.

Yet international efforts to establish a cooperative regime among groundwater-basin states have centered on just a few transnational aquifers, such as the Nubian Sandstone Aquifer in the Sahara Desert and the Guaraní Aquifer, the world's third largest, shared by Argentina, Brazil, Paraguay, and Uruguay. An agreement reached in the 1990s among Chad, Libya, and Egypt to set up a joint management authority for the Nubian Aquifer has done little to halt unsustainable practices. Some of the world's largest transnational aquifers are actually located in South America and North Africa.

If water locked up in polar ice caps and glaciers is excluded from calculation, groundwater represents the bulk of the world's freshwater resources. Some aquifers receive much of their recharge on one side of an international border, but their discharge through extraction lies largely on the other side. Sound management of shared groundwater resources demands the delineation of interstate borders of all hydraulically interconnected aquifers, as well as the establishment of rules on sustainable water withdrawals from them.

Strengthening International Rules

To promote the reasonable use and protect the integrity of shared waters, it is essential to establish the rule of law throughout the global riparian community. Weak international law on transnational water resources remains a major handicap, providing encouragement for countries to voraciously appropriate shared waters and compound the challenge of setting up a rules-based international water order. In the absence of clear international rules and enforcement mechanisms, powerful riparians will continue to impose unilateral solutions. To regulate water competition and forestall violent conflict, the building of international rules must take priority.

Despite the wide extent of shared freshwater resources, there is no universal treaty in force to govern the resources of transnational watercourses. The 1997 United Nations Convention on the Law of the Non-Navigational Uses of International Watercourses (hereafter the "UN Convention") was designed to serve as the global water law. It was passed by the United Nations General Assembly in the teeth of stiff opposition by the two countries that have fashioned hydro-hegemony in their respective regions: China and Turkey. Yet years later, the UN Convention has still to take effect because it has not been ratified by the required minimum number of countries.

There is no internationally available mechanism to prevent a country from materially altering the cross-border flow of a transnational watercourse. Moreover, despite the existence of a customary principle that seeks to prohibit causing palpable harm to a co-riparian through deprivation of water rights, pollution, or other means, international law on transnational water pollution remains feeble and poorly codified.

International law rests primarily on treaties and customary (or unwritten) law. Decisions of international courts and arbitral tribunals serve as subsidiary sources for determining international law. International legal customs are norms that have evolved from the practice of states in the absence of formal agreements. No practice, however, can be considered a principle of customary international law unless it is widely followed and deemed by states to be obligatory. Within the jurisprudence of customary law are major decisions of the International Court of Justice and its forerunner, the Permanent Court of International Justice. Today, because no written international water law is in force, the principles of customary water law remain dominant.

However, the development of comprehensive, widely accepted international rules on transboundary watercourses, backed by enforcement mechanisms, continues to prove elusive. Even customary international water law remains institutionally underdeveloped to be able to manage transboundary water disputes or conflicts. Yet the principles and guidelines incorporated in the UN Convention are increasingly being invoked in international forums because they reflect customary international law, although the convention's entry into force is still not within sight.

Truth be told, several of the key principles of customary international water law overlap each other or are in conflict with one another. Nations over the years have conveniently chosen to rely on the specific customary principle that would support their claims to or actions over shared waters. The prevailing power equations and other factors in a basin have also influenced which principle a nation has sought to embrace. Here is a summary of the basic customary principles, including the unresolved issues surrounding them:

The principle of absolute territorial sovereignty (the upstream state rules) versus the principle of absolute territorial integrity (the downriver state's interests are shielded) in a river basin. Under the first principle, an upper riparian has the right to assert absolute territorial sovereignty over transboundary waters on its side of an international boundary—the right to divert as much water as it wishes for its developmental needs, irrespective of the effects on a co-riparian. The maximalist principle of absolute territorial sovereignty, which extends to all natural resources within a country's borders, was embodied in the now-discredited "Harmon Doctrine" in the United States and appears to guide Chinese and Turkish riparian conduct today.

Turkey's assertion of absolute territorial sovereignty over the upstream waters of the Tigris and the Euphrates, regardless of the impact on the downriver states' interests, was publicly summed up on July 25, 1992, at the opening of the Atatürk Dam, by Turkish leader Süleyman Demirel, who served as prime minister seven times before becoming president: "Neither Syria nor Iraq can lay claim to Turkey's rivers any more than Ankara could claim their oil. This is a matter of sovereignty. We have a right to do anything we like. The water resources are Turkey's, the oil resources are theirs. We don't say we share their oil resources, and they can't say they share our water resources."[9]

China, for its part, justified its opposition to the adoption of the UN Convention on the grounds that the "text did not reflect the principle of territorial sovereignty of a watercourse state. Such a state had indisputable sovereignty over a watercourse which flowed through its territory. There was also an imbalance [in the convention] between the rights and obligations of the upstream and downstream states."[10] China's opposition, as one analyst put it, represents "a significant modern reassertion of the absolute territorial sovereignty approach."[11]

In stark contrast, the principle of absolute territorial integrity (tacitly invoked by Egypt to protect its utilization of the lion's share of the Nile waters) is founded on the doctrine of restricted sovereignty that seeks to compel the upstream state to forgo uses of a river that would harm the downstream state. It thus severely limits the rights of the upper riparian so that the river flows down undiminished to another state.

The 1960 Indus Waters Treaty—which, in a very methodical way, circumscribes India's sovereignty over the basin's three uppermost rivers in order to reserve them for Pakistan's use—is the only existing intercountry water-sharing agreement embodying the doctrine of restricted sovereignty. In contrast, the 1909 U.S.-Canada Boundary Waters Treaty—at U.S. insistence—was founded on the principle that both parties enjoy absolute sovereign rights over shared waters within their territories, although the pact reflects important American concessions on other issues, including legal remedies for material injury and joint commission powers.

The plain fact is that the principles of absolute territorial sovereignty and absolute territorial integrity are inflexible and antithetical to each other and to the present-day imperative for collaborative interriparian relationships. No new water treaty is likely to materialize without the parties seeking some middle ground between these diametrically opposed principles.

A balance between the doctrine of "equitable and reasonable utilization" and the principle of causing no significant harm. This balance of rights and obligations, to be achieved on the basis of fairness and give-and-take, has been laid down to help promote harmonious relations between co-basin states. Although an upper riparian has the first right to exploit basin resources, all riparians have the same right to equitable and reasonable utilization. The right of any state, however, comes with the obligation not to cause significant harm to another party over a shared resource.

However, customary international law lacks clarity on how to strike a balance between the right of "equitable" and "reasonable" utilization (which itself is not clearly spelled out) and the corresponding obligation not to cause palpable harm. The key unresolved issue is what is primary—the right to utilize shared water resources, or the duty not to cause appreciable harm.

The doctrine of prior appropriation. A priority right falls on the first user of river waters. This favors the upper-riparian state or any first appropriator of water (called the "senior appropriator"). A priority right thus can fall on the downstream state if it is the first user of river waters. The right of the "senior appropriator" to meet its water requirements takes precedence over the entitlement of a "junior appropriator" to draw resources from a waterway.

The central element in prior appropriation is the diversion of water from a watercourse for "beneficial" applications, including irrigation, industrial or mining purposes, electric power generation, and municipal supply. The definition of beneficial application of water has expanded since the 1990s to include environmental protection uses.

The Colorado Doctrine. Named after the U.S. Supreme Court case *Wyoming v. Colorado*, this principle recognizes the evolution of the doctrine of prior appropriation in the American West as a means of determining the right to use scarce waters from rivers, streams, springs, lakes, and ponds.[12] Legal issues related to water rights in the American West date back to the precious-metal mining boom of the nineteenth century and still remain contentious.

The Colorado Doctrine, like the broader doctrine of prior appropriation, rests on the customary dictum "First in time, first in right." The first user or appropriator of water acquires the priority to its future use, as against later users or appropriators. Also, in keeping with the maxim "Use it or lose it," water rights can be lost due to nonuse.

The law of riparian rights. This doctrine confers proprietary water rights to the owner of land contiguous to a river. After all, the term "riparian" comes from the Latin *ripa*, meaning "the bank of a stream," with the riparian landowner in customary law being the one whose continuous ownership title extends beyond a riverbank.

The law of riparian rights permits an upper riparian to secure "reasonable" water share, yet it does not define that term. Whereas the doctrine of prior appropriation allows an appropriator to remove the water from its source and put it to beneficial use at another location, including through interbasin transfer, the law of riparian rights tends to impose a geographical limitation.

The law of reason. This nascent, still-evolving principle seeks to link modern water norms and rules with ecosystem values, rather than merely with human demand and competition.

Water rights equal property rights. Within some nations or subnational regions, water rights have evolved in such a way that they have become analogous to property rights. Water rights thus can be exercised, mortgaged, or transferred independently of the land on which the water originates or on which it is intended to be used. National or state-level legislation in some countries, by decoupling water rights from land rights, has allowed water to be traded on an open market.

The UN Convention, Helsinki Rules, and Berlin Rules

The 1997 UN Convention "codifies," or writes down, several principles of customary water law. Although the United Nations has 193 member states today, the convention's entry-into-force was pegged to just thirty-five ratifications.[13] Yet getting even that number of countries to ratify the convention has proved difficult thus far.

The convention was the product of a long process that traced its origin to the adoption in the 1960s of a document designed to help draft the principles of international water law. That document—the declaration of the so-called Helsinki Rules—collated the different principles of customary international water law.[14] The 1966 Helsinki Rules, adopted by the nongovernmental International Law Association (ILA), were subsequently complemented in 1986 by the ILA-approved Groundwater Rules, creating norms for any transboundary basin with surface or underground waters.[15]

Although the Helsinki Rules were not accepted by the United States due to sovereignty-related and other concerns, their adoption greatly swayed the international discourse and created a slow but steady movement toward an international water convention. After all, those rules represented the first international effort to establish common standards for nations on shared waters. The Helsinki Rules, of course, lacked a binding effect, yet they were treated as authoritatively mirroring the principles of customary international law. They thus helped to shape the drafting of the provisions of the UN Convention, whose final text drew heavily on them.

It was the Helsinki Rules that outlined the principle of "reasonable and equitable share" in the waters of an international basin and the obligation not to cause "substantial injury" to a co-basin state through either overutilization of resources or water pollution.[16] The Helsinki Rules sought to overcome the problem of clearly defining "reasonable and equitable" apportionment of shared waters by listing eleven factors to be "considered together and a conclusion reached on the basis of the whole"—a formulation that shaped the UN Convention's articles 5 and 6. In addition, the Helsinki Rules attempted to establish a general framework for dispute settlement by listing the measures the feuding states could invoke, including creating a joint commission of inquiry or submitting the dispute to an arbitral tribunal or the International Court of Justice.[17]

The Helsinki Rules were developed further in the 1970s and 1980s by the International Law Commission (ILC), which was created by the United Nations to focus on specific international legal issues and is distinct from the London-based ILA, whose membership ranges from lawyers and academics to representatives of commercial organizations and chambers of commerce. The ILC—made up of thirty-four international lawyers serving in their individual capacity and representing the world's major legal systems—designed the 1989 Bellagio draft treaty to help set international norms on shared underground waters.[18] It was the ILC that was tasked by the United Nations to draft the UN Convention.

The ILA, for its part, helped draft the 2004 "Berlin Rules" broadening, amplifying, and superseding the Helsinki Rules.[19] The detailed Berlin Rules, consisting of seventy-three articles divided into fourteen chapters, are contentious because—even before an international water law has entered into force—these rules seek to apply modern customary norms to both national and international waters. For example, they require states to adopt integrated and sustainable management of wholly domestic waters.[20] And to overcome the UN Convention's failure to address environmental concerns, the Berlin Rules also extend the customary principles to the aquatic environment.

Because the sphere of domestic law traditionally has been viewed as beyond the reach of international rules, there was dissent within the ILA over the expansive scope of the Berlin Rules.[21] Although the Berlin Rules claim that most of the incorporated principles are "firmly based on generally recognized customary international law," some experts have openly questioned the wisdom of extending international norms to waters that are exclusively domestic, even before an international law on transnational waters has taken effect.

More important, the Berlin Rules emphasize each riparian state's primary *duty* (to "manage" the shared watercourse in an equitable and reasonable manner, with "due regard for the obligation not to cause significant harm to other basin states"), in sharp contrast to the UN Convention's enunciation of a *right* (to "utilize" shared waters in an equitable and reasonable manner, "taking into account the interests of the watercourse states concerned"). The Berlin Rules actually seek to codify not just the established principles of customary international law but also the emerging principles, thus opening to debate what rules should or should not govern interriparian relations. A more charitable view of the twenty-first-century Berlin Rules—anchored in the progressive evolution of international norms—is that they represent a forward-thinking approach to help forestall violent water conflicts in an era where water scarcity is beginning to plague many parts of the world.

The travails of the UN Convention should be seen against this background. It took twenty-three years for the ILC to complete its assigned task to draft the UN Convention.[22] The United Nations Sixth Committee (the Legal Committee) then deliberated on the draft law for almost three years, before the UN General Assembly approved the convention by an overwhelming vote at a specially convened session on May 21, 1997, with 103 nations in favor, 3 against, 27 abstaining, and 52 absent. Tellingly, China and Turkey, with their viselike grip on regional water sources, voted against the convention, roping in on their side a small water-stressed country, Burundi, which was then convulsed by civil war.

There are important gaps, admittedly, in the UN Convention.[23] It offers few clearly defined guidelines for water sharing, a highly contentious issue at the heart of most water conflicts. It seeks to establish, without spelling out in practical terms, two key principles that are to guide the conduct of nations that share watercourses: "equitable and reasonable utilization" (articles 5 and 6), and "the obligation not to cause significant harm" to another party over a shared waterway (Article 7).

Whereas equal co-riparian access to transboundary resources is implied, equal share is not, because the convention identifies the following seven issues relevant to determining equitable and reasonable utilization: (1) geographic, hydrographic, hydrological, climatic, ecological, and other factors of a natural character; (2) the social and economic needs of the concerned watercourse states; (3) the size of the populations dependent on the watercourse in each co-basin state; (4) the effects of utilization of shared waters on another watercourse state; (5) existing and plausible patterns of utilization; (6) the conservation, protection, development, and economy of use of watercourse resources, as well as the costs of measures undertaken for these purposes; and (7) any comparable alternatives to an existing or planned use.[24]

Yet, having set out the relevant factors for consideration "together," the convention leaves it to the watercourse-sharing parties to define the equitable-and-reasonable-use principle in practice. On balance, the convention does not prioritize or grade these seven factors in terms of relative importance; rather it states that "no use of an international watercourse enjoys inherent priority over other uses."[25]

The fact is that nothing a priori delineates the weighting among the seven factors, which themselves are open to different interpretations—a recipe for potential discord between parties. By recommending that "all relevant factors are to be considered together and a conclusion reached on the basis of the whole," the convention has of course sought to accommodate different needs and prospective uses.

The convention, in its Article 7, laconically defines the "obligation not to cause significant harm"—that the state utilizing shared waters shall take "all appropriate measures" to preclude "significant harm" to co-basin states. And that in case significant harm has nevertheless been caused, the culpable state shall take "all appropriate measures, having due regard for the provisions of articles 5 and 6, in consultation with the affected state, to eliminate or mitigate such harm and, where appropriate, to discuss the question of compensation."[26] Article 7 does not include a requirement that an upstream state halt any activity that harms a downstream state.

The loose formulation in the convention implies that causing harm may be tolerable if it is followed by mitigation, including possible compensation.[27] Such leeway is tantamount to granting a carte blanche to the upriver state, or to the country overexploiting a transnational aquifer.

In fact, this provision's reference to articles 5 and 6 (which deal with equitable and reasonable utilization) suggests that the no-significant-harm obligation is subordinate to the equitable-utilization rule. The implicit subordination happened after a number of countries forcefully argued in the Sixth Committee deliberations that an unencumbered obligation not to cause significant harm would effectively undercut the right of equitable utilization.

Still, the main reason for the convention's slow progress toward entry into force is the way it seeks to address a hot-button issue—the relationship between equitable and reasonable utilization and the obligation not to cause significant harm. In trying to match these two key but seemingly incompatible provisions, the negotiators settled for the middle ground. Yet the compromise has continued to evoke concern among many states because of the somewhat ambiguous language in the provisions and the lack of clarity as to which of the two principles holds primacy in international law.

The UN Convention, to its credit, has a detailed dispute-settlement provision, which provides that, if the concerned parties fail to mutually resolve a dispute, they "may jointly seek the good offices of, or request mediation or conciliation by, a third party, or make use, as appropriate, of any joint watercourse institutions that may have been established by them or agree to submit the dispute to arbitration or to the International Court of Justice."[28] Although the convention permits some flexibility on how a dispute is handled in its earliest phase, it mandates "impartial" fact-finding if it remains unresolved.

A fact-finding commission, set up at the request of any of the parties, is to investigate and report its findings and recommendations for "an equitable solution of the dispute, which the parties concerned *shall consider in good faith*" (emphasis added).[29] Even closely aligned states have expressed diametrically opposite views on the dispute-settlement provision. For example,

China said it cannot "support provisions on the mandatory settlement of disputes," while its ally Pakistan complained that the mechanism was "not binding" and sought "obligatory and binding settlement procedures."[30] By and large, downstream states want effective means to resolve disputes, while upper riparians are wary of a binding process involving any outside role.

The convention, even if imperfect in some respects, represents a vital step toward evolving a body of international law with strong rules to avert diversion or rapacious overexploitation of shared water resources and to forestall water wars. The convention indeed embodies some important safeguards. For example, given that upstream dams, reservoirs, and barrages can potentially be used as a weapon to trigger flash floods downstream or become a target of military strike, the convention's Article 29 stipulates, "International watercourses and related installations, facilities, and other works shall enjoy the protection accorded by the principles and rules of international law applicable in international and noninternational armed conflict and shall not be used in violation of those principles and rules."

This is actually a "framework" convention that will need a follow-up international protocol with regulatory and enforcement mechanisms, just as the 1997 Kyoto Protocol followed the 1992 United Nations Framework Convention on Climate Change. Once the convention has entered into force, efforts can begin to make it effective on the environmental and other fronts and to remove some of its infirmities through new measures as well as basin-level accords. The fact is that the convention is widely seen as the most current and respected statement of the law on shared waterways.

The convention's entry into force, of course, would bind only the acceding countries to its rules. Yet an enforceable convention with a protocol would greatly aid efforts to establish the rule of law on shared water resources by putting pressure on recalcitrant states.

Still, developing clear international rules on shared groundwater has proved a particularly complex and vexed task. Given that many aquifers extend across state boundaries, groundwater has emerged as a critical transnational resource. But the drafters of the convention shied away from bringing within its rules the deep, fossil-groundwater resources that are cut off from any significant recharge.[31] As a result, several of the convention's key provisions do not extend to shared fossil aquifers, including the right to reasonable and equitable utilization, the obligation not to cause significant harm, the requirement to exchange on a regular basis available data and information, and the duty to notify co-basin states before undertaking any measure that may carry transboundary effects.

The convention restrictively defines a watercourse as "a system of surface waters and groundwater constituting by virtue of their physical relationship a unitary whole and normally flowing into a common terminus."[32] The definition thus includes only underground water that is either tributary to, or sharing a common terminus with, any surface watercourse. In other words, the convention's applicability is limited to aquifers that contribute water to, or receive water from, a surface-water basin. Indeed, they must cross a high scientific bar: if the groundwater supply does not form a "unitary whole" with surface waters, then it would not constitute an international watercourse. The convention's exclusion of some major types of groundwater resources potentially crimps the development of a legally inclusive and integrated approach to transnational water-resource management.

To help comprehensively clarify and build international rules on all types of transboundary aquifers, the ILC has drawn up a separate draft law, which was welcomed by the UN General Assembly in late 2008.[33] Some countries, however, have opposed the draft articles from being fashioned into a global groundwater convention on the grounds that it would cover some transnational waters already within the scope of the 1997 UN Convention and thus create potential confusion in the form of two overlapping framework conventions.

The United States, for example, has contended that regional arrangements, as opposed to a global convention, offer the best way to ease pressures on transnational aquifers.[34] Several important countries also insist that the principle of inalienable sovereignty over natural wealth and resources—embodied in a 1962 UN General Assembly resolution[35]—must extend to any portion of a transboundary aquifer located within a state's territory. The future of the draft articles on the law of transboundary aquifers thus remains uncertain.

The Role of the International Court of Justice

Clarity on international rules for how co-basin states must divide and manage transboundary watercourse resources is imperative. But it is also important to enforce compliance with rules. If a rule is not enforceable, it may be repeatedly violated by the more powerful states. Even if the UN Convention took effect and a dispute arose between two of its state parties, the matter could be submitted for international arbitration, or adjudication by the International Court of Justice (ICJ), only with the consent of both, unless an applicable agreement between them provided for the appointment of a mediator or court of arbitration at the request of one side.

The fifteen-judge ICJ, also known as "the World Court," is the principal judicial organ of the United Nations and plays two roles: it decides disputes of a legal nature between nations that agree to submit to the court's jurisdiction, and it issues advisory opinions on important legal questions posed by the United Nations or its agencies.[36] Although its decisions are binding only on the parties and not on others, they nonetheless carry precedential value because they contribute to the evolution of customary law.

The ICJ, however, lacks any practical mechanism to enforce its rulings. If one state fails to abide by the court's ruling, the recourse for the other party is to take the matter to the hotbed of great-power politics—the United Nations Security Council, where any action demands unanimity among the permanent members, thereby often precluding enforcement against any such noncompliant nation, especially if it has close ties with one or more of the permanent members.

In truth, the ICJ has contributed little to clarifying international rules on shared waters. Since the ICJ was established in 1946, it has decided only one case involving the management of an international watercourse—a dispute between Hungary and Slovakia over the Gabčíkovo-Nagymaros barrage system, a bilateral project designed under the 1977 Budapest Treaty to prevent severe flooding, improve navigability of the Danube River, and generate hydropower through dam building. But the 1997 ICJ decision in that case did little to clarify international rules. In fact, Slovakia approached the court again in 1998 for a further decision because of Hungary's reluctance to implement the original ruling.

The case centered on Hungary's 1989 unilateral abandonment of its part of the joint works on grounds that the binational project entailed serious risks to the Budapest municipal water supply and the natural environment. Slovakia, while insisting that Hungary meet its treaty obligations, proceeded to build on its territory an alternative project with transboundary implications for Hungary. The ICJ, in its 1997 decision, held that both Hungary and Slovakia had breached their legal obligations and thus each was entitled to receive compensation from the other, although it declined to specify the value of damages.[37]

The court's decision implicitly endorsed the primacy of the right of equitable and reasonable utilization vis-à-vis the obligation not to cause significant harm. The ruling referred to the right of equitable and reasonable utilization but did not mention the no-significant-harm principle, although Hungary had placed heavy reliance on that norm in its pleadings.

The decision implied that the prevention of significant harm may matter only if a particular use is adjudged not equitable or if the harm is "grave and imminent." The court said a nation, under certain circumstances, could cite "environmental necessity" to ignore its treaty obligations but only if it convinced the court that it faced "grave and imminent" harm and that it had no choice other than to suspend actions required under the treaty. This set a high bar.

The point is that the ICJ decision in the case neither helped to end the dispute nor provided a clear definition of the precise standards that should govern the application of even the principle of equitable and reasonable utilization. The court, however, underscored the idea that a cross-border river represents a "community of interests"—an expression used by the Permanent Court of International Justice in 1929 in a dispute over navigation rights on the Oder River in Central Europe.

Any "community of interests" will benefit from stronger and clearer international rules on shared freshwaters. The World Water Council, which calls itself an "international multi-stakeholder platform" and is best known for organizing the World Water Forum every three years, proposed abortively in 2000 that a World Commission on Water, Peace, and Security be established to offer third-party mediation on transboundary water disputes. Third-party mediation can be effective only if the rules on shared resources are clear at the international or basin level. The evolution of the principles of international law is a continuous process, and the global water crisis is likely to increase pressure for clearer rules and enforcement mechanisms.

BUILDING BILATERAL OR BASINWIDE INSTITUTIONS

Because of an infirm, still-evolving international water regime, it has become important to develop rules-based water cooperation at the regional level through bilateral or basinwide water institutions. Arrangements at the bilateral or regional level, of course, can draw on the work already accomplished at the global level. Context-specific arrangements, for example, can be built by taking into account the provisions of the 1997 UN Convention and the draft articles on the law of transboundary aquifers.

New basin-level arrangements can also draw on existing bilateral or regional water treaties in other basins, so as to incorporate provisions that have proved valuable and resilient in practice, while steering clear of the weak elements. None of the water-sharing or institutionalized cooperation treaties currently in force at the basin or subregional level serves as a model by itself, yet some of their strong features are worth embracing or improving upon. For example, the U.S.-Canada joint commission, set up under the bilateral water treaty, has evolved into an effective dispute-settlement mechanism, thanks to its structure, although its lack of sufficient supranational powers has hampered a holistic, ecosystem-protective approach to water management.

Transboundary "No Harm" Principle in Practice

Institutionalized, basin-level cooperation in the developing world can benefit by emulating the clarity developed in Europe by the United Nations Economic Commission for Europe (UNECE)—one of the five regional commissions of the UN—on the principle of causing no significant transboundary harm to others.[38] The 1992 Convention on the Protection and Use of Transboundary Watercourses and International Lakes (known as the "Helsinki I Convention") is the core UNECE water-related agreement. Two other UNECE conventions—the Espoo Convention on Environmental Impact Assessment in a Transboundary Context, and the Helsinki II Convention on the Transboundary Effects of Industrial Accidents—also have a bearing on transboundary water management.

The Helsinki I Convention, with the goal of a holistic, sustainable development path, was designed to promote the protection of transboundary waters by preventing, controlling, and reducing pollution. In addition, it calls for the management and sharing of such waters in an equitable, eco-friendly manner, as well as the conservation and restoration of ecosystems.[39]

It was amended in 2003 to allow accession by countries outside the UNECE region, thereby inviting other nations to use the convention's legal framework and benefit from its institutional experience.[40] In fact, three of the principles elaborated by this convention have already gained wide international respect and acceptability.

The precautionary principle. Building on the "no harm" norm, the precautionary principle imposes the burden on a government or another entity to ensure that any planned works on or along an international watercourse (such as the construction of a dam or pulp mill) will not seriously damage the larger environment. In addition to casting the onus on the authority behind a project, the principle demands that action to prevent transnational environmental impact must not be postponed on any ground, even if there is no proven causal link between an activity and a transboundary effect. Several international declarations that fall in the category of "soft law" have endorsed the precautionary principle.

The "polluter pays" principle. The costs of pollution reduction and control measures must be borne by the polluter. The polluter's liability, in effect, will be absolute. The "polluter pays" principle also found mention in the Rio Declaration on Environment and Development, adopted at the United Nations–sponsored "Earth Summit" held just three months after the Helsinki I Convention came into being. Principle 16 of the Rio Declaration stated, "National authorities should endeavor to promote the internalization of environmental costs and the use of economic instruments, taking into account the approach that the polluter should, in principle, bear the cost of pollution, with due regard to the public interest and without distorting international trade and investment."[41]

The sustainability principle. Water resources must be managed so that the needs of the present generation are met without compromising the ability of future generations to satisfy their requirements—the same principle at the core of the sustainable-development concept.

With its long rainy seasons and temperate climate, Europe has had few intercountry water-sharing disputes. Even today, water sharing is not an important intercountry issue in Europe. Since the 1986 toxic chemical spill in Switzerland—one of Europe's worst post–World War II environmental disasters that turned the Rhine red—the European emphasis has been on developing and implementing water-quality criteria for the purpose of forestalling any cross-border pollution impact. State parties to the Helsinki I Convention, for example, are required to check watercourse contamination by subjecting municipal wastewater to at least biological treatment and by adopting the best available technology and practices to help reduce nutrient pollution from industrial and other sources, besides developing contingency planning for accidental contamination.

Water sharing, however, is at the heart of basin competition and discord in the developing world. The usefulness of the UNECE-developed principles elsewhere is largely limited to preventing or minimizing transboundary pollution and encouraging sustainable water-resource management, including the application of the ecosystem approach. These are important objectives, yet the intercountry race in the developing world to dam international rivers and pump the resources of transboundary aquifers means that broader norms and goals must be set.

Moderating the Problematic Doctrine of Prior Appropriation

Regional water institutions must be designed to rein in the "dam racing" being witnessed in several international basins. National projects on transnational rivers are often pursued unilaterally, with little regard for the potential cross-border effects. Consequently, dam building is becoming an increasing source of intercountry friction, underscoring the need for rules and transparency on building projects that can trigger a chain of environmental reactions affecting downstream interests.

Dam-building competition actually draws encouragement from the doctrine of prior appropriation, which legitimizes the principle "First in time, first in right." Given that the first user of waters under this doctrine acquires a priority right to the utilization of river resources, this creates a strong incentive to be an early mover to commandeer shared resources.

In Asia, for example, China has taken the lead to appropriate the resources of shared rivers so as to set itself as the prime user of the waters of international watercourses. It wants to present itself as the country whose water rights are protected under the doctrine of prior appropriation vis-à-vis lower riparians, including Kazakhstan, India, Nepal, Burma, Laos, Vietnam, and Russia. India's own moves in recent years to initiate dam projects in the state of Arunachal Pradesh, along its northeastern border with Tibet, have seemingly been driven by the intent to assert a priority right as the first user of river waters, although India is located downstream. Grassroots protests and red tape, however, have delayed the Indian projects.

The doctrine of prior appropriation has also influenced Pakistan's riparian moves. Pakistan, under the Indus treaty's dispute-settlement provisions, instituted international arbitration proceedings against India in 2010 over its run-of-river hydropower project on a small Indus tributary, the Kishenganga (also known as Neelum). A seven-member court of arbitration was set up, with the Hague-based Permanent Court of Arbitration—an intergovernmental arbitral institution—acting as its secretariat.

But much before the court of arbitration heard arguments and delivered a partial award in February 2013, Pakistan succeeded in persuading the court to order India in 2011 to suspend work on the project. India was enjoined not to construct, until the court rendered its award, any permanent works on or above the riverbed "that may inhibit the restoration of the full flow of that river to its natural channel."[42] In the meantime, Pakistan put the construction of its own three-times-larger hydropower plant on the same Himalayan stream on the fast track to gain a priority right on river-water use under the doctrine of prior appropriation. Indeed, it told the court that the older Indian project threatened the hydrological viability of its newer project. Its Chinese-aided, $2.16 billion project—formally known as the Neelum-Jhelum Hydropower Plant—is located downstream from the Indian project at a nearby border site.

The plain fact is that the doctrine of prior appropriation has proven problematic in the intercountry context by encouraging the trend currently being witnessed in a number of international basins—a dam-building competition between co-riparians. Because one of the key roles of dams is to manage resource variability, including controlling flooding and storing water for the dry season, climate shifts are likely to only promote greater dam building.[43]

Dam racing, like arms racing, has the potential to deepen mutual distrust and discord, sharpen territorial and resource feuds, and trigger greater regional instability by stoking dangerous tensions. The doctrine of prior appropriation must not be allowed to become a legitimating credo for commandeering shared water resources. That is why clear and enforceable rules are needed to prevent resource grabs. There must also be little international tolerance of moves to start projects by stealth and then present them as a fait accompli, or to pass them off as old schemes supposedly under development for years.

To stem the growing geopolitical risks arising from dam racing, it is essential to develop strong norms on dam building so that such projects are pursued transparently and without significantly altering cross-border flows. The only concrete way to moderate dam racing is by building treaty-based water institutions. Such institutions cannot completely eliminate dam-building competition, as the experience in the Mekong, Indus, and other basins illustrates.

Yet institutional arrangements, backed by clear rules, can help restrain or slow rival plans through mechanisms that provide for, among other things, constructive dialogue and dispute settlement. Any project proposal with potential cross-border implications should be the subject of joint consultations and independent scrutiny, with the process proceeding on the basis that each riparian would pay due regard to the rights and legitimate interests of the other.

Broader Parameters to Alleviate Conflict

Inclusive water institutions at the basin level—backed by a comprehensive set of rules, regulatory and oversight standards, and cooperation mechanisms—can regulate interstate water competition, help balance the rights and obligations of co-riparians, and seek to promote sustainable practices. In the absence of institutionalized cooperation, a downriver state will be in a precarious position vis-à-vis an upstream nation that seeks to siphon off cross-border flows in the name of meeting its thirst for water and energy. If the upstream country in such a case was also the more powerful state, the downriver nation would be able to do little more than plead with and seek to cajole the upper riparian to be more considerate. International law, after all, offers little remedy for any internationally wrongful act.[44]

Basin arrangements, however, can be formed only if the stronger riparians believe that structured cooperation will further their interests. If a strategically placed riparian is to agree to a rules-based basin community and eschew the arbitrary exercise of power in its own interest, it must have confidence that significant benefits would accrue. Basin arrangements, by fostering greater regional cooperation and integration, can yield broader economic and political benefits, thereby aiding the interests of all parties, including the more powerful riparians, and underpinning mutually beneficial relationships.

Despite the immense challenge involved in setting up institutional arrangements in any international basin, a successfully established collaboration is likely to create a mutual stake in strategic stability, resource sustainability, economic growth, and environmental protection. The following fifteen principles sum up the key elements that can make a basin regime robust and forward looking.

1. Manage shared water resources as part of a broader effort to build regional peace and stability and environmental security, with watershed management being a joint priority to achieve clearly defined objectives.

2. Incorporate mechanisms that promote constructive dialogue and joint collaboration, strengthen flood management, support climate-change adaptation planning, and permit monitoring and compliance verification. Mechanisms to provide compensation for the transfer of local water rights to help
augment environmental flows or for causing cross-border economic loss
through unintentional actions may also be useful.

3. Pursue optimal yet rational and sustainable utilization of shared water
resources on the basis of transboundary transparency and unhindered flow
of data on the hydrological, meteorological, environmental, and other
relevant aspects. The information sharing must occur in a compatible and
integrated manner and include the proficient collection, processing, and
dissemination of data.

4. Establish an instrument for prior consultations on any water project with
potentially significant transboundary implications, with the exchange of
information extending to project-related technical studies and environmental impact assessment. The state seeking to build the project must
allow the other party to independently conduct an expert appraisal and, if
necessary, be ready to modify the proposed works or accept international
arbitration in order to allay its co-riparian's concerns.

5. Agree to a reasonable sharing of basin waters, whether on a proportional
basis, or by specifying minimum cross-border flows, or by reserving full
or partial flows from specific subbasins for the downstream party. Because toothless sharing arrangements only allow disagreements to fester
and even engender new discord, a pact must provide for mutually agreed
mechanisms to sort out differences or disputes over resource division
and utilization.

6. Emphasize equitable water distribution as much as equitable water allocations. Distributing water-use benefits opens up "positive-sum agreements, whereas dividing the water itself only allows for winners and
losers."[45] An equitable distribution of water, to the extent possible, can
help bring about an equitable spreading of benefits through energy and
food production, fisheries, economic development, public health, and
ecological preservation. For example, the link between water and public
health is such that combating poverty, disease, and pollution demands
improved water distribution, including the provision of safe drinking
water and sanitation services.

7. Permit built-in flexibility in water-allocation formulas and institutional
arrangements so as to account for hydrological fluctuations and new
understandings of basin dynamics, environmental impacts, and changing
needs. Elasticity to deal with new issues will also facilitate adoption of
new technologies and practices.

8. Develop joint, basin-level contingency plans to deal with exigent circumstances with transboundary implications, including severe flooding, drought, or a chemical spill into a river. A state must be obligated to immediately notify co-riparians of any emergency situation arising out of natural or human-caused events, such as a dam burst, upstream flooding or landslides, earthquake-triggered damage to a major water project, or an industrial accident that threatens to contaminate shared waters. Joint disaster-related intervention can help to contain loss of life and economic and environmental damage.

9. Commit to avoiding actions that may be damaging to a downriver state's interests, such as discharging highly saline drainage from irrigation into a cross-border waterway or abruptly releasing water from a rain-swollen storage structure located close to an international boundary.

10. Prevent pollution by taking effective measures and adopting the best available technologies and practices to protect transboundary waters (both surface and underground), including by accepting the precautionary principle and the "polluter pays" principle.

11. Embrace binding dispute settlement as an integral element of basin arrangements. If a dispute cannot be resolved through consultations, good offices, mediation, and conciliation, the mandated next step must be independent fact-finding, arbitration, or adjudication. As interstate water disputes often arise from disagreements over questions of fact, an impartial opinion on disputed facts resulting from independent fact-finding by an expert panel may help to moderate, mute, or settle a dispute before any binding process becomes necessary. A basin-level treaty must also provide for the appointment of international arbitrators if a dispute remains unresolved. It could also expressly grant jurisdiction to the ICJ to resolve disputes.

12. Shield biodiversity and freshwater ecosystems through enforceable water-quality standards, conservation, and forest cover in upper catchment regions. Preserving the integrity of ecosystems and protecting freshwater-living species demands control over human activities—ranging from deforestation and elimination of grasslands to mining and manufacturing pollution—that are degrading watershed areas, leading to excess watercourse sedimentation, deterioration in water and habitat quality, high rates of loss of genetic variability, and extinction of species. Given that exotic fast-growing plant species usually consume more water than native forests and cause other environmental impacts, there must be a commitment not to introduce alien species that can upset ecosystem processes and degrade biological diversity.

13. Establish policy linkages between water and land use on both sides of a transboundary basin with the aim of achieving greater productivity in

the utilization of water and land resources and controlling soil erosion in catchment regions or above aquifers. Such an approach demands the participation in planning of all key agencies across the basin.

14. Work to harmonize each other's water-resource strategies and action programs relating to a transnational watercourse and its catchment regions. Such joint efforts must seek to shore up the most vulnerable subbasins and sustain ecosystem balance.

15. Integrate water with other transboundary issues holistically so that projects related to irrigation, hydropower, flood control, navigation, fishing, and reforestation can be bunched together in a sustainable development framework.

These recommendations are not meant to be exhaustive, yet they point to the elements essential for creating basin arrangements that can advance genuine collaboration and sustainable practices and forestall conflict in an era of growing water-supply constraints. No existing water treaty in any region comes close to this ideal, although some arrangements have successfully managed transboundary issues. With cross-border water issues becoming increasingly linked with regional security issues, cooperative and integrated management of international watersheds is assuming importance for peace and resource sustainability.[46]

To be effective, a basin regime must include all riparians with an important stake. The absence of one key riparian can easily stymie the formation of a meaningful basin community. For example, the nonparticipation of China in the Mekong regime has blighted prospects for a basinwide community and undercut the 1995 Mekong treaty's goal of sustainable practices.

The interstate sharing of hydrological data on transboundary basins is another important area where regional and international norms need to be strengthened. Sharing data with a lower riparian on upstream precipitation and river levels should not be a political favor but a normative obligation discharged on a regular, uninterrupted basis. Regular cross-border information flows on the hydrological, meteorological, hydrogeological, and other relevant conditions in the upper basin are essential to help instill a cooperative and collaborative spirit.

Only robust norms can thwart attempts to employ data sharing as a tool of leverage. In some basins, data sharing has even been turned into a lucrative commercial enterprise, with the extent of upstream data transfer pegged to the fees the lower riparian is willing to pay. This runs counter to the UN Convention's directive that only while supplying data or information that was "not readily available" may a state seek to recover "reasonable costs" from a co-riparian.[47] However, several upstream countries provide hydrological data free to co-basin states.

There must be transparency on the collection, processing, and timely dissemination of data so that it inspires confidence in its reliability. After all, an upriver state can fob off poor or even misleading data to a downstream nation.

In this regard, the international sharing of high-resolution satellite imagery can serve as a means to encourage transparency on the harnessing of shared waters. The use of remote sensing and even geographic information system (GIS) technology—capable of capturing, managing, analyzing, and displaying all forms of geographically referenced information—in the management of water resources indeed offers a new scientific tool that transcends political frontiers and controls.[48] Radar, infrared, and photographic imagery can be helpful in developing cooperative and environmentally sound water management among co-basin states.

The more information that becomes available, the greater will be the level of transparency. Collection of information and its dissemination among key stakeholders is vital to the process of preventing or resolving water conflicts. The role of the private sector, civil society, international institutions, and even Google Earth is important in ensuring information flow. Regional and international data hubs providing harmonized data on different aspects of water resources and their nexus with energy- and food-production issues will be very useful. Employing Internet technology, a virtual database (VDB) can be created for each transboundary subbasin and the issues of concern or contention there.

The "democratization" of water-related information, data, and pictures is becoming an essential tool in the development of more sustainable, peacebuilding approaches to shared water resources. Plugging information gaps can by itself serve as a conflict-prevention mechanism, helping to defuse claims and counterclaims. After all, factual disagreements often stem from conditions where the parties are relying on completely different sets of information or data. Conclusions actually depend on how different pieces of information are used and what weight is assigned to each piece, making objective processing and analysis of information critical.

In the absence of strong and enforceable rules, however, the present situation, as the United Nations acknowledges, offers states "clear incentives to capture and use water before it goes beyond their political control," especially because there is "no immediate incentive to conserve or protect supplies for users beyond" national borders.[49] In many countries, the public discourse on transboundary water resources tends to be very political because it is guided more by nationalistic than larger cross-basin considerations. As a result, data on water resources is closely guarded.

In this light, the limited incidence of water-related armed conflict in history offers little consolation. The sharpening of interstate and intrastate water competition augurs a more conflict-ridden trajectory for the future. That is why robust basin-level institutions with normative standards and rules have become a must for conflict prevention and management.

PATHWAYS TO HELP CONTAIN THE RISKS

There is no silver bullet or single pathway to solve all the world's water problems. But if the various available options are pursued in conjunction, they can together help to ease the global water crisis and alleviate the potential for conflict. In fact, the water crisis is opening new opportunities for political cooperation, financial investment, improved water management, and technological innovation, which, if seized, can help to mitigate the economic and security risks. With diminished water supplies already becoming a reality in several regions, the world, however, does not have the luxury of time.

A serious challenge confronting the world is how to assure fair and adequate access to water to meet the needs of growing populations; to support the rising demands of the energy, agricultural, manufacturing, and services sectors; and to ensure the unharmed provision of water-related ecosystem goods and services. If this challenge is to be addressed, governments must safeguard the integrity of the water cycle, or else the interests of future generations would stand compromised. In partnership with important stakeholders, governments need to develop equitable and efficient water-allocation policies and ensure that water withdrawals are within the limits of sustainability.[50]

The holistic management of water resources has become essential to prevent their depletion or degradation. Translating this imperative into policy, however, is proving difficult. Nowhere is this truer than in the developing world, where the rate of abstraction in many countries already exceeds nature's freshwater-renewable capacity. But even in some areas within the United States, tomorrow's water is being used to meet today's needs.

Although the world's urban population is growing at an exponential pace, the rate of increased freshwater utilization in economic production will continue to outstrip that by urban households. The fast-growing industrial and agricultural production, after all, is largely geared to meet the rising per capita consumption levels in the cities, especially in emerging and other developing economies.

Water has long been tomorrow's investment. But with adequate availability of potable water coming under increasing strain, water is becoming the investment option for today, despite the continuing global focus on other issues, such as energy, food, sovereign debt, and unemployment. Dealing with declining water security demands forward-looking approaches that break free from business as usual and seek to bring on board all key stakeholders. Because water has largely been supplied at a heavily subsidized price (and, in some areas, even free)—with the continued availability of this resource taken for granted—the crisis demands fundamental shifts in policies and practices.

The key to containing mounting water challenges includes integrated management of watersheds; aggressive conservation strategies; greater water efficiency; tapping nonconventional water sources (including rainwater capture, reuse of water, utilization of brackish water, desalination, and artificially recharged groundwater); and employing new clean-water technologies. The cost competiveness of clean-water technologies, of course, will be greatly aided by environmentally friendly production of electricity. Technological breakthroughs on the renewable-energy front, such as the availability of cheap solar power, are likely to brighten the water outlook by making desalination and wastewater recycling cost attractive and carbon light.

There is, however, one critical uncertainty: whether new cost-effective and environmentally friendly clean-water and energy technologies will be developed and commercialized in time to forestall a water-scarcity-driven slowdown in economic growth—a slowdown that could potentially keep the world's poorest states on the margins of globalization and stall the continued rise of some emerging economies (figure 5.1). As a U.S. National Intelligence

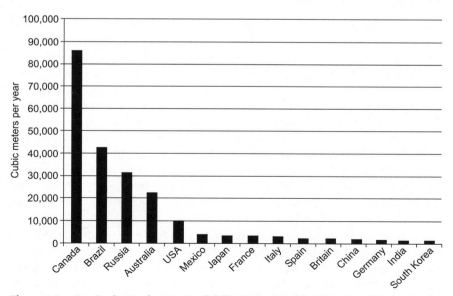

Figure 5.1. Per Capita Freshwater Availability in the World's Fifteen Leading Economies
Source: Based on Aquastat online data, 2013.

Council report has cautioned, "A world in which shortages predominate could trigger behaviors different from one in which scarcities are overcome through technology or other means."[51]

Historically, new technologies, especially in the energy sector, have had a long "adoption lag." But it may be possible for new clean-water and energy technologies to be developed and adopted within shorter time frames owing to the high economic and social costs of continued water and energy shortages. Technological breakthroughs could also open up diverse sources of water supply.

The imperative at present is to move from purely supply-side approaches to demand management and improved water quality. The ever-increasing use of waterways for irrigation, power, navigation, flood control, drinking-water supply, industrial fishing, and recreation is being accompanied by a fall in water quality. When freshwater quality is maintained, paucity of water can be better managed.

Water is expected to serve as an instigator of economic, social, and technological change in the world. If the spatial distribution of water becomes more lopsided, it could influence population settlement patterns, including migration and refugee flows. Acute water stress, however, could also bring about a shift from consumption-oriented values to more sustainable practices.

Stemming the Security Risks

History attests that the relative scarcity of natural resources can contribute to political conflict. Water feuds today are more common at the intrastate than interstate level, yet the risks of intercountry conflict are growing and carry greater potential costs, as highlighted by the dam-building or pumping race between some riparian neighbors to gain relative advantage. The competition to appropriate resources of shared rivers or aquifers is engendering mutual distrust and discord. For example, an important state's heavy dependence on transboundary river flows controlled by an upper riparian through dam projects is a recipe for major vulnerability and regional instability.

So, what can be done to prevent the sharpening struggle for water resources from becoming a tipping point for overt conflict? Harmonious ties between riparians depend on each side exercising its water rights in such a way as not to affect the rights of the other party.

Only basin-level water institutions, along with preventive hydrodiplomacy, can lower the geopolitical risks, promote better management of shared resources, and forestall disputes from flaring into open conflict. River-basin organizations (RBOs) have been established even between adversarial states that have fought wars or remain locked in hostility. Such institutions have helped moderate competition, although their record in the developing world in rationally managing transnational resources has been anything but flattering.

Their structural limitations and other shortcomings notwithstanding, the intercountry arrangements established in the Amazon, Ganges, Indus, Jordan, La Plata, Mekong, Niger, and Senegal basins serve as conflict-avoidance mechanisms. So do the arrangements in North America, as symbolized by the two omnibus water authorities—the Canada-U.S. International Joint Commission and the Mexico-U.S. International Boundary and Water Commission.

The only peace pact in the world with water provisions is the 1994 U.S.-brokered Treaty of Peace between Israel and Jordan. This treaty openly links water with peace, yet its water allocations and water-cooperative elements are modest.[52] The 1967 war, of course, radically transformed Israel's water-related strategic position: instead of having to protect its cross-border inflows from any Arab diversion, Israel now has a hold over the subregional flows and thus is able to extend—in return for peace—asymmetric water cooperation to Jordon (and the Palestinians, even as it takes the lion's share of the waters from the three West Bank aquifers).

A World Bank report found that Palestinians have access to only a fifth of the West Bank's water supply, while Israel, which controls the region, takes the rest.[53] After the 1967 war, Israel abrogated the Palestinians' water rights and blocked their access to Jordan River waters. The 1995 Israeli-Palestinian Interim Agreement on the West Bank and the Gaza Strip ("Oslo II") deferred the issue of water rights until the "final status" negotiations on other thorny issues such as Jerusalem's political future. But, while estimating the "future needs" of the West Bank Palestinians at between 70 and 80 million cubic meters yearly, it recognized the "necessity to make available to the Palestinians during the interim period a total quantity of 28.6 million cubic meters per year."[54]

For Israel, water has become an important instrument to underpin its regional preeminence and leverage. Although it has a small population compared with its Arab neighbors—Egypt, Jordan, Syria, and Lebanon—Israel outshines their combined military strength in number of regular troops and reservists, quality of conventional weapons, and defense spending.[55] In addition, it enjoys a nuclear-weapons monopoly in the Middle East.

The core challenges in the world's transboundary basins vary significantly. They range from the need to limit fast-growing downstream water demand in the Nile and Indus basins as well as upstream diversions in the Tigris-Euphrates and Mekong basins, to controlling pollution in the Niger, the Jordan, and the Amu and Syr Darya.

Downriver Egypt and Pakistan use 75 percent and 85 percent of the total Nile and Indus waters respectively, much of them for irrigated farming, although irrigation has caused serious problems of waterlogging and soil salinity in both countries.[56] Such is the low level of water-use efficiency in Egypt that it uses more than five times the per capita water of Switzerland.[57] The crops cultivated in Egypt—a vast desert plateau other than the 4 percent of land that comprises the Nile Basin—can be grown with greater water efficiency in the more fertile soils of upstream countries like Sudan and Ethiopia. Pakistan's water productivity is even lower than Egypt's, although it is less arid. For example, Pakistan's average wheat crop yield of 2.24 tons per hectare is two and a half times lower than Egypt's.[58]

The intercountry water arrangements in Asia, the Middle East, and Africa, despite their continuing usefulness, have not succeeded in exorcising the specter of conflict, with disputes between parties still rife. This is largely because of institutional and other deficiencies in the arrangements, whose scope in most cases is limited. A modest cooperative framework, of course, is still better than there being no arrangement at all in a basin.

Although South America is the world's richest continent in per capita freshwater resources, a significant part of its population and economic activity is concentrated in areas where water availability is limited. The La Plata Basin, for example, is the freshwater source for nearly half the populations of Argentina, Bolivia, Brazil, Paraguay, and Uruguay and helps to generate an estimated 70 percent of their gross domestic product (GDP). The 1969 umbrella basin agreement for cooperation has spawned multilateral and bilateral accords on hydropower development, transportation, and investment.[59] The basin's strategic importance, however, has made conflict avoidance and wider cooperation critical to regional security.

The list of intercountry freshwater agreements in the world is deceptively long. Most such accords relate more to mundane issues than to core issues like sharing waters or collaboratively managing transboundary basin resources through sustainable practices. Masquerading as water agreements are commercial contracts, such as to sell hydrological data to a downstream nation; joint research initiatives; flood-control projects; accords on mutual use of river islands, hydropower development, water for industrial uses, fishing, recreation, or preservation of freshwater species; and nonbinding memorandums of understanding.

For example, the otherwise useful, oft-cited Transboundary Freshwater Dispute Database lumps together all sorts of accords under the heading of international freshwater treaties, including those relating to protection of the marine environment, cooperation in the sphere of geological exploration, establishment of a technical commission, resolution of boundary differences (if there is a mere reference to freshwater), commercial navigation, prevention of floods, wastewater pumping and disposal, environmental protection, salmon farming, cross-border visits of water officials, basin cooperation on meteorological and hydrological surveys, construction and maintenance of a bridge across a river, joint studies, bilateral cooperation between two ministries, and funding of a basin commission.[60]

Clumping all these issues under a single title creates a false impression that transboundary water resources generate cooperation and little conflict. Indeed, no distinction is often drawn between treaties and nonbinding agreements-in-principle or ad hoc accords that incorporate no abiding commitments. A treaty, under the Vienna Convention on the Law of Treaties, is a *binding* agreement between two or more nations (whether called a convention, protocol, charter, pact, or other name) that is usually ratified by the lawmaking authority of each state party.[61] The existence of just eighteen genuine water-sharing treaties with clearly specified national allocations is particularly striking—and disturbing. Recognizing the gaps and inadequacies in many of the existing intercountry water arrangements is important so as to promote efforts to improve and strengthen them.

Building institutional arrangements in any transnational basin is never an easy task because of the intricate factors at play, including asymmetrical power equations, relative riparian advantage or disadvantage, mismatched levels of economic development, the dissimilar contributions of the states to the aggregate basin waters, the size of the population dependent on the basin waters in each country, existing water-use patterns, and the unilateral diversion of shared waters by one or more co-riparians. Yet, to reduce the risk of overt water conflict and build a mutual stake in the sustainability of basin resources, rules-based cooperation is vital in all basins currently lacking any meaningful collaborative mechanisms.

RBOs and other institutions provide for constructive dialogue and structured cooperation between riparian neighbors. Their legal, institutional, and consultative mechanisms act like a safety valve in a boiler to release pent-up pressure created by popular passions and whipped-up apprehensions. A robust institutional framework, however, can be built only with the participation of the headwaters-controlling state, high-level political commitment, well-defined rules and functions, transparency, tangible benefits for each party, conflict-resolution mechanisms, and flexibility to deal with hydrological variability. A forward-looking arrangement should have the structural elasticity to adapt to change and be open to improvements as a result of the experience gained and new understandings gleaned about basin dynamics.

Any water institution demands an organizational structure, financing, and technical capacity. To carry out its tasks competently, it should have the support of high-quality technical experts and the powers to consult, investigate, and arrive at decisions. Its commissioners must act not as agents of their respective countries (as they do, for example, in the hamstrung Permanent

Indus Commission, Senegal Basin Permanent Water Commission, and Administrative Commission of the River Uruguay) but as professionals, basing their decisions on technical assessments and other specialized inputs from experts.

A focus on clear, pressing objectives (such as water sharing and joint management) is important because broad, diffused agendas tend to obscure the core challenges. Water sharing must involve a clearly set allocation formula so as to minimize disputes. To prevent or combat degradation of basin resources, parties should be obligated to pay attention to water quality, as mandated, for example, by the 1975 Argentinean-Uruguayan pact on the Uruguay River and the Israeli-Jordanian agreement.

Meeting water demand is usually a function of both quantity and quality, although water competition and discord tends to blur the line between available quantity (or access to resources) and diminishing quality (which may be caused by agricultural runoff, municipal sewage, industrial effluence, or aquatic weeds and invasive species). Generally, the more populous a subregion, nation, or substate area, the greater its water challenges tend to be, with water stress often being accompanied by a fall in water quality. When water quality is maintained, water stress can be better managed, thereby reducing the conflict potential, whereas extensive water pollution is likely to precipitate or intensify a water crisis and raise security risks. Many existing intercountry arrangements, however, call for no concrete action to prop up water quality.

Considering the frequency with which interstate disputes arise even when a treaty is in force, as in the six-nation Rhine River Basin and the Indus Basin, an effective institution must have clear and detailed conflict-resolution procedures, which can include independent fact-finding, neutral-expert assessment, arbitration, and adjudication. For example, the bitter political row between Argentina and Uruguay over the latter's decision to set up two pulp mills along the shared Uruguay River—an action Argentina contended violated Uruguay's no-pollution pledge—was settled by a 2010 ICJ judgment, which was followed by a political accord between the two neighbors on a joint oversight of activities.[62]

The involvement of a mutually agreed third party is usually important in creating and nurturing an institutional arrangement. The third party could be the United Nations or one of its agencies like the United Nations Development Programme, or a multilateral financial institution such as the World Bank. The third party typically provides financial aid in what is a lengthy process to institutionalize cooperation and implement the arrangement. The third party also serves as a facilitator or intermediary to help settle any difference or dispute through conciliation, fact-finding, or referral to an impartial tribunal for a binding decision.

In the absence of conflict-resolution mechanisms or a role for a third party, the majority of existing intercountry arrangements remain weak or even toothless. Less than 10 percent of the existing water treaties, according to United Nations data, permit fact-finding, arbitration, or adjudication. In fact, many agreements allow no monitoring of any kind, including information sharing, and most that do permit monitoring contain very rudimentary elements. Few agreements incorporate any provision for scrutinizing compliance.

Since the 1950s, military conflicts over water in the Nile, Niger, Senegal, Jordan, Amu Darya, Syr Darya, and Indus basins, among others, were averted largely because of external interventions. In one case—the Ganges Basin—the offer of outside help, however, was rebuffed, yet the parties managed to resolve their protracted political conflict by entering into a treaty in 1996 that almost equally divides the downstream water flows between India and Bangladesh. President Jimmy Carter in 1978 proposed a Ganges version of the 1960 Indus treaty, with the U.S. government and the World Bank to play the lead role again, but the idea was politely spurned by the Indians, still smarting from their assessment that they "lost out to Pakistan in water-sharing arrangements under the Indus treaty."[63]

In several basins, conflict situations were defused only through strong external political involvement, backed at times by international funding of projects that were set up as part of the new conflict-avoidance arrangements.[64] The role of the World Bank and UN agencies indeed has been critical in aiding the birth and advancement of new water institutions. The World Bank, for example, has supported initiatives ranging from the 1980 formation of the Niger Basin Authority covering Africa's third-longest river to the 1960 establishment of the now-defunct multilateral Indus Basin Development Fund to underwrite Pakistan's construction of its first major dams and other hydroengineering projects.

The changing international lending and aid dynamics—underscored by China's overtaking of the World Bank as the biggest global lender—suggest that multilateral financial institutions may now no longer have the same clout as before. In a spectacular reversal of fortunes, the emerging economies, with their large foreign-currency holdings, now finance the mounting deficits of the wealthy economies, whose accumulation of sovereign debt and other structural problems are fostering an uncertain environment for global economic growth. China, the global leader in dam building at home and abroad, is the top funder of overseas dam projects. Having voted against the 1997 UN Convention and remaining opposed to bilateral or multilateral water institutions with its own neighbors, China has emerged as an important obstacle to the building of international water rules and mechanisms. These developments have a bearing on the ability of the World Bank and UN agencies to promote and sponsor new water institutions.

However, at a time when intercountry water competition—coupled with internal and regional instabilities—has turned some basins into potential flash points for serious conflict, multilateral financial institutions and the United Nations still have an important role to play in providing technical know-how, aiding information collection and sharing, and sponsoring cooperative initiatives and projects. They can encourage the development of a partnership among key actors in the transnational basins where no cooperative framework currently exists.

Many of the existing cooperative arrangements also need to be strengthened to help buttress regional stability and to make them more capable of handling the challenges arising from soaring water demand, sharpening basin hydropolitics, increasing strain on already fragile water reserves, and global warming. Some of the present intercountry arrangements actually face a test whether they will survive and thrive or gradually become dormant and wither away.

This test is a reminder that the challenge of negotiating and concluding a basin treaty is distinct from the challenge of faithfully and successfully implementing the terms of such a pact. Often the latter challenge has proved to be more onerous. The implementation process can proceed productively only if it is kept largely insulated from the pressures of hydropolitics.

To contain the security risks, co-basin states have little choice but to invest in institutional water mechanisms designed to underpin strategic stability, protect continued economic growth, and promote environmental sustainability. Only basin communities tied together by institutionalized cooperation and involving all important co-riparians can manage transnational watercourses in a holistic and peaceful manner.

Such management is critical to the promotion of a dynamic, multisectoral approach that integrates technological, socioeconomic, environmental, and human-health considerations. To plan for the rational, sustainable utilization of shared resources, harmonization of national strategies across a transboundary basin is usually necessary, even if joint initiatives are concentrated only at the subbasin level. None of these steps will be easy to undertake, yet such action is essential.

The international community's ability to prevent water wars will depend on its success in building wider and improved cooperative management of transboundary water resources. Balancing competing demands on shared resources is central to the prevention of conflict. To help fragile, water-scarce states win enduring peace after prolonged civil strife, the postconflict management of water resources must be integrated into broader relief and recovery efforts. International assistance, in general, can serve as an important facilitator of a more cooperative intranational and intercountry approach to shared water resources.

Mitigating the Economic Risks

Farsighted decisions can turn the global water crisis into an opportunity for innovation and profit for investors. The international economic risks indeed come with important new business and innovation opportunities. The business opportunities are expected to draw massive amounts of public and private capital into the modernization and upgrading of water systems.

The fact that water remains a greatly underappreciated and underpriced resource despite being central to life and livelihoods has imbued it with qualities appealing to investors in a water-stressed world. Because it is becoming difficult to supply water at a heavily subsidized price, investors are exploring the long-term business opportunities that this resource holds. Some investors and private equity funds in the United States, for example, are buying real estate in order to acquire water rights, while some others are purchasing just water entitlements in places at home and abroad (such as Australia and Chile) where water rights have been separated from land ownership.

In fact, the water crisis is bringing governments under increasing pressure to change their business-as-usual approach, lest the water gap between demand and supply widen further. The key to mitigating the economic risks lies in greater investments in water treatment and storage as well as in enhancing water-use efficiency in agriculture, energy, and manufacturing, along with improved water management (including better allocation, distribution, and pricing). Better water management must also include the slashing of distortive electricity-for-water-pump subsidies in many developing economies.

The risks associated with unmitigated water problems are so high that governments must start addressing them in earnest. If left unaddressed, the water problems, coupled with environmental degradation, are likely to lead to economic and social disruptions. Depletion of groundwater resources, for example, threatens food security and rural livelihoods and carries the potential to unleash serious socioeconomic consequences in the developing world, where nearly half the labor force is engaged in farming-related activities. Water-availability constraints will also impinge on the expansion of electric power generation, resource mining and processing, and manufacturing.

To be sure, the ability of states to lessen water-paucity-related risks very much depends on their political and economic capabilities. A nation's institutional capacity and financial resources are critical even to deal with the potential impact of climate change, which could prove a threat multiplier, compounding water shortages and exacerbating water competition.

States with good governance and ample financial resources are expected to deal with their water challenges in a much better way than cash-strapped nations racked by internal turmoil, rising extremism and lawlessness, rampant corruption, and corroding governance. For example, South Korea and Singapore are better placed to address their water shortages than water-scarce Somalia or the Maldives—a group of strategically located islands in the Indian Ocean where the first democratically elected president was forced at gunpoint to resign in 2012. In fact, the risks of state failure are high when water and environmental stresses are compounded by explosive population growth, internal civil war, and weak political authority.

The more-capable emerging economies will seek to engineer potential solutions or mitigation measures for their water challenges. By contrast, containing water distress may not be within the capacity of weak states, which will need international assistance. For example, in impoverished, conflict-torn Yemen—the first country in the world expected to run out of water—desalination is not an affordable supply option because most citizens live at elevations of one to two kilometers above sea level, making pumping desalinated water prohibitively expensive, even as the rampant overexploitation of groundwater has promoted saltwater intrusion into its coastal plains.[65]

More fundamentally, the real choice for water-stressed countries is not between supply-side and demand-side options but to find ways to amalgamate the two holistically so as to create potential win-win solutions. Indeed, multilateral financial institutions, in a shift from their earlier stance, are now increasingly reluctant to fund large supply-side projects without their integration with demand-side initiatives in national policy.[66]

The policy options at the national and multilateral levels may be limited, but delayed action to stem the economic and other risks of water stress are likely to exact a heavy price.[67] After all, water is a resource that will determine whether the future of humankind will be peaceful or perilous.[68] The economic risks, in fact, are serious enough to call into question the ability of water-constrained economies to continue to register high GDP growth rates. For example, China, which is in danger of slipping into the category of water-stressed countries, estimates that growing water shortages in its semi-arid north are costing the national economy roughly $28 billion in lost annual industrial output.[69]

What is needed internationally is a new strategic approach centered on water conservation, efficiency and productivity gains, and integrated water resources management (IWRM) combining economic, social, and environmental objectives across sectors. Collaborative paths embracing all key stakeholders can help to open up crisis-mitigation pathways. Most countries, however, have yet to go beyond paying lip service to the IWRM concept by taking concrete steps to implement it in policy so as to sustainably manage both demand and supply.

Water-sector reforms have also become imperative. Public authorities in many countries often are regulators, utility owners, and service providers, all rolled into one. For better oversight and management, and to encourage private industry to finance, build, and operate water-service facilities, it is useful to disaggregate these roles. Granting management autonomy to water utilities, for example, will allow public authorities to concentrate on playing a more effective role as regulators. Only by adopting good policies and practices for long-term, strategic management of water resources can governments hope to ease their water challenges.

Any crisis presents opportunities for investment, innovation, and improved management. In fact, international experience indicates that societies do not deal with a problem unless faced with a crisis. The water problem is now turning into a water crisis in several parts of the world. The new opportunities being opened up for investment, improved water management, and technological innovation are principally in three broad areas, which can help unlock solutions.

These areas center on achieving greater water efficiency and productivity gains; using clean-water technologies to open up new supply sources, including ocean and brackish waters and recycled wastewater; and expanding and enhancing the water infrastructure to build distribution efficiency, to correct spatial and seasonal imbalances in water availability, and to harvest rainwater, which can serve as yet another new supply source to ease local water shortages. Boosting water supplies essentially demands tapping unconventional sources and adopting nontraditional approaches, including in the management of resources.

Traditional supply-side measures, besides facing a steep cost escalation, are in any event running into natural limitations because water resources in many regions are already overexploited. Such measures, at best, can augment water supplies in a small way, especially in countries where water abstraction rates are relatively high and water-resource depletion is already a problem. The easy-to-get water has largely been tapped. That is why developing nontraditional supply sources and management methods has become imperative. Each of the three broad areas, in turn, can be divided into three subcategories.

1. Water Efficiency and Productivity Gains

One key area centers on securing higher water efficiency and productivity gains. The annual rate of efficiency improvement in water use for economic production lags the corresponding rate for energy use. Greater water-use efficiency will translate into not just major water savings but also considerable energy savings. Water stress dictates that users be much smarter about every drop. Paying more for water and wasting less is becoming unavoidable. Three subcategories are pivotal to securing greater water-efficiency gains.

The centrality of making water savings in agriculture. Because agriculture uses approximately 70 percent of the world's freshwater supply, the greatest potential for easing the water crisis is through technology and practices that cut the amount of water channeled for food production. This principally underscores the imperative of shifting from the old, gravity-flow irrigation systems (whose water profligacy is flagrant) to microirrigation systems, like drip-feed irrigation—so called because water is conveyed under pressure through a pipe system to the fields, where it drips slowly onto the soil through emitters or drippers. Drip irrigation, which directs the water flow straight to the root zone of plants, can cut agricultural water use by 50 to 70 percent compared with gravity irrigation, and by 10 to 20 percent in comparison with sprinkler irrigation.

Only a tiny minority of the world's farms currently employ microirrigation systems, with the portion of the global irrigated acreage under microirrigation estimated to be just over 3 percent at the end of 2012. This in large part is because few farmers have adopted such systems in Asia, which boasts 72 percent of the world's 310.3 million hectares of land equipped for irrigation.[70] In fact, just South Asia, China, and Southeast Asia account for more than half of the world's total irrigated land.

The limited prevalence of microirrigation calls attention to the immense business opportunities in an important area where a number of governments are beginning to offer important subsidies to encourage farmers to switch to such modern systems so as to help save water and energy. To ensure that power outages—common in many developing economies—do not undermine the appeal of microirrigation methods, some manufacturers, in fact, are marketing drip irrigation with a solar-powered system that can provide continuous electricity to farmers to water their fields.

Because most farms in Asia, Latin America, Africa, and the Middle East are relatively small by Western standards, farmers often have to bank on state support to make capital investments, including in water-efficient technologies and crops. Therefore, forward-thinking state policies and support are necessary to encourage the shift to less wasteful practices on farms. After all, traditional irrigation practices, through the steady accumulation of soluble salts in soil, have contributed to soil and water degradation.

Microirrigation systems prevent soil salinization or degradation by efficiently applying water and even fertilizers and herbicides. Efficient application can prevent the runoff of fertilizers (particularly nitrates) from contributing to the chemical contamination of drinking-water supplies and the spawning of blooms of algae in waterways that deplete oxygen and leave "dead zones" in their wake.[71]

The upfront capital costs of drip irrigation (estimated at \$4,000 per hectare in South Africa, for example) are more than offset by the savings it brings in water, fertilizer, and energy use and by the higher crop yields, averaging 45 percent.[72] Yield improvement, of course, varies considerably by area and crop. Drip irrigation, however, has little utility for the cultivation of paddy rice, which demands flooding of fields. But even in Asia, where much of the world's rice is grown and consumed, rice accounts for barely one-third of the yearly planted acreage. This leaves significant room for adoption of microirrigation to cultivate other crops.

Agricultural water-productivity measures, of course, must seek to go beyond improved efficiency of water application to achieve even greater water gains through a focus on crop-yield enhancement. To secure "crop per drop," dramatic improvements in agricultural water productivity have to come from an approach that seeks not only to overhaul antediluvian irrigation systems but also to develop new grain varieties that are more tolerant of drought and flooding yet boost yield.

Given the exceptionally high (and unsustainable) water withdrawals for farming, which are putting more strain on already stretched water resources, savings will need to come primarily through the aid of technology and improved practices that promote water conservation and efficiency, permitting greater food production with less water use. Policies that encourage sustainable water use in agriculture are central to fashioning a more stable water future.[73] The greater the water savings in agriculture, the greater will be the water available for manufacturing, energy production, and urban households.

Improving industrial water-use efficiency. Industrial water prices have been steadily rising across the world in response to the increasing costs of abstraction, treatment, and transport of water. Industrial water-use efficiency levels vary significantly across the world. Yet even in states with higher efficiency levels, there is ample room for securing greater gains. The major industrial users of water (the metals, mining, manufacturing, and energy companies) need to reduce their water footprints. This task can extend from achieving improved water efficiency in the power sector to adopting specific methods such as coke dry quenching in steel production and paste technology for tailings disposal.

Speeding up technology upgrades is the single most important factor in enhancing industrial water-efficiency levels. Given that the industrial sector has been the main culprit behind the problem of water pollution in many parts of the world, greater investments to upgrade technologies currently in use are important to prevent such contamination. Corporations now seem willing to be part of the solution because their own business future hinges on mitigation of the water crisis.

Adopting intelligent water management. Better water-management technologies and practices offer new opportunities to help cut operating costs for businesses, especially in mining activities, energy-related operations, and manufacturing processes. If the available quantity of water is more intelligently managed, it can be put to greater productive use and at lower cost.

Australia's "Every Drop Counts" program has shown that an astute water-management strategy that aggressively emphasizes water efficiency can bring big water savings. For example, Sydney's $30 million investments between 1999 and 2003 in demand-side management enabled the water-stressed city to stabilize its 2003 water demand at 1983 levels, despite a population increase of almost 1 million during the intervening period.[74]

Increasing the diversity of sources of water supply for agriculture and energy so as to address the problems of overdrawing and degradation of resources must be a key goal of improved water management. The power sector's role in contributing to water stress, for example, can be curtailed by utilizing non-freshwater sources for cooling, to the extent possible. Such utilization is critical if the world is to produce much more power, given that nearly 20 percent of the global population still lacks access to electricity.[75]

In order to reap freshwater savings, seawater and impaired groundwater or surface water should increasingly be used in energy extraction and processing and electric power generation. The greater use of seawater as a coolant, however, may be constrained by the difficulty of finding sites for new power plants in coastal areas, which developers often regard as prime real estate. Wastewater reclamation (discussed later in this section) can also be a valuable new source of water supply for agricultural and energy-related activities.

New technologies and practices, extending from treatment solutions to "smart" water meters, are likely to offer a promising market to help meet the demand for innovative water management. Indeed, they could facilitate the development of responsible management policies designed to boost clean-water supplies for human needs and essential ecosystem functions.

2. Clean-Water Technologies

Another promising area centers on clean-water technologies to purify degraded or contaminated water, treat wastewater in a way that turns it into reusable water, and desalinate seawater or brackish water. Purification and desalination technologies are critical to managing global water challenges. Treatment can turn any water potable. Consequently, the demand for water-treatment systems and chemicals is likely to spiral in the coming years.

Clean-water technologies actually hold the key to boosting water availability through two parallel, nontraditional supply-side paths. One path is to tap ocean and brackish waters and make them potable, and the other is to chemically treat and reuse wastewater as well as to cleanse polluted surface waters, whose sheer volume in many regions drastically reduces the overall clean-water availability. Such unconventional sources of water supply, however, will come at a much higher cost, even as the price of conventional water is set to be driven up by scarcity.

Clean-water technologies fall mainly into two categories: distillation technologies, principally multistage flash (MSF) and multi-effect distillation (MED), and membrane technologies, such as reverse osmosis (RO), nanofiltration (NF), and electrodialysis reversal (EDR). MSF (also known as the thermal process) and RO are usually applied for treating high-salinity ocean water, with RO emerging as the fastest-growing method of seawater desalination since the 1990s. In the drinking-water segment, the common technologies employed for purification include membrane filters and ultraviolet radiation, as well as disinfection with ozone, chlorine, or chlorine dioxide. There are also new hybrid technologies, including membrane bioreactor (MBR), which combines biological treatment and membrane separation processes to effectively treat wastewater.

Energy intensity is the principal downside of the currently available clean-water technologies. Yet in water-scarce regions, these technologies offer the only hope for mitigating water distress. Nations and corporations that are the first to develop and deploy more cost- and energy-efficient clean-water technologies could gain a huge geopolitical or commercial advantage.[76]

Desalinated water. Given the vast coastlines of the various continents and the concentration of many economic-boom zones along the coasts, the desalination option is attractive. Desalination indeed offers a pathway to ease the human impacts on the environment by generating additional quantities of potable water from an unconventional source and thereby helping to reduce the degradation and depletion of freshwater sources. Large-scale desalination may also help to ease the rise of ocean levels. Desalination actually presents hope of meeting global water needs in the future in an economical and sustainable manner.[77]

Desalination costs are falling, and a number of new commercial plants are coming up, especially in the oil-rich Arab states, which have bounteous energy resources to fuel such plants. Desalination, however, remains at present an expensive, energy-intensive, and greenhouse-gas-emitting way to secure freshwater supplies. Because of the chemicals employed, the desalination process also creates toxic residues, whose safe disposal through deep-well injection or other means imposes considerable costs.[78] This may explain why a number of nations thus far have taken only modest steps to commercialize the desalination option.

Like any new technology, commercial-scale desalination depends on major state subsidies. Still, new innovations could make it cost effective. Solar-powered desalination, in particular, could potentially emerge as the silver-bullet answer to the global water crisis.

Several nations, as well as the Vienna-based International Atomic Energy Agency, have been working on nuclear desalination—that is, producing potable water in a complex where the desalination plant is integrated with a nuclear power plant so that excess heat from the reactor is used to evaporate seawater and condense the pure water. Given that energy input accounts for up to 45 percent of the total cost of producing desalinated water through RO and as high as 75 percent through the conventional thermal process, nuclear evaporative desalination presents important cost savings by employing the free heat energy available as a by-product of generating nuclear electricity.[79]

Desalination of brackish groundwater or brackish coastal water offers another unconventional source of water supply. Brackish water—less salty than seawater yet undrinkable—is found underground (including in deep, ancient aquifers) or in places where freshwater and seawater meet, such as estuaries, deltas, and mangrove swamps. In fact, a considerable part of the world's subterranean water reserves are brackish or saline.

According to one scientific estimate, the energy requirements for desalinating inland brackish water are up to three times lower than for seawater desalination.[80] Desalinating brackish water thus is an appealing option for opening a new source of supply to address water shortages and halt the overexploitation of conventional freshwater sources.

The production of desalinated ocean or brackish waters at economical rates promises to fundamentally change the geopolitics of water, as if rivers magically reversed the direction of their flows to run from downstream to upstream. In this scenario, the upstream riparian's geopolitical advantage through the control of river headwaters will dissipate.

Indeed, the riparian advantage would move from the mountains to the valleys and coastal plains, and from the basin headwaters to the oceans. And many downstream states, currently dependent on inflows from their upstream neighbors, would—with their proximity to the sea—become important sources of freshwater supply, including potentially across their frontiers to an upriver nation.[81] Even within countries, provinces with a shoreline would gain advantage over inland regions.

Recycled ("reclaimed") water. Wastewater recycled as new water after treatment is another unconventional supply option whose large-scale commercialization is likely to help moderate the international water crisis. Such reclaimed water is already in use in several cities, although the aggregate global quantity of recycled water remains modest. Water-scarce Singapore, which has taken the lead in Asia to treat wastewater for everyday reuse, has discovered that such recycling is less expensive than desalinating seawater.

Actually, the relative costs of wastewater recycling and desalination depend on the type of technology employed and the intended final use of the treated water.[82] For agricultural use, for example, recycled wastewater is less expensive than desalinated water.[83] Minimal freshwater quality is adequate for growing crops, and therefore it is easier and less expensive to produce agriculture-grade water from wastewater treatment than through desalination. However, it is possible to bunch large desalination and wastewater-treatment plants together to provide economies of scale and a hybrid solution.

According to a 2012 U.S. National Academy of Sciences report, the risk from potable reuse—or supplementing streamflows or reservoirs with reclaimed water—"does not appear to be any higher, and may be orders of magnitude lower" than any risk from the conventional treatment of drinking water.[84] Municipal wastewater treatment, for example, is a three-step process involving filtration through microfibers, then the use of RO technology to remove salt ions and other impurities, and finally the disinfectant exposure to ultraviolet light. The World Health Organization has set international standards for recycled water for potable uses.

Still, much of the reclaimed water globally is channeled for nonpotable uses, such as irrigation and industrial purposes. Some is utilized for drinking and for replenishment of aquifers, rivers, and reservoirs. In the coming years, recycled water could also be used for ecological purposes, such as restoring or enhancing wetlands and riparian habitats.[85]

To be sure, desalinated water has greater public acceptability than reusing treated wastewater. Recycling wastewater also demands government subsidies. Yet some fast-growing cities in the world are grudgingly embracing the toilet-to-tap concept—already long in operational use on manned spacecrafts—by mixing treated wastewater with conventional water in the city supply system.

Such mixing is designed to overcome the "yuck factor" that keeps many people from accepting reclaimed water. In cities as varied as San Diego, London, and Singapore, recycled water now makes up a portion of the municipal supply.[86] With many urban areas increasingly searching for additional water resources to tap, more metropolises are likely to go in for recycling of treated wastewater so as to augment their municipal supplies. At present, however, just a fraction of 1 percent of municipal wastewater is internationally recycled into local supplies.

Cleaning up contaminated water. Given the wide extent of pollution of surface waters, the cleansing of such waters offers an important source of additional freshwater supplies to water-stressed communities. In fact, cleaning up polluted freshwater—unless it has been heavily contaminated with untreated industrial wastewater or potentially toxic elements from sewage sludge—is usually less energy intensive than the treatment of wastewater or seawater for potable uses.

A fall in water quality usually carries consequences as serious as those resulting from a decline in water quantity. Water-quality improvements are thus an important means of alleviating water distress. In cases where water quality and productivity have both appreciably increased, conditions of water stress tend to perceptibly lessen. China, for example, has a slightly lower water abstraction ratio than Japan as a percentage of its water resources.[87] Yet, because of the wide extent of its water-pollution problem, China is worse off than Japan in being able to meet its water requirements.

Consider another example: South Korea and Pakistan are both densely populated and water stressed, with an identical per capita availability of freshwater—about 1,400 cubic meters per year.[88] Yet the world hears about Pakistan's water crisis but not about South Korea's because the latter has focused on improving water quality and productivity through massive investments, while the former has done little to arrest its deteriorating water quality.

These examples underscore the importance of preventive measures and investments in clean-water technologies for improving quality and easing supply constraints. If water-stressed economies are to appreciably raise their water-productivity levels, they must begin by increasing their water quality.

3. Expanded and Enhanced Water Infrastructure

A final area offering promising business opportunities to help reduce water shortages centers on the growing demand for expanded and enhanced water infrastructure. This includes storage and supply structures, distribution networks, and pipelines to deliver clean water and transport water within countries from regions with abundant resources (or from desalination or wastewater-treatment plants) to areas seared by scarcity.

The water sector globally is plagued by inadequate infrastructure and poor system maintenance as well as by financially strapped utilities. Owing to leaks and system inefficiencies, a sizable portion of the international water supply is lost before reaching users.

Large public and private investments are needed to upgrade infrastructure so as to increase storage and supply, control water pollution, boost water-use efficiency, and promote wastewater reclamation. Capital expenditures by governments on water infrastructure indeed are growing in response to water shortages. The Chinese government, for example, has unveiled an annual $63.5 billion spending on water infrastructure between 2012 and 2022.[89]

Storage and supply. To deal with water shortages, nations will need to increase their capacity to store water in large reservoirs during periods of relative water glut, such as the seasonal snowpack melt and main rainy time of year, so as to be able to augment supplies during the dry periods. The water for storage can be taken from rivers and streams overflowing in the water-surplus season or can be directly captured from rainfall.

Storing rain and melt runoff helps to moderate major variations in the seasonal cycles of water supply and demand. This task entails major investments in storage and supply infrastructure. In some reservoir-dotted basins, however, the potential for tapping water for storage may already have been fully exploited.

Poorer countries, with their limited financial resources, have by and large much lower surface-water storage capacities than the richer nations. But as figure 5.2 shows, emerging economies such as Brazil and China have relatively impressive water-storage capacities.

Water naturally recycles, but nature's finite replenishment capacity limits the world's total renewable freshwater resources to about 42,370 billion cubic meters per year—the maximum theoretical amount of water available under natural conditions, excluding human influence and the effects of climate change. However, 813 millimeters of precipitation falls annually, measuring

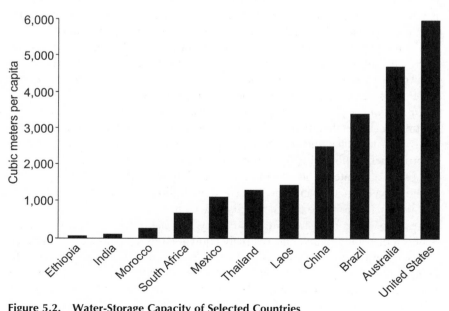

Figure 5.2.   Water-Storage Capacity of Selected Countries
Source: World Bank, 2005; and United Nations, 2010.

108,831 billion cubic meters in volume, or more than two and a half times the world's annual freshwater resources.[90]

Much of the rain falls on the oceans, or the rainfall is carried to the seas by rivers and other natural drainage. If just a small percentage of the rain that falls on the land surface were captured, without upsetting the rainfall's ecological uses and recharge of aquifers, it would significantly augment the world's usable water resources. A capture of 5 percent of the total annual precipitation would translate into a 12.8 percent increase in the world's aggregate water resources.

Climate-change models indicate a likely shift in rainfall patterns, with the projected long-term decrease in river flows (largely because of the accelerated thawing of glaciers and mountain snows) being somewhat offset by greater rainfall in the wet season, especially in the tropics, as discussed in chapter 3. Higher average surface temperatures are expected to strengthen monsoon circulation.[91] To compensate for decreased river flows, rainwater capture on a large scale will likely become critical to meet some of the additional annual freshwater demand of 64 billion cubic meters being imposed by a human population growing by about 80 million a year.[92]

Rainwater harvesting, in fact, is an ancient, relatively low-cost technique dating back to the ninth or tenth century, and its global revival is making it one of the most promising frontiers to ease local water shortages.[93] Rainwater capture is already being practiced in cities ranging from Singapore to the southern Indian metropolises of Bangalore and Chennai, where new apartment complexes and commercial buildings are required to have a rainwater-harvesting system. For many rural communities in Australia, rainwater harvesting serves as the only source of freshwater. The heavy rain during the wet season in several water-stressed regions makes it easier to capture rainwater for dry-season use.

Upgrading and strengthening distribution. Greater public and private investments are needed to upgrade and maintain water-distribution networks so as to plug leakages and prevent recontamination of water. The aim of such improvements should be the building of distribution efficiency.

Losses of treated water from leaky distribution in Asia, according to the Asian Development Bank, are valued at $9 billion—a conservative estimate that perhaps hides more than it reveals.[94] Even in a rich country like the United States, the water infrastructure is so leaky that nationwide losses of potable water in distribution are conservatively equivalent to the combined water consumption of the country's ten largest cities.[95] According to other estimates, the aging pipes lose up to 25 percent of the water. A U.S. Senate committee was told in 2011 that it would take $335 billion to fix the country's water systems. Although water pipes in the United States are being changed at an average of eighteen kilometers a year, replacing the water system's entire 2,080 kilometers of old pipes is projected to take more than one hundred years.[96]

Better infrastructure is also needed to prevent microbial contamination of treated water supplies during storage and transport. It is not uncommon in many developing and some developed economies for water, although free of contamination when it leaves the municipal water-treatment plant, to become recontaminated by the time it reaches the household tap.

Building pipelines to transport water. Intercountry water pipeline projects have often failed to materialize because of hydronationalism and fears of disruption of supplies, but there are usually no political impediments to building pipelines within a nation to take water from a surplus area to a deficient area. Pipelines are also essential for building a desalinated-water grid and for recycling water by first transporting the wastewater for treatment and then conveying the reclaimed water to users.

To be sure, pipelines are expensive to build and maintain. And like oil and gas, the transport of water by pipeline, unless based on gravity flow, demands the use of pumps along the route. The diameter of water pipelines, however, is typically larger than that of oil and gas pipelines. The pumps, too, are larger and more energy intensive because water is about 20 percent heavier than oil. So piping water across long distances is indeed expensive. Yet in an increasingly water-stressed world, water-pipeline networks in the near future are likely to surpass the carrying capacity of oil and gas pipelines today.[97]

None of the measures outlined above can be a solution by itself to the international water crisis. But if these measures were pursued in tandem so as to complement and strengthen each other, they would likely help contain the crisis and its economic and security risks.

Addressing global water challenges fundamentally demands new skills, technologies, and management practices, as well as innovative approaches, including building demand-side efficiency and tapping nontraditional water sources. In such a framework of integrated management designed to help cut environmental costs, virtual-water trade will hold important but limited benefits for both importing and exporting countries.

A key unresolved issue is pricing water appropriately within countries to prevent wastage and conserve supplies. Working out a fair price to charge for water is very complicated, especially because the value of water and the price of water are often not one and the same thing. Water as an economic good must be balanced against water as a social good by taking into account different variables, including income levels, usage patterns, environmental protection, and social rights of access. Some benefits, such as water for recreation, may not lend themselves to charges.

Countries with different levels of economic and social development and institutional capacity predictably have different approaches to water pricing. Water tariffs often vary significantly even within nations because the pricing is greatly influenced by substate subsidy policies and by the value locally assigned to different uses of water by various authorities and categories of consumers.

Water tariffs in most countries are usually inadequate to cover expenditures on system management, operation, and maintenance or, in many cases, to even discourage profligate practices. With water being pumped at greater depths and transferred across longer distances, the growing costs of water abstraction and transport have in many regions not been passed on to end users.

Even the concept of full-cost pricing includes charges for everything except the water itself. The bottled-water industry often pays a pittance or nothing for the resource it extracts, whether it is high-end glacier and mountain-spring water or plain groundwater.

Municipal water supply in many wealthy countries comes with direct or indirect subsidies. In the United States, for example, most households do not pay the real cost of the water they use, enjoying an unacknowledged subsidy.[98] In many developing economies, well-to-do citizens tend to be the main beneficiaries of subsidized municipal-water supply, while slum dwellers often depend on water tankers.[99] Creative pricing policies are required.

Subsidies need to be specifically targeted at the poor, including marginal farmers, and in a way that helps to incentivize water conservation and efficiency. South Africa has set a good example through its "Working for Water" program, under which the old flat rate for water gave way to a system in which people who save water pay less. The program's pricing formula, with the tariff and tax rising steeply as consumption increases, gives poor people access to water at a low cost, while high-end users pay more per unit of water because they use far more water.

The social and political constraints on introducing full-cost pricing for all end users mean that most governments will continue to subsidize water supply, especially to low-income households and farmers. Even in the thirty-four rich member states of the Organization for Economic Cooperation and Development, agricultural water use remains heavily subsidized.[100] Underpriced water has long been misallocated to powerful lobbies, such as rich farmers.

But now the introduction of rationalized and calibrated agricultural water tariffs—along with the gradual phasing out of power subsidies, where they still exist—is becoming unavoidable in order to deter wasteful practices on farms. In Thailand's southern Chao Phraya area, the mere levying of water charges for pumping led to an 80 to 90 percent increase in water-use efficiency.[101] In contrast, some farmers in California—where irrigation subsidies remain a contentious political issue—still find it more profitable to sell their quota of subsidized water and leave their fields fallow than to plant crops.

Businesses, for their part, must get ready to pay more for the water they use. Even if new energy- and cost-efficient clean-water technologies emerge in the coming years, the supply costs are still bound to escalate because of the treatment expenses and additional infrastructure and maintenance needs.

The cost of water supply thus is likely to mount. Just as many consumers have overcome an aversion to pay for the water they drink by switching to bottled water, businesses, farms, and households will adjust to what seems inevitable—a sharp reduction in government subsidies so as to promote movement toward an economic price for water.

With public spending on water-related projects rising, the world is also set to see a big expansion of investment in the water sector. Revenues accruing from higher water charges and pollution taxes are likely to be increasingly reinvested in water-resource management and infrastructure development. Moreover, to help attract the required level of private investment and technology in the water sector and to build public-private partnerships, an increasing number of governments are likely to offer appropriate incentives for businesses.

Water, as the new oil, must lead to a similar response as oil has done— conservation and efficiency. More than four-fifths of the countries in the world have already carried out some water reforms, including overhauling antediluvian water laws, with the aim of improving water management and drinking-water access. However, progress on water efficiency continues to lag behind other water-management enhancements, with the majority of reforms not addressing ways to achieve improved water productivity.[102] A greater focus on water efficiency and conservation seems unavoidable in the next phase of reforms, given the limited options policy makers have to tackle the water crisis.

A global market for bulk freshwater, however, may not emerge in the coming years, although water trading within nations is likely to become more prevalent. Technical, economic, and political constraints make a globally integrated market for bulk-water trade difficult to establish. Water can emerge as an important physical-commodity-based asset class internationally only if bulk water becomes a widely traded global commodity. Of course, if bulk water does become a major exportable commodity, water-rich states like Brazil, Canada, and Russia would reap commodity-driven wealth while running the risk of falling victim to a "resource curse."

WHY COOPERATIVE POLITICS HOLDS THE KEY

The success or failure of an institutional arrangement to alleviate water insecurity and tensions in any basin hinges not on just its legal terms but also on the political equation between its parties. If relations between co-basin states are strained, any treaty arrangement between them may also come under pressure, especially if mutual obligations are not faithfully discharged.

Make no mistake: a genuinely cooperative and collaborative approach cannot develop without a politically conducive environment. Improving regional geopolitics is thus usually indispensable to building better hydropolitics. International water law, in any event, remains too weak to avert or manage conflicts, with a rules-based international water order still distant. This situation only serves as an incentive for riparian powers to try to control transnational water resources. Even booming trade and economic engagement between neighbors can do little to improve their hydropolitics.

The experience in several parts of the world, including North America, Europe, Southern Africa, and even the Middle East, indicates that bilateral or regional cooperation on shared water resources is more likely to develop when it is embedded in a larger framework of political collaboration. This is a reminder that in basins where watercourses ignore political boundaries, politics holds the key to building intercountry collaboration on their resources. Without riparian neighbors demonstrating the political will and investing requisite political capital, it will be hard to address regional water challenges.

Some pundits romantically saw the Cold War's end as heralding a new age in which geoeconomics would dictate geopolitics—a thesis reminiscent of the nineteenth-century liberal belief that growing trade and financial interdependence would make war obsolete. That trade is not constrained by political differences in today's market-driven world, with rival countries boasting fast-increasing bilateral trade, has not stopped powers from playing on the grand geopolitical chessboard.

Indeed, as underscored by the grating hydropolitics in several basins and the recrudescence of Cold War–era territorial disputes, booming trade is no guarantee of moderation or restraint between states. The growing economic stakes in an increasingly interdependent world have only sharpened geopolitical rivalries, even as economic powers like Germany and Japan have struggled to carve out a bigger role for themselves in world politics.

Economic interdependence helps to raise the costs of political miscalculation, yet economics alone cannot solve politics or avert armed conflict. Economic forces, for example, have failed to rein in geopolitical competition over natural resources or to open up autocratic political systems thriving on market capitalism. Despite a greater role for economic power in international relations, politics continues to drive economics, with political risk, for example, dominating the financial markets.

The widening gulf between economics and politics in some regions where interriparian relations are already under pressure could even set off political tremors. Unless co-basin states fix their political relations, economics alone will not be enough to build riparian collaboration between them. The water competition, if anything, is a reminder that the interplay between economic and political risks to riparian peace must be addressed.

Law versus Politics

If the politics between two co-basin states are bad, even a fine water treaty, with clearly defined water allocations and binding dispute-settlement provisions, is unlikely to prove adequate, as the parties will have a tendency to engage in charges and countercharges over the observance of the pact's terms, leading to recurrent frictions on alleged noncompliance and recourse to conflict-resolution mechanisms. But if the politics are stable, with the bilateral relationship bolstered by a peace, friendship, or regional-cooperation treaty, even a tentative water arrangement between the two parties is likely to carry an aura of success.

An example from this second category is the Israel-Jordan water agreement, while the India-Pakistan Indus treaty—one of the world's most detailed pacts—typifies the first kind. Annexes are critical to both of these agreements of unlimited duration: whereas Israel and Jordan have set out their water arrangements not in the main body of their peace treaty but in its Annex 2, much of the meat of the Indus treaty, after the partitioning of the six rivers has been defined, is in its voluminous annexes.[103] It is this treaty's multiple annexes—examined by few studies—that spell out a series of curbs on India to preclude any Indian control over the quantum and timing of transboundary flows of the rivers reserved for Pakistan. No other existing water treaty in the world places such intricate fetters on an upper riparian.

The important point is that political trust is central to building genuine collaboration and forward-looking management of basin resources extending beyond a mere sharing or division of waters between two co-riparian states. Without cooperative politics, attempts to build water collaboration between countries (or even provinces) cannot go very far. If water tensions and conflicts within and between nations are to be averted or minimized, shared resources need to be protected and judiciously managed in a cooperative spirit as part of a progressive political deal.

The likely effects of unregulated competition also undergird the critical importance of cooperative politics. Water-resource depletion and pollution, cropland degradation, and other environmental and climatic impacts can foster conditions that undermine state capacity to deal with such stresses and lead to grave civil strife. Dysfunctional states, as present-day examples illustrate, are the breeding grounds for extremism, fundamentalism, and terrorism. Intrastate effects, by compounding water distress and triggering refugee flows and other problems, would likely assume transnational dimensions over time and destabilize a region. Water security in a transnational basin cannot be truly achieved just on any side of the jurisdictional boundaries, in isolation of the developments across the frontiers.

To be sure, a well-structured treaty, with clear aims and objectives, delineation of rights and responsibilities, and monitoring mechanisms, is vital to build water cooperation. But focusing merely on the legal terms of cooperation and ignoring ways to improve basin politics is likely to progressively undermine an institutional arrangement in a basin where the parties have yet to reconcile their geopolitical interests and fully normalize their relationship. Transboundary water cooperation—and any treaty intended to underpin it—cannot be expected to last if the co-riparians do not get along and remain estranged for good.

It is also true that the mutual benefits any institutional arrangement yields should be such that all parties have a stake in its continuation. The comparative benefits and burdens from any arrangement must clearly weigh in favor of the former, or else the party that sees itself as a loser will in due course stop extending cooperation or faithfully complying with its obligations. If that party is the dominant riparian, the arrangement's future would be doomed.

In a river basin, the lower riparian, for example, cannot seek to undermine the security of the upriver country through the use of terrorist proxies or other covert means and still expect the latter to devotedly share waters with it under an existing pact. Good neighborly ties, after all, cannot be on a piecemeal basis, with one party seeking good relations on water sharing but blocking or undermining cooperation on larger bilateral issues. There are simple ways for states to be good neighbors, and they all start with the observance of international norms on good neighborly relations.

In the absence of improved regional geopolitics, even a water treaty of indefinite duration can offer little reason for smug satisfaction. Such a treaty's future can hardly be secure if the geopolitics stay messy. After all, under the Vienna Convention on the Law of Treaties, "indefinite" does not necessarily mean "permanent."

In fact, world history attests that treaties are not automatically permanent, even if they were intended to be of indefinite duration. As the obligations of states in international law arise only from their consent, no party in principle can be stopped from withdrawing its consent due to fundamentally changed circumstances or national-security exigencies. A water treaty cannot be seen as imposing obligations imbued with immutability and permanence, irrespective of a changed circumstance. In fact, without faithful discharge of obligations by both sides, no bilateral pact of any term or type can survive.

Some international treaties of unlimited duration expressly permit withdrawal as long as procedures of notification are followed; some others are silent on the issue of withdrawal by either party while allowing the pact to be modified by mutual accord. Modifying a treaty of unlimited duration by mutual agreement to help create a more level playing field between two parties is unlikely to happen because each would seek to change the pact to its advantage.

Yet, in such a situation, nothing can stop one party from regarding the treaty as holding little utility. When Russia said it would oppose any modification of the 1972 Anti-Ballistic Missile Treaty as sought by Washington, the United States first unveiled a national missile defense program that openly violated the pact's terms, before President George W. Bush's administration announced America's unilateral withdrawal from the bilateral treaty.[104]

International law recognizes that a party may withdraw from its treaty obligations in the event of substantially changed circumstances. Titled "Fundamental Change of Circumstances," Article 62 of the Vienna Convention states that a party may invoke a fundamental change of circumstances to terminate or withdraw from the treaty, even absent an express provision, if such change undermined the "essential basis" of consent by the party concerned and if it radically transformed the extent of obligations still to be performed.[105] For example, the use of state-reared terrorist groups by one party against another state with which it has a water pact could possibly be invoked by the injured party as representing fundamentally changed circumstances that both undermine the essential basis of its original consent and significantly alter the balance of obligations, thus constituting reasonable grounds for its unilateral withdrawal from the treaty.

The ICJ has upheld the applicability of the general principle that a treaty may be dissolved by reason of a fundamental change of circumstances. In a fisheries jurisdiction case between West Germany and Iceland in the 1970s, the court, while finding no fundamental change of circumstances that radically increased the burden of obligations on Iceland (which was not represented in the proceedings), endorsed the doctrine of changed circumstances in the following words:

> International law admits that a fundamental change in the circumstances which determined a party to accept a treaty, if it has resulted in a radical transformation of the extent of the obligations imposed by it, may, under certain conditions, afford the party affected a ground for invoking the termination or suspension of the treaty. This principle, and the conditions and exceptions to which it is subject, have been embodied in Article 62 of the Vienna Convention on the Law of Treaties, which may in many respects be considered as a codification of existing customary law on the subject of termination of a treaty relationship on account

of change of circumstances. One of the basic requirements embodied in that Article is that the change of circumstances must have been a fundamental one.[106]

Significantly, Article 62 of the Vienna Convention states that if "a party may invoke a fundamental change of circumstances as a ground for terminating or withdrawing from a treaty, it may also invoke the change as a ground for suspending the operation of the treaty." So the withdrawing party could suspend the operation of a water treaty from the day it announced its withdrawal, although the withdrawal itself would take effect only once the notice period was over. In case a treaty contains no provision regarding its termination, a "party shall give not less than 12 months' notice of its intention to denounce or withdraw from a treaty," according to the Vienna Convention's Article 56.

In this light, the future of any water treaty really hinges on its parties' readiness to pursue good neighborly relations and to refrain from actions that breach international norms. One party, for example, cannot expect to continuously derive benefits from a binational treaty by holding the other party to its terms while it blithely flouts fundamental canons of interstate relations. A scofflaw cannot simply compel the other party to abide by the treaty, especially if it refuses to reform its conduct.

Investing in Water Peace

To help build basin peace and alleviate water insecurity, joint hydroengineering projects between two or more riparian states can go a long way to advance mutually beneficial cooperation, including by developing new sources of supply and protecting existing ones. Such projects of peace, aimed at deriving joint benefits from hydrological science and engineering, are important for building political trust and shared interests in the effective management of common water resources. Only by developing ground-level political cooperation among co-basin states can the resources of transnational watercourses be sustainably harnessed and water shortages be potentially eased through national and binational projects.

The apparent difficulties of getting competing nations to cooperate with each other on joint projects or to insulate the subregional water dynamics from the larger geopolitics notwithstanding, innovative ideas and schemes are vital to address the growing water insecurities. Transboundary waters must be dealt with in their totality, including through joint research and development, pollution control, conservation, and mutual assistance in the alleviation of water shortages.

Cross-border investments in water infrastructure can serve as an important means of optimizing basin-level collaboration and moderating competition. Such investments, in the form of joint ventures involving private- and public-sector players, may also help to plug water losses by opening commercial avenues to upgrade run-down infrastructure. An open flow of transboundary investment will help bring new technologies and best practices to boost water efficiency and productivity and to tap nontraditional sources of water supply.

By fostering interdependence and facilitating a resolution of political differences, collaborative water projects are likely to yield significant peace dividends. Tangible benefits from joint investment and resource development could even spur political cooperation in other fields. Cross-border water investments and joint projects, by helping to reduce the risks of unilateral action by any side, may actually contribute to building regional crisis stability.

A fragmented or compartmentalized approach to water issues—at the national, basin, and regional levels—must be shed in favor of an institutionalized, holistic framework that integrates the agricultural, manufacturing, energy, household, and environmental aspects of water withdrawals and seeks to shield water resources from degradation and depletion. Co-basin states, for example, can productively conjoin the development and management of both underground and surface water resources. Even simple water sharing between co-riparians, underpinned by stable interstate relations, serves as an important conflict-avoidance arrangement.

The global water challenges, although formidable, should serve not as a pretext for inaction or a ground for pessimism but as a call for urgent action by moving beyond business as usual and recognizing the link between water security and regional security. Action at the regional, national, and substate levels must co-opt principal stakeholders through transparent institutions that holistically combine surface water, groundwater, and the ecosystems through which water flows in a policy framework. Such efforts, while seeking to sustainably optimize resource availability, could make pollution-control policies more effective by enforcing the "polluter pays" principle and implementing liability and compensation rules for economic and public-health impacts from water degradation.

At a time when the world is at a defining moment in its history, the process has begun, even if slowly, to reform several existing international institutions and explore setting up new institutions to deal with the emerging challenges. Some have called for the formation of a United Nations Environment Agency—backed by sufficient legal mandate and financial resources—to coordinate a concerted international response to the growing water and environmental stresses. International assistance to the least-developed economies to help them cope with the water, food, energy, and environmental challenges must extend to technology transfer and capacity building.

Although weak international institutions may have dimmed hopes of more effective action on the freshwater front, the initiatives taken in the past couple of decades have helped raise global awareness on the threat to the already fragile water resources and the need for their integrated management. The actions include the codification of international water principles, the establishment of international water organizations, the adoption of nonbinding multilateral declarations, the holding of an annual World Water Day, and the release of a World Water Development Report every three years at the World Water Forum by UN-Water, the new name for an odd acronym—ACC-SCWR, or the Subcommittee on Water Resources of the United Nations Agency Coordinating Committee. A lot more international work needs to be done, including translating agreed principles on water cooperation and sustainable management into action.

A central aim of international scientific cooperation on water resources must be the development of more cost- and energy-efficient clean-water technologies. A global public-private alliance initiative is needed to help fund international research into more advanced water technologies.

Such research, utilizing the unique chemical and physical properties of nanoparticles and nanofibers, could build on existing technological progress on separation and purification methods. Advances in membrane bioreactors—actively employed for municipal and industrial wastewater treatment—and materials substitution promise new technological breakthroughs. Recent developments in membrane materials, including low-cost filters, membrane modification, and dynamic membranes, actually signal a growing role for membrane technology to produce quality water.[107]

New clean-water technologies, by permitting low-cost, energy-efficient treatment of wastewater, brackish water, seawater, and polluted freshwater, could help to significantly expand the availability of water resources and create sustainable and diverse sources of water supply. Different qualities of water could be reserved for different purposes—for drinking, for irrigation, for thermoelectric cooling, and for manufacturing. Innovations that help to significantly boost water-use efficiency in industry and agriculture would offer another means to address the gap between water demand and supply.

Developing sustainable and affordable supplies of clean water is an issue tied to the future of international peace and stability. Success on this front holds the key to producing adequate food to feed the world's growing population and meeting the soaring global electricity demand.

In an age of diminishing water security, the world indeed needs a second green revolution to meet the projected 70 percent increase in food demand by 2050. Thanks to the first green revolution, food production rose dramatically in the period since 1960, with the largest production gains concentrated in the developed world, which was quick to apply the latest technologies for high yields. Irrigation expansion and the green revolution also transformed agriculture in many developing economies, best illustrated by Asia's rise as an important food-exporting continent.

Yet the problem of malnourishment and occasional famines in the developing world—the scene of much of the population growth—and the fast-increasing international food demand underscore the imperative for technological breakthroughs to usher in a new green revolution. Such is the inequality of international food distribution at present that the developed world, with barely a quarter of the global population, consumes almost half of the world's agricultural products, partly because it converts a lot of its grain into meat. Significant increases in global food production in an era of water-supply constraints demand net water-productivity gains through higher crop yields per unit of irrigation water applied—that is, increasing "crop per drop." They also demand the use of superior germplasm or other new seed developments, more efficient utilization of nitrogen fertilizers and agrochemicals, and innovative crop-protection methods.

But unlike the first green revolution that set in motion the degradation of water and land resources, the new green revolution must be a doubly green revolution—a "green green revolution," anchored in the principles and practices of environmental sustainability.[108] The challenge, after all, is to grow more food with less water, less land, less energy, and less carbon and nitrogen in the atmosphere. Whereas the overuse of nitrogen fertilizers has contributed to water pollution, increased soil acidity, and atmospheric deposition of nitrogen, the acceleration of urbanization, coupled with land degradation, is reducing global farmland by 5 to 7 million hectares annually.[109]

More fundamentally, there is much ground to cover between rhetoric and reality on the imperative to break the link between economic development and the depletion of natural resources. The concept of development is broad based and signifies far more than GDP growth rate. The benchmarks of comprehensive development include protection of the biological and physical environment, public health, low income disparity, social equity, resource conservation, and environmental sustainability. National progress must be measured not merely in GDP terms but by how well human needs are being met through other measures as well, including resource efficiency and the protection of natural ecosystems.

The world has made the mistake of overemphasizing GDP growth—which demands more and more consumption—to the virtual exclusion in many countries of other indices of development. As a consequence, environmental degradation and resource depletion are turning into serious problems, even as many societies are becoming more unequal and facing popular discontent.

If the rest of the world were to aspire to U.S. levels of consumption, the world would likely confront an environmental catastrophe. The imprudent overuse of water, energy, land, minerals, and biological resources is already exacting increasing environmental costs and contributing to climate change. The growing gap between near-term development objectives and long-term human aspirations means that the costs of development are being shifted onto future generations.

There is hope, however, that the world will live in peace over water and other natural resources. The human mind is innovative. History is a testament to how human civilization has successfully overcome dire situations, including through a combination of political will, new technological methods, and an enduring capacity to adapt. Water shortages in the past have served as an engine of human ingenuity, motivating and compelling societies to find ways to alleviate the scarcity, including by controlling environmentally harmful traditional practices.

Water insecurity—as a pressing cross-sectoral challenge demanding cooperative politics, integrated policies, adaptation, and new technological solutions—is now challenging the human ability to innovate and live in harmony with nature. Market mechanisms, public-private partnerships, new practices and technologies, conservation, and astute water management can help to advance adaptation and affordable solutions, thereby potentially opening the path to a more sustainable and peaceful world.

Appendix A

Web Links to International Water Norms

The 1966 Helsinki Rules on the Uses of the Waters of International Rivers; location of the text: Max Planck Institute, http://goo.gl/WmTTU.

The 1992 Convention on the Protection and Use of Transboundary Watercourses and International Lakes (the so-called Helsinki I Convention); location of the text: United Nations Economic Commission for Europe, http://goo.gl/0ZVjx.

The 1997 United Nations Convention on the Law of the Non-Navigational Uses of International Watercourses; location of the text: United Nations Treaty Collection, http://goo.gl/ZNmNV.

The 2004 Berlin Rules on Water Resources; location of the text: International Water Law Project, http://goo.gl/UQOa0.

The 2008 Draft Articles on the Law of Transboundary Aquifers; location of the text: Internationally Shared Aquifer Resources Management initiative, http://goo.gl/Iqo6O.

Note: None of these rules, conventions, or articles is legally binding at the global level. The 1997 United Nations Convention has not yet come into force, while the 1992 Helsinki I Convention was designed with a regional scope by the United Nations Economic Commission for Europe (UNECE), although the convention was amended in 2003 to allow accession by countries outside the UNECE region. The 1966 Helsinki Rules and the 2004 Berlin Rules were framed by the nongovernmental International Law Association. The fate of the 2008 groundwater-related draft articles remains uncertain.

Appendix B

Genuine Intercountry Water-Sharing Agreements Currently in Effect

Date	Basin	Signatories	Accord
January 11, 1909	Great Lakes–St. Lawrence system and other shared watercourses	U.S., Great Britain (on behalf of Canada)	The Boundary Waters Treaty defines principles to demarcate joint surface-water resources and establishes the International Joint Commission to help resolve or prevent disputes.
November 14, 1944	Colorado	U.S., Mexico	The treaty, with effect from 1950, guarantees Mexico 1.85 billion cubic meters. The then-existing International Boundary Commission is turned into a new International Boundary and Water Commission, with a mandate extending to other shared surface waters.
August 11, 1957	Araks, Atrek	USSR, Iran	Agreement for joint utilization (on a 50–50 sharing basis) of the frontier sections of the Araks and Atrek rivers for irrigation and hydropower.
November 8, 1959	Nile	Egypt, Sudan	Agreement on the "full utilization" of the Nile waters says Egypt's "acquired right" is 48 billion cubic meters per year, as measured at Aswan, and Sudan's is 4 billion cubic meters per year. The two parties to equally share increased river yield resulting from new projects to cut Nile water losses in the swamps in Sudan.
December 4, 1959	Gandak	Nepal, India	Agreement on the Gandak provides for India to construct western and eastern canals to supply irrigation water to Nepal. India secures similar share, plus one-third of electricity output from joint project.
September 19, 1960	Indus	India, Pakistan, World Bank	Treaty partitions the six rivers of the Indus system flowing downstream from India, giving Pakistan the largest three rivers (or 80.52 percent of the basin waters aggregating to 167.2 billion cubic meters yearly) and India the small rivers making up 19.48 percent of the waters. It remains the most generous sharing treaty in any basin.
December 15, 1971	Waters of frontier rivers and lakes	Finland, Sweden	Agreement states that in "frontier rivers with branches, each party shall be entitled to an equal share of the water volume."

Date	Water body	Parties	Description
December 26, 1975	Frontier rivers	Iraq, Iran	Agreement on watercourses that follow or intersect the boundary states that Bnava Suta, Qurahtu, and Gangir river waters will be divided equally, while division of Alvend, Kanjan Cham, Tib (Mehmeh), and Duverij waters will be based on a 1914 arrangement and customary pattern.
October 24, 1986	Senqu/Orange River	South Africa, Lesotho	Treaty is signed to increasingly channel Senqu/Orange River waters to South Africa (with the diversion to reach 2.2 billion cubic meters per year by 2020 or later) through the new Lesotho Highlands Water Project. Lesotho, in return, to get hydropower.
April 17, 1989	Euphrates	Syria, Iraq	Iraq's Law No. 14 of 1990 ratifies the joint minutes concerning the provisional division of the Euphrates River waters (Syria to limit its utilization to 42 percent of the Euphrates waters and keep the remaining 58 percent for downstream Iraq).
March 13, 1992	Komati River Basin	South Africa, Swaziland	Treaty is signed giving upstream South Africa the larger water share of the Komati River Basin. Treaty sets up an authority to build and maintain the Driekoppes Dam in South Africa and the Maguga Dam in Swaziland.
September 20, 1994	Al-Asi/Orontes	Lebanon, Syria	Bilateral agreement on the division of the waters of the Al-Asi/Orontes (Syria gets the major share; Lebanon's share quantified at 80 million cubic meters per year if the river's flow inside Lebanon is 400 million cubic meters or more, but if the flow falls below 400 million cubic meters, Lebanon's share is to be proportionately reduced).
October 26, 1994	Jordan River Basin, freshwater springs, saline springs	Israel, Jordan	Bilateral peace treaty carries Annex II titled, "Water and Related Matters." Under it, Israel to pump 45 million cubic meters per year from Yarmouk and, in return, transfer to Jordan 20 million cubic meters from Jordan River every summer. Jordan gets to store 20 million cubic meters and secures access to 10 million cubic meters of desalinated spring waters.
September 28, 1995	Groundwater	Israel, Palestine Liberation Organization	Interim Agreement on the West Bank and the Gaza Strip establishes a Joint Water Committee, with Israel agreeing to grant West Bank Palestinians access to 28.6 million cubic meters of water per year on an interim basis.

(continued)

Date	Basin	Signatories	Accord
December 22, 1995	Mesta/Nestos	Bulgaria, Greece	Agreement on the use of the waters of the Mesta/Nestos River obligates Bulgaria to leave 29 percent of the annual flow for Greece.
February 12, 1996	Mahakali	India, Nepal	Treaty on the integrated development of the Mahakali River provides for equitable water sharing but with Nepal's requirements to be "given prime consideration." Nepal also secures specified allocations from India's Sarada and Tanakpur Barrages.
December 12, 1996	Ganges	India, Bangladesh	Treaty on sharing of the Ganges River waters divides downstream flows equally in the March–May dry season, with Bangladesh and India guaranteed a minimum of 35,000 cusecs each. In the other seasons when total flows surpass 80,000 cusecs, Bangladesh's share is larger than India's.
May 6, 1997	Quarai/Cuareim	Uruguay, Brazil	River flows equally split under complementary settlement to the agreement of cooperation for the use of natural resources and basin development.
April 20, 2002	El-Kaber	Syria, Lebanon	Agreement is reached on sharing the waters of the small border river El-Kaber and jointly building a dam (60 percent of the river flow, estimated at 76 million cubic meters per year, reserved for Syria and 40 percent for Lebanon).

Note: This list includes only intercountry agreements with specific allocations or a formula dividing the waters of shared watercourses that are recognized by their parties as still mutually binding. It excludes project-specific accords lacking a defined watercourse-sharing arrangement.

Notes

INTRODUCTION: OUR MOST
PRECIOUS RESOURCE UNDER THREAT

1. See United Nations Educational, Scientific and Cultural Organization (UNESCO), *Water Security and Peace: A Synthesis of Studies Prepared Under the PCCP-Water for Peace Process*, comp. William J. Cosgrove (Paris: UNESCO, 2003); Peter Gleick, "Water and Conflict: Freshwater Resources and International Security," *International Security* 18, no. 1 (Summer 1993); and T. Le-Huu, "Potential Water Conflicts and Sustainable Management of International Water Resources Systems," *Water Resources Journal of United Nations Economic and Social Commission for Asia and the Pacific* (2001), Paper No. ST/ESCAP/SER.C/210.

2. Among the analysts to warn of water wars are Fred Pearce, *When the Rivers Run Dry: Water—the Defining Crisis of the Twenty-First Century* (Boston: Beacon Press, 2007); Diane Raines Ward, *Water Wars: Drought, Flood, Folly, and the Politics of Thirst* (New York: Riverhead Books, 2002); Brahma Chellaney, *Water: Asia's New Battleground* (Washington, DC: Georgetown University Press, 2011); John K. Cooley, "The War over Water," *Foreign Policy*, no. 54 (Spring 1984): 3–26; John Bulloch and Adel Darwish, *Water Wars: Coming Conflicts in the Middle East* (London: Victor Gollancz, 1993); Joyce R. Starr, "Water Wars," *Foreign Policy*, no. 82 (Spring 1991): 17–36; Thomas F. Homer-Dixon, *Environment, Scarcity, and Violence* (Princeton, NJ: Princeton University Press, 1999); Arnon Soffer, *Rivers of Fire: The Conflict over Water in the Middle East* (Lanham, MD: Rowman & Little-field, 1999); and Vandana Shiva, *Water Wars: Privatization, Pollution, and Profit* (Cambridge, MA: South End Press, 2002).

3. United Nations, *Managing Water under Uncertainty and Risk*, vol. 1 of the World Water Development Report 4 (Paris: United Nations Educational, Scientific and Cultural Organization and UN-Water, March 2012).

4. United Nations Environment Program, *Environmental Degradation Triggering Tensions and Conflict in Sudan* (Nairobi: UNEP, 2007).

5. Sarah C Walpole, David Prieto-Merino, Phil Edwards, John Cleland, Gretchen Stevens, and Ian Roberts, "The Weight of Nations: An Estimation of Adult Human Biomass," *BMC Public Health* 12, no. 439 (2012); World Health Organization, "Obesity and Overweight," Fact-Sheet No. 311 (Paris: WHO, May 2012); Phil Edwards and Ian Roberts, "Population Adiposity and Climate Change," *International Journal of Epidemiology* 38, no. 4 (April 2009).

6. U.S. National Intelligence, *Global Water Security*, Intelligence Community Assessment ICA 2012-08, February 2, 2012 (Washington, DC: Office of the Director of National Intelligence, March 22, 2012), 3–4.

7. UNESCO, *Water Security and Peace*, 50.

8. See Lin Noueihed and Alex Warren, *The Battle for the Arab Spring: Revolution, Counter-Revolution and the Making of a New Era* (New Haven, CT: Yale University Press, 2012); John R. Bradley, *After the Arab Spring: How Islamists Hijacked the Middle East Revolts* (New York: Palgrave Macmillan, 2012); James L. Gelvin, *The Arab Uprisings: What Everyone Needs to Know* (New York: Oxford University Press, 2012); and Marc Lynch, *The Arab Uprising: The Unfinished Revolutions of the New Middle East* (New York: PublicAffairs, 2012).

9. The steam engine and the water turbine were invented a century apart, but each invention marked a turning point in history. While American Lester Allan Pelton's efficient form of impulse water turbine in 1878 opened the path to hydropower generation, James Watt's improved steam engine earlier "helped transform Britain into the world's dominant economy with a steam-and-iron navy that lorded over a colonial empire spanning a quarter of the globe." Steven Solomon, *Water: The Epic Struggle for Wealth, Power, and Civilization* (New York: HarperCollins, 2010), prologue.

10. Food and Agriculture Organization, *FAO Statistical Yearbook 2012* (Rome: FAO, 2012).

11. Donella H. Meadows, Dennis L. Meadows, Jorgen Randers, and William W. Behrens III, *The Limits to Growth: A Report for the Club of Rome's Project on the Predicament of Mankind* (New York: Universe Books, 1972), "Conclusions." *The Limits to Growth* was published barely four years after another book, *The Population Bomb*, predicted that population growth would outstrip food production, leaving "hundreds of millions of people" to starve to death as early as the 1970s. John R. Ehrlich, *The Population Bomb* (Cutchogue, NY: Buccaneer Books, 1968).

12. Richard Dobbs, Jeremy Oppenheim, Fraser Thompson, Marcel Brinkman, and Marc Zornes, *Resource Revolution: Meeting the World's Energy, Materials, Food, and Water Needs* (McKinsey Global Institute and McKinsey Sustainability & Resource Productivity Practice, November 2011), 1–4.

13. World Bank and infoDev, *Maximizing Mobile* (Washington, DC: World Bank, July 2012); United Nations, *Managing Water under Uncertainty and Risk*.

14. Examples of what grassroots opposition can do include the multiyear delays in the plans of the Luxembourg-based ArcelorMittal and Posco of South Korea to set up large multibillion-dollar steel plants in India's water-stressed iron ore belt.

15. Sanctity of borders, paradoxically, has allowed the emergence of weak states, whose internal wars spill over and create wider regional tensions and insecurities.

Boaz Atzili, *Good Fences, Bad Neighbors: Border Fixity and International Conflict* (Chicago: University of Chicago Press, 2012).

16. Kevin Watkins and Anders Berntell, "A Global Problem: How to Avoid War over Water," *International Herald Tribune*, August 23, 2006.

CHAPTER 1: THE SPECTER OF WATER WARS

1. Citigroup Global Markets, *Global Themes Strategy*, July 20, 2011, 18–24.

2. U.S. National Intelligence, *Global Water Security*, Intelligence Community Assessment ICA 2012-08, February 2, 2012 (Washington, DC: Office of the Director of National Intelligence, March 22, 2012), 4.

3. Catherine B. Asher, *The New Cambridge History of India: Architecture of Mughal India* (Cambridge, UK: Cambridge University Press, 1992); Percy Brown, *Indian Architecture* (Bombay: Taraporevala, 1959); Christopher Tadgell, *The History of Architecture in India* (London: Phaidon Press, 1994).

4. Christopher Boucek, "Yemen: Avoiding a Downward Spiral," Carnegie Papers, No. 102 (Washington, DC: Carnegie Endowment for International Peace, September 2009), 6.

5. Food and Agriculture Organization, *Irrigation in Africa in Figures—Aquastat Survey 2005*, Water Report 29 (Rome: FAO, 2006), sec. "Libyan Arab Jamahiriya."

6. Sharlene Leurig, *The Ripple Effect: Water Risk in the Municipal Bond Market* (Boston: Ceres and New York: Water Asset Management), 44–46; Tim P. Barnett and David W. Pierce, "When Will Lake Mead Go Dry?" *Water Resources Research* 44 (2008), doi:10.1029/2007WR006704.

7. Natural Resources Defense Council, *Climate Change, Water, and Risk: Current Water Demands Are Not Sustainable* (New York: NDRC, July 2010), 4. Also see Randy Stapilus, *The Water Gates: Water Rights, Water Wars in the 50 States* (Carlton, OR: Ridenbaugh Press, 2010); Marc Reisner, *Cadillac Desert: The American West and Its Disappearing Water* (New York: Penguin, 1993).

8. National Research Council of the National Academies, *Water Reuse: Potential for Expanding the Nation's Water Supply through Reuse of Municipal Wastewater* (Washington, DC: National Academies Press, 2012), 4.

9. Peter Annin, *The Great Lakes Water Wars* (Washington, DC: Island Press, 2009).

10. For details, see Jared Diamond, *Collapse: How Societies Choose to Fail or Succeed* (New York: Viking, 2011), 157–177. According to one study, climate change also likely played a role in the collapse of the Maya civilization. Douglas J. Kennett, Sebastian F. M. Breitenbach, Valorie V. Aquino, Yemane Asmerom, Jaime Awe, James U. L. Baldini, Patrick Bartlein, Brendan J. Culleton, Claire Ebert, Christopher Jazwa, Martha J. Macri, Norbert Marwan, Victor Polyak, Keith M. Prufer, Harriet E. Ridley, Harald Sodemann, Bruce Winterhalder, and Gerald H. Haug, "Development and Disintegration of Maya Political Systems in Response to Climate Change," *Science* 338, no. 6108 (November 9, 2012): 788–791, doi:10.1126/science.1226299.

11. Sharon Squassoni, "Hanging Questions," *IAEA Bulletin* 50, no. 2 (May 2009): 53–54.

12. United Nations secretary general Kofi Annan, question and answer session after statement at the Federation of Indian Chambers of Commerce and Industry in New Delhi, March 15, 2001 (New York: United Nations, SG/SM/7742).

13. United Nations secretary general Kofi Annan, message on occasion of World Water Day 2002 (New York: UN, March 22, 2002).

14. Cited in United Nations, *Water in a Changing World*, United Nations World Water Development Report (Colombella, Italy: United Nations World Water Assessment Program, March 2009).

15. Michael Blackhurst, Chris Hendrickson, and Jordi Vidal, "Direct and Indirect Water Withdrawals for U.S. Industrial Sectors," *Environmental Science & Technology* 44, no. 6 (2010): 2126–2130, doi:10.1021/es903147k.

16. The term "water stress" was popularized by Swedish hydrologist Malin Falkenmark, who in 1989 developed the Water Stress Index, which divided the volume of available freshwater resources by the national population and factored in food-production water requirements. A per capita annual water availability of 1,000 to 1,666 cubic meters made a country water stressed; availability below that level indicated chronically water-stressed or water-scarce conditions. Because Falkenmark's index does not account for spatial variability of water resources within nation-states, it cannot capture the intrastate distinctions between water-deficient and water-rich areas—important in large states like the United States and China. Leif Ohlsson developed a rival Social Water Stress Index to introduce social and political "adaptive capacity," but that index too failed to take into account spatial variability. Peter H. Gleick, for his part, has identified four broad indices of a nation's vulnerability—external water-dependency ratio, per capita water availability, ratio of water supply to demand, and dependence on hydropower as a fraction of total national electricity supply.

17. Food and Agriculture Organization of the United Nations (FAO), Aquastat online data, http://goo.gl/83tfb.

18. S. Siebert, J. Burke, J. M. Faures, K. Frenken, J. Hoogeveen, P. Doll, and F. T. Portmann, Groundwater Use for Irrigation—A Global Inventory, *Hydrology and Earth System Sciences Discussions* 7 (2010): 3977–4021, doi:10.5194/hessd-7-3977-2010.

19. See I. S. Zekster and L. G. Everett, eds., *Groundwater Resources of the World and Their Use*, IHP-VI, Series on Groundwater No. 6 (Paris: United Nations Educational, Scientific and Cultural Organization, 2004).

20. Groundwater extraction jumped from 312 to 734 billion cubic meters per year between 1960 and 2000. Yoshihide Wada, Ludovicus P. H. van Beek, Cheryl M. van Kempen, Josef W. T. M. Reckman, Slavek Vasak, and Marc F. P. Bierkens, "Global Depletion of Groundwater Resources," *Geophysical Research Letters* 37 (2010), doi:10.1029/2010GL044571.

21. Joan F. Kenny, Nancy L. Barber, Susan S. Hutson, Kristin S. Linsey, John K. Lovelace, and Molly A. Maupin, *Estimated Use of Water in the United States in 2005*, U.S. Geological Survey Circular 1344 (Reston, VA: U.S. Geological Survey, 2009), 1.

22. FAO, Aquastat online data; UN, *Water in a Changing World.*

23. World Bank and infoDev, *Maximizing Mobile* (Washington, DC: World Bank, July 2012).

24. See European Environment Agency, *The European Environment: State and Outlook 2010* (Copenhagen: EEA, December 2010).

25. Carbon Disclosure Project, *CDP Water Disclosure 2010 Global Report* (London: Carbon Disclosure Project, 2010), 5.

26. The 2030 Water Resources Group, *Charting Our Water Future*, Barilla Group, Coca-Cola Company, International Finance Corporation, McKinsey & Company, Nestlé S.A., New Holland Agriculture, SABMiller, Standard Chartered Bank, and Syngenta AG (2030 Water Resources Group, 2009), 40–44.

27. "Warning about the Water," MarketWatch.com, February 28, 2009.

28. Bridget R. Scanlon, Claudia C. Faunt, Laurent Longuevergne, Robert C. Reedy, William M. Alley, Virginia L. McGuire, and Peter B. McMahon, "Groundwater Depletion and Sustainability of Irrigation in the U.S. High Plains and Central Valley," *Proceedings of the National Academy of Sciences* 109, no. 24 (2012): 9320–9325.

29. Global International Waters Assessment, *Challenges to International Waters: Regional Assessments in a Global Perspective* (Kalmar, Sweden: GIWA, 2007); Bernadette McDonald, ed., *Whose Water Is It? The Unquenchable Thirst of a Water-Hungry World* (Washington, DC: National Geographic Society Press, 2003); Julie Stauffer, *The Freshwater Crisis: Constructing Solutions to Freshwater Pollution* (Montreal: Black Rose Books, 1999); Paul Simon, *Tapped Out: The Coming World Crisis in Water and What We Can Do about It* (New York: Welcome Rain Publishers, 1998).

30. United Nations Development Programme, *Human Development Index*, http://hdr.undp.org/en/statistics; Fund for Peace and *Foreign Policy* journal, *The Failed States Index*, http://goo.gl/xx7oT.

31. United States Senate, *Avoiding Water Wars: Water Scarcity and Central Asia's Growing Importance for Stability in Afghanistan and Pakistan*, Majority Staff Report, Prepared for the Use of the Committee on Foreign Relations (Washington, DC: U.S. Government Printing Office, February 22, 2011), 1.

32. See, for example, Victoria Clark, *Yemen: Dancing on the Heads of Snakes* (New Haven, CT: Yale University Press, 2011).

33. FAO, Aquastat online data.

34. In 2012, freshwater availability in Iceland was 537,975 cubic meters per person per year; in Guyana 320,053; in Suriname 234,615; in Gabon 110,961; in Canada 86,177; and in Norway 79,024. FAO, Aquastat online data.

35. FAO, Aquastat online data.

36. *The Water Poverty Index: International Comparisons* (Wallingford, UK: Centre for Ecology and Hydrology, Natural Environment Research Council, 2002).

37. U.S. Geological Survey, "Water Science for Schools," https://bitly.com/bV3lIC; United Nations Development Programme, *Human Development Report 2006—Beyond Scarcity: Power, Poverty, and the Global Water Crises* (New York: UNDP, 2006).

38. United Nations Development Programme, *Human Development Report*; Water Footprint Network, online database, http://goo.gl/T7NXY.

39. The Dublin Statement emerged from the 1992 expert-level International Conference on Water and the Environment in Dublin. Full text of the "Dublin Statement on Water and Sustainable Development," including its four guiding principles, at http://goo.gl/n7wug.

40. United Nations Population Fund, *The State of World Population 2009* (New York: UNFPA, November 2009).

41. Charles Fishman, *The Big Thirst: The Secret Life and Turbulent Future of Water* (New York: Free Press, 2011), 250.

42. United Nations, *World Water Development Report—Water for People, Water for Life* (Oxford, UK: Berghahn Books, 2003), 4.

43. UN-Water, *Status Report on the Application of Integrated Approaches to Water Resources Management* (New York: United Nations, 2012).

44. Article 25 of the Universal Declaration of Human Rights. United Nations, *The Universal Declaration of Human Rights*, adopted by the UN General Assembly on December 10, 1948, http://www.un.org/en/documents/udhr.

45. Food and Agriculture Organization (FAO) of the United Nations, *Water at a Glance*, http://goo.gl/l9tCz.

46. UN-Water and World Health Organization, *GLAAS 2012 Report: The Challenging of Extending and Sustaining Services* (New York: United Nations, April 2012).

47. World Health Organization, *Emerging Issues in Water and Infectious Diseases* (Paris: WHO, 2003).

48. Water for Life, 2005–2015, http://goo.gl/G5cl4.

49. The Millennium Development Goals (MDGs) are eight goals that were to be achieved by 2015. Water and sanitation targets fall in Goal 7, "Ensure Environmental Sustainability." The MDGs were drawn from the Millennium Declaration that was adopted by 189 nations, and signed by 147 heads of state or government, during the UN Millennium Summit in September 2000.

50. *Plan of Implementation of the World Summit on Sustainable Development*, World Summit on Sustainable Development, Johannesburg, August 26–September 4, 2002, http://www.un-documents.net/jburgpln.htm.

51. World Health Organization and United Nations Children's Fund, *Progress on Drinking Water and Sanitation 2012*, report of WHO/UNICEF Joint Monitoring Program for Water Supply and Sanitation (Geneva and New York: WHO and UNICEF, March 2012).

52. The Global Water Partnership was established in 1996 by the World Bank, the United Nations Development Programme, and the Swedish International Development Agency to foster integrated water resources management. See International Network of Basin Organizations and Global Water Partnership, *The Handbook for Integrated Water Resources in Transboundary Basins of Rivers, Lakes, and Aquifers* (Paris: INBO and GWP, March 2012).

53. M. Falkenmark and J. Rockström, "The New Blue and Green Water Paradigm: Breaking New Ground for Water Resources Planning and Management," *Journal of Water Resources Planning and Management*, May/June 2006, 132.

54. Jill Boberg, *Liquid Assets: How Demographic Changes and Water Development Policies Affect Freshwater Resources* (Santa Monica, CA: Rand Corporation, 2005).

55. Peter M. Vitousek, Harold A. Mooney, Jane Lubchenco, and Jerry M. Melillo, "Human Domination of Earth's Ecosystems," *Science* 277, no. 5325 (July 25, 1997): 494–499.

56. Clive Ponting, *A New Green History of the World: The Environment and the Collapse of Great Civilizations* (New York: Penguin, 2007).

57. Vitousek et al., "Human Domination of Earth's Ecosystems."

58. Millennium Ecosystem Assessment, *Ecosystems and Human Wellbeing: A Framework for Assessment* (Washington, DC: Island Press, 2005).

59. Paul M. Kennedy, *The Rise and Fall of the Great Powers: Economic Change and Military Conflict, 1500–2000* (New York: Vintage, 1989), 143–157.

60. Roland H. Worth Jr., *No Choice But War: The United States Embargo against Japan and the Eruption of War in the Pacific* (Jefferson, NC: McFarland, 1995), 100–135.

61. Niall Ferguson, *The War of the World: Twentieth-Century Conflict and the Descent of the West* (New York: Penguin, 2006), xix.

62. Brahma Chellaney, *Water: Asia's New Battleground* (Washington, DC: Georgetown University Press, 2011), 61–67.

63. Samuel P. Huntington, "The Clash of Civilizations?" *Foreign Affairs* 72, no. 3 (Summer 1993): 22–49; Samuel P. Huntington, *Clash of Civilizations and the Remaking of World Order* (New York: Simon & Schuster, 1998).

64. The 1982 World Charter for Nature was adopted by the United Nations General Assembly with 111 votes in favor, 1 against (the United States), and 18 abstentions. Full text in Scott S. Olson, *International Environmental Standards Handbook* (Boca Raton, FL: CRC Press, 1999), 39.

65. Michael T. Klare, *The Race for What's Left: The Global Scramble for the World's Last Resources* (New York: Metropolitan Books, 2012); Michael T. Klare, *Resource Wars: The New Landscape of Global Conflict* (New York: Holt Paperbacks, 2002).

66. Hussein Amery, "Water, War, and Peace in the Middle East: Comments on Peter Beaumont," *Arab World Geographer* 4, no. 1 (2001): 51; Thomas F. Homer-Dixon, *Environment, Scarcity, and Violence* (Princeton, NJ: Princeton University Press, 1999).

67. Philip Andrews-Speed, Raimund Bleischwitz, Geoffrey Kemp, Stacy VanDeveer, Tim Boersma, and Corey Johnson, *The Global Resource Nexus: The Struggles for Land, Energy, Food, Water, and Minerals* (Washington, DC: Transatlantic Academy, May 2012).

68. North Atlantic Treaty Organization, "Environmental Security," http://bit.ly /JuNzR2.

69. Institute for Security Studies, *The New Global Puzzle: What World for the EU in 2025* (Paris: Institute for Security Studies, 2006), 190.

70. Nazli Choucri, "Perspectives on Population and Conflict," in *Multidisciplinary Perspectives on Population and Conflict*, ed. Nazli Choucri (Syracuse, NY: Syracuse University Press, 1984), 23–24.

71. United Nations Population Division, *World Population Prospects: The 2010 Revision* (New York: United Nations, 2011).

72. Margaret C. Hogan, Kyle J. Foreman, Mohsen Naghavi, Stephanie Ahn, Mengru Wang, Susanna Makela, Alan Lopez, Rafael Lozano, and Christopher Murray, "Maternal Mortality for 181 Countries, 1980–2008: A Systematic Analysis of Progress toward Millennium Development Goal 5," *Lancet*, April 12, 2010.

73. UN Population Division, *World Population Prospects.*

74. See UN Population Division, *World Population Prospects*, and U.S. Census Bureau, *Statistical Abstract of the United States: 2012* (Washington, DC: U.S. Census Bureau, 2011), table 1328. The UN Population Division has also put out for 2100 a "high variant" projection (exceeding 15 billion) and a "low variant" (6.2 billion). However, there are no reliable techniques to project long-term population growth.

75. Daniel Wild, Carl-Johan Francke, Pierin Menzli and Urs Schön, *Water: A Market of the Future* (Zurich, Switzerland: Sustainable Asset Management, December 2007), 10.

76. United Nations, *World Urbanization Prospects: The 2011 Revision* (New York: UN, April 2012).

77. UN Population Division, *World Population Prospects.*

78. World Health Organization, *World Health Statistics 2012* (Paris: WHO, 2012), 36.

79. National Heart Lung and Blood Institute, U.S. Department of Health & Public Services, "How Are Overweight and Obesity Diagnosed?" http://bit.ly/MS9hja.

80. Institute of Medicine of National Academies, *Accelerating Progress in Obesity Prevention: Solving the Weight of the Nation* (Washington, DC: Institute of Medicine, May 2012).

81. Phil Edwards and Ian Roberts, "Population Adiposity and Climate Change," *International Journal of Epidemiology* 38, no. 4 (April 2009): 1137–1140.

82. Institute of Medicine, *Accelerating Progress in Obesity Prevention.*

83. Obesity is prevalent across the United States, affecting every group and income level, although some racial and ethnic groups are disproportionately obese.

84. Sarah C. Walpole, David Prieto-Merino, Phil Edwards, John Cleland, Gretchen Stevens, and Ian Roberts, "The Weight of Nations: An Estimation of Adult Human Biomass," *BMC Public Health* 12, no. 439 (2012), doi:10.1186/1471-2458-12-439. According to the study, "One ton of human biomass corresponds to approximately 12 adults in North America and 17 adults in Asia." The average BMI in Japan is 22.9, while in the United States it is 29.7.

85. Paul Harrison and Fred Pearce, *AAAS Atlas of Population & Environment* (Washington, DC: American Association for the Advancement of Science, 2007), "Part 1—Overview: Population and Consumption Trends."

86. UN Population Division, *World Population Prospects.*

87. FAO, Aquastat online data.

88. Bjørn Lomborg, "Environmental Alarmism, Then and Now," *Foreign Affairs* 91, no. 4 (July/August 2012).

89. See United Nations Environment Program (UNEP), *Challenges to International Waters: Regional Assessments in a Global Perspective* (Nairobi: UNEP, 2006);

United Nations Educational, Scientific and Cultural Organization (UNESCO), *Water Security and Peace: A Synthesis of Studies Prepared Under the PCCP–Water for Peace Process*, comp. William J. Cosgrove (Paris: UNESCO's International Hydrological Program, 2003); United Nations, *Water for People, Water for Life*, World Water Development Report (New York: United Nations, 2003); Fekri A. Hassan, Martin Reuss, Julie Trottier, Christoph Bernhardt, Aaron T. Wolf, Jennifer Mohamed-Katerere, and Pieter van der Zaag, *History and Future of Shared Water Resources* (Paris: UNESCO's International Hydrological Program, 2003); and C. W. Sadoff and David Grey, "Cooperation on International Rivers: A Continuum for Securing and Sharing Benefits," *Water International* 30, no. 4 (2005): 420–427.

90. Peter H. Gleick, "Water and Conflict: Freshwater Resources and International Security," *International Security* 18, no. 1 (Summer 1993): 79.

91. Dan Smith and Janani Vivekananda, *A Climate of Conflict: The Links between Climate Change, Peace and War* (London: International Alert, 2007), 13.

92. Richard H. Ullman, "Redefining Security," *International Security* 8, no. 1 (Summer 1983): 153.

93. See, for example, Salman M. A. Salman, and Laurence Boisson de Charzoures, eds., *International Watercourses: Enhancing Cooperation and Managing Conflict* (Washington, DC: World Bank, 1998); Arun P. Elhance, *Hydropolitics in the Third World: Conflict and Cooperation in International River Basins* (Washington, DC: U.S. Institute for Peace, 1999); and Heather L. Beach, Jesse Hammer, J. Joseph Hewitt, Edy Kaufman, Anja Kurki, Joe A. Oppenheimer, and Aaron T. Wolf, *Transboundary Freshwater Dispute Resolution: Theory, Practice, and Annotated References* (New York: United Nations University Press, 2000).

94. Cited in U.S. Senate, *Avoiding Water Wars*, 11.

95. Mikhail Gorbachev and Jean-Michel Severino, "Climate Change Raises Threat of Water Wars," *Japan Times*, June 9, 2007.

96. See Kennedy Graham, ed., *The Planetary Interest: A New Concept for the Global Age* (London: UCL Press, 1999).

97. United Nations, "Water without Borders," Backgrounder, Water for Life 2005–2015 project (New York: United Nations), 1.

98. Food and Agriculture Organization (FAO) of the United Nations, "FAO Water," http://www.fao.org/nr/water.

99. To help lend support to the "water peace" school, UNESCO in 2000 launched a program, "From Potential Conflict to Cooperation Potential (PCCP)," which has stressed that water spawns cooperation more than conflict.

100. Wendy Barnaby, "Do Nations Go to War over Water?" *Nature* 458, no. 7236 (March 19, 2009): 282–283.

101. Aaron Wolf, "Water Wars and Water Reality: Conflict and Cooperation along International Waterways," in *Environmental Change, Adaptation and Human Security*, ed. S. C. Lonergan (Berlin: Springer, 1999), 251–266.

102. Kent Hughes Butts, "The Strategic Importance of Water," *Parameters*, Spring 1997, 65–66.

103. See, for example, Arthur H. Westing, ed., *Global Resources and International Conflict: Environmental Factors in Strategic Policy and Action* (New York:

Oxford University Press, 1986); and Norman Myers, *Ultimate Security: The Environmental Basis of Political Stability* (New York: Norton, 1993).

104. See, for example, Paul Ekins, *A New World Order: Grassroots Movements for Global Change* (London: Routledge, 1992), 24–44.

105. Aaron T. Wolf, Annika Kramer, Alexander Carius, and Geoffrey D. Dabelko, "Water Can Be a Pathway to Peace, Not War," a brief (Washington, DC: Worldwatch Institute, June 1, 2005).

106. The Indus River case cited by one author to argue that "cooperation is water rational" actually undercuts her thesis because Pakistan, far from being at peace in this basin, has waged a protracted war by proxy against the dominant riparian, India, with cross-border terrorism being publicly justified by militant Pakistani elements in the name of national water interests. Undala Z. Alam, "Questioning the Water Wars Rationale: A Case Study of the Indus Waters Treaty," *Geographical Journal* 168, no. 4 (December 2002): 341–353.

107. Martin Reuss, "Historical Explanation and Water Issues," in *History and Future of Shared Water Resources*, by Fekri A. Hassan, Martin Reuss, Julie Trottier, Christoph Bernhardt, Aaron T. Wolf, Jennifer Mohamed-Katerere, and Pieter van der Zaag (Paris: UNESCO's International Hydrological Program, 2003), 12.

108. U.S. National Intelligence Council, *Global Trends 2025: A Transformed World* (Washington, DC: US National Intelligence Council, 2008), 51.

109. National Intelligence Council, *Global Trends 2025*.

110. Thomas F. Homer-Dixon, "Environmental Scarcities and Violent Conflict: Evidence from Cases," *International Security* 19, no. 1 (Summer 1994): 5.

111. United Nations, *UN-Water Factsheet* (New York: United Nations Department of Economic and Social Affairs, 2009).

112. Meredith A. Giordano and Aaron T. Wolf, "Transboundary Water Treaties," Science and Issues, *Water Encyclopedia*, http://goo.gl/5Wkq8.

113. U.S. Senate, *Avoiding Water Wars*, 7.

114. Article I of Annex II on "Water and Related Matters," in the 1994 Treaty of Peace between the State of Israel and the Hashemite Kingdom of Jordan, http://bit.ly/afYao2.

115. Pakistan Ministry of Water and Power, *Pakistan Water Sector Strategy*, Detailed Strategy Formulation, vol. 4 (Islamabad: Ministry of Water and Power, October 2002), 82; Food and Agriculture Organization, "Country Profile: Pakistan (2010)," http://goo.gl/xaEHx.

116. National Geographic Education, "Germ-Killing Ganges," encyclopedic entry, http://goo.gl/C9GNn.

117. Ismail Serageldin, "Water: Conflicts Set to Arise within as Well as between States," *Nature* 459, no. 7244 (May 14, 2009): 163.

118. Figure cited in Food and Agriculture Organization, "Country Profile: Lesotho," Aquastat online data.

119. Pakistani military ruler Mohammad Ayub Khan subsequently published his autobiography. But it did not cover the 1965 war. Yet it provided useful insights on how Pakistan had linked the Kashmir territorial dispute to water. Mohammad Ayub Khan, *Friends Not Masters: A Political Autobiography* (Oxford, UK: Clarendon Press, 1967).

120. "Operation Gibraltar" was detected by India on August 5, 1965. It was, however, not until early September 1965 that Pakistan declared open war, believing it was a "now or never" opportunity for it to wrest control of Indian Kashmir at a time when India—still reeling from its 1962 humiliating rout by China—was in dire straits, with domestic political discord and food shortages growing.

121. See, for example, Mark Zeitoun, *Power and Water in the Middle East* (London: I. B. Tauris, 2008); Myers, *Ultimate Security*; and Avi Shlaim, *The Iron Wall: Israel and the Arab World 1948–1998* (London: Penguin, 2001).

122. Ariel Sharon with David Chanoff, *Warrior: An Autobiography* (New York: Simon & Schuster, 2001), 165–167.

123. David Ben-Gurion, "Why I Retired to the Desert," *New York Times Magazine*, March 28, 1954, 47.

124. Levi Eshkol, in fact, was involved in the 1937 founding of the Mekorot water company and served as its chief executive until 1951.

125. Shlaim, *The Iron Wall*, 218–282.

126. Efraim Inbar, "Israeli Control of the Golan Heights: High Strategic and Moral Ground for Israel," Mideast Security and Policy Studies Paper No. 90, Begin-Sadat Center for Strategic Studies, Bar-Ilan University, September 2011, 12.

127. Food and Agriculture Organization, "Israel," Aquastat, http://goo.gl/VoOeX.

128. Lenard Milich and Robert G. Varady, "Openness, Sustainability and Public Participation in Transboundary River-Basin Institutions—Part I: The Scientific-Technical Paradigm of River Basin Management," *Arid Lands Newsletter*, no. 44 (Fall–Winter 1998); Sharif S. Elmusa, *Water Conflict: Economics, Politics, Law and Palestinian-Israeli Water Resources* (Washington, DC: Institute for Palestine Studies, 1998); Mostafa Dolatyar and Tim S. Gray, *Water Politics in the Middle East: A Context for Conflict or Cooperation?* (New York: Palgrave Macmillan, 1999).

129. Food and Agriculture Organization, *Irrigation in the Middle East Region in Figures—Aquastat Survey 2008*, Water Report 34 (Rome: FAO, 2009), sec. "Israel."

130. Kim Murphy, "Old Feud over Lebanese River Takes New Turn," *Los Angeles Times*, August 10, 2006.

131. See, for example, Miriam R. Lowi, "Bridging the Divide: Transboundary Resource Disputes and the Case of West Bank Water," *International Security* 18, no. 1 (Summer 1993): 113–138; and Joseph W. Dellapenna, *Middle East Water: The Potential and Limits of Law* (The Hague: Kluwer Press, 2002).

132. Adel Darwish, "Middle East Water Wars," BBC online analysis, May 30, 2003.

133. Masahiro Murakami, *Managing Water for Peace in the Middle East: Alternative Strategies* (Tokyo: United Nations University Press, 1995), sec. 2.5, "The Jordan River."

134. John C. Cooley, "The War over Water," *Foreign Policy*, no. 54 (Spring 1984): 3.

135. J. Eliseo da Rosa, "Economics, Politics, and Hydroelectric Power: The Parana River Basin," *Latin American Research Review* 18, no. 3 (1983): 77–107; Andrew Nickson, "Brazilian Colonization of the Eastern Border Region of Paraguay," *Journal of Latin American Studies* 13 (1981): 111–131.

136. "Water Dispute May Trigger Indo-Pak War: Assef," *The Nation* (Pakistan), January 3, 2010, http://ow.ly/SivD.

137. See, for example, Marq de Villiers, *Water: Is the World's Water Running Out?* (Boston: First Mariner Books, 2001), 216–231.

138. UNESCO, *Water Security and Peace*, 14.

139. Pacific Institute, *Water Conflict Chronology*, http://goo.gl/Xt2zy.

140. Peter Gleick, "Water Conflict Chronology," September 2000, http://goo.gl /gt4VA.

141. Ameur Zemmali, "The Protection of Water in Times of Armed Conflict," *International Review of the Red Cross*, no. 308 (October 31, 1995): 550–564; Frederick Lorenz, *The Protection of Water Facilities under International Law* (Paris: UNESCO's International Hydrological Program, 2003).

142. Full text of the 1982 World Charter for Nature, which proclaimed five "principles of conservation by which all human conduct affecting nature is to be guided and judged," at http://goo.gl/vtahH.

143. Zbigniew W. Kundzewicz and Piotr Kowalczak, "The Potential for Water Conflict is on the Rise," *Nature* 459, no. 7243 (May 7, 2009): 31.

CHAPTER 2: THE POWER OF WATER

1. For a list of drinking-water contaminants, see U.S. Environmental Protection Agency, *List of Contaminants and Their MCLs* (Washington, DC: EPA, undated), http://www.epa.gov/safewater/contaminants/index.html.

2. Mikhail Gorbachev, "People: Water Rights," introductory article as guest editor, *Civilization* (the magazine of the U.S. Library of Congress), October–November 2000.

3. Stefanie Rost, Dieter Gerten, Alberte Bondeau, Wolfgang Lucht, Janine Rohwer, and Sibyll Schaphoff, "Agricultural Green and Blue Water Consumption and Its Influence on the Global Water System," *Water Resources Research* 44 (2008), doi:10.1029/2007WR006331.

4. Jan Lundqvist and Eliel Steen, "The Contribution of Blue Water and Green Water to the Multifunctional Character of Agriculture and Land," in *Cultivating our Futures*, background papers prepared for FAO/Netherlands conference on "The Multifunctional Character of Agriculture and Land" (Rome: FAO, 1999).

5. M. Falkenmark and M. Lannerstad, "Consumptive Water Use to Feed Humanity—Curing a Blind Spot," *Hydrology and Earth System Sciences* 9 (2005): 18–19; M. Falkenmark and J. Rockström, "The New Blue and Green Water Paradigm: Breaking New Ground for Water Resources Planning and Management," *Journal of Water Resources Planning and Management*, May/June 2006, 129–132.

6. Rost et al., "Agricultural Green and Blue Water Consumption," 12.

7. Timothy B. Sulser, Claudia Ringler, Tingju Zhu, Siwa Msangi, Elizabeth Bryan, and Mark W. Rosegrant, "Green and Blue Water Accounting in the Limpopo and Nile Basins," discussion paper, International Food Policy Research Institute, November 2009.

8. The 1996 Rome Declaration on World Food Security said it was "unacceptable" that 800 million people still did "not have enough food to meet their basic nutritional needs."

9. See Food and Agriculture Organization, *The State of Food Insecurity in the World* (Rome: FAO, October 2010).

10. Dana Gunders, "Wasted: How America Is Losing Up to 40 Percent of Its Food from Farm to Fork to Landfill," issue paper (Washington, DC: Natural Resources Defense Council, August 2012), 4.

11. See FAO Food Price Index, http://bit.ly/gC6TGF.

12. Lester R. Brown, *Outgrowing the Earth: The Food Security Challenge in an Age of Falling Water Tables and Rising Temperatures* (New York: Earthscan, 2005), 90–92.

13. Robert Glennon, "Our Water Supply, Down the Drain," *Washington Post*, August 23, 2009.

14. The 2030 Water Resources Group, *Charting Our Water Future*, Barilla Group, Coca-Cola Company, International Finance Corporation, McKinsey & Company, Nestlé S.A., New Holland Agriculture, SABMiller, Standard Chartered Bank, and Syngenta AG (2030 Water Resources Group, 2009), 6.

15. United Nations, *World Urbanization Prospects: The 2011 Revision* (New York: UN Department of Economic and Social Affairs, Population Division, 2012), executive summary.

16. See, for example, FAO, *State of Food Insecurity in the World*; and Alan Wild, *Soils, Land, and Food: Managing the Land during the 21st Century* (Cambridge, UK: Cambridge University Press, 2004).

17. FAOSTAT, Food Balance Sheets, http://goo.gl/wzvz1.

18. Food and Agriculture Organization, Aquastat online data, http://faostat.fao.org.

19. Susan Payne, founder and chief executive of Emergent Asset Management, quoted in Diana B. Henriques, "Food Is Gold, So Billions Invested in Farming," *New York Times*, June 5, 2008.

20. National Academy of Sciences, *The Impact of Genetically Engineered Crops on Farm Sustainability in the United States* (Washington, DC: National Academies Press, 2010).

21. See, for example, Ivan B. T. Lima, Fernando M. Ramos, Luis A. W. Bambace, and Reinaldo R. Rosa, "Methane Emissions from Large Dams as Renewable Energy Resources: A Developing Nation Perspective," *Mitigation and Adaptation Strategies for Climate Change* 13, no. 2 (2007): 193–206, doi:10.1007/s11027-007-9086-5.

22. United Nations, *Managing Water under Uncertainty and Risk*, vol. 1 of the World Water Development Report 4 (Paris: United Nations Educational, Scientific and Cultural Organization and UN-Water, March 2012), chapter 2.

23. FAO, Aquastat online data.

24. FAO, Aquastat online data.

25. Africa's water withdrawal for agriculture is 82 percent, followed by Asia's 81 percent, Oceania's 60 percent, North and South America's 49 percent, and Europe's 22 percent, according to the FAO.

26. R. Lal, "Enhancing Eco-Efficiency in Agro-Ecosystems through Soil Carbon Sequestration," *Crop Science* 50 (March–April 2010): S-121, doi:10.2135/crop sci2010.01.0012.

27. International Bank for Reconstruction and Development and World Bank, *World Development Report 2008: Agriculture for Development* (Washington, DC: World Bank, 2007).

28. FAO, Aquastat online data.

29. David Pimentel, "Soil Erosion: A Food and Environmental Threat," *Environment, Development, and Sustainability* 8, no. 1 (2006): 119, doi:10.1007/s10668-005-1262-8.

30. Anne Simon Moffat, "Ecology: Global Nitrogen Overload Problem Grows Critical," *Science* 279, no. 5353 (February 13, 1998): 988–989.

31. Hugh Turral, Jacob Burke, and Jean-Marc Faurès, *Climate Change, Water and Food Security* (Rome: Food and Agriculture Organization, 2011), xvii.

32. Dabo Guan and Klaus Hbacek, "Lifestyle Changes and Its Influences on Energy and Water Consumption in China," in Leeds Institute of Environment, School of the Environment, University of Leeds, *Proceedings for the International Workshop on Driving Forces for and Barriers to Sustainable Consumption* (Leeds, UK: University of Leeds, 2004), 385, 389.

33. The Food and Drug Administration released *voluntary* guidelines in April 2012 advising the "judicious" use of farm antibiotics and identifying the use of antibiotics for promoting animal growth or feed efficiency as not judicious.

34. FAO, *Water at a Glance.*

35. Akifumi Ogino, Hideki Orito, Kazuhiro Shimada, and Hiroyuki Hirooka, "Evaluating Environmental Impacts of the Japanese Beef Cow–Calf System by the Life Cycle Assessment Method," *Animal Science Journal* 78, no. 4 (August 2007): 424–432, doi:10.1111/j.1740-0929.2007.00457.x.

36. Daniel Wild, Carl-Johan Francke, Pierin Menzli, and Urs Schön, *Water: A Market of the Future* (Zurich, Switzerland: Sustainable Asset Management, December 2007), 8.

37. See Igor A. Shiklomanov, *World's Water Resources and Their Use* (Paris: IHP, UNESCO, 1999); Falkenmark and Lannerstad, "Consumptive Water Use to Feed Humanity," 15–28; H. Yang, P. Reichert, K. Abbaspour, and A. J. B. Zehnder, "A Water Resources Threshold and Its Implications for Food Security," *Environmental Science and Technology* 37, no. 14 (2003): 3048–3054; Peter H. Gleick, "Basic Water Requirements for Human Activities: Meeting Basic Needs," *Water International* 21, no. 2 (1996): 83–92; and A. J. B. Zehnder, H. Yang, and R. Schertenleib, "Water Issues: The Need for Actions at Different Levels," *Aquatic Sciences* 65 (2003): 1–20.

38. Joel E. Cohen, "Seven Billion," *New York Times*, October 23, 2011.

39. See, for example, A. Jägerskog and T. Jønch Clausen, eds., *Feeding a Thirsty World: Challenges and Opportunities for a Water and Food Secure World* (Stockholm: Stockholm International Water Institute, 2012).

40. Jim Galloway, Marshall Burke, Eric Bradford, Rosamond L. Naylor, Walter P. Falcon, Harold A. Mooney, Joanne Gaskell, Kirsten Oleson, Ellen McCollough, and

Henning Steinfeld, "International Trade in Meat—The Tip of the Pork Chop," *Ambio* 36, no. 8 (December 2007): 622; FAOSTAT data, http://faostat.fao.org.

41. Henning Steinfeld, Pierre Gerber, Tom Wassenaar, Vincent Castel, and Mauricio Rosales, *Livestock's Long Shadow: Environmental Issues and Options* (Rome: Food and Agriculture Organization, 2006), xx.

42. United Nations, *Sustainable Agriculture and Food Security in Asia and the Pacific* (Bangkok, Thailand: United Nations Economic and Social Commission for Asia and the Pacific, or ESCAP, 2009), 63.

43. U.S. Census Bureau, *The 2011 Statistical Abstract*, International Statistics, http://goo.gl/64IGB.

44. Sandra Postel, "Water for Food Production: Will There Be Enough in 2025?" *BioScience* 48, no. 8 (August 1998): 636.

45. United Nations Population Division, *World Population Prospects: The 2010 Revision* (New York: UNPD, 2011).

46. Paul Harrison and Fred Pearce, *AAAS Atlas of Population & Environment* (Washington, DC: American Association for the Advancement of Science, 2007), "Part 2—Natural Resources: Introduction"; U.S. Census Bureau, *The 2011 Statistical Abstract*.

47. The global average, assessed by the Food and Agriculture Organization, is for 2007.

48. U.S. Census Bureau, *The 2011 Statistical Abstract*, table 1377.

49. Per capita meat consumption in the United States fell from the 2004 high point of 83.5 kilograms to 77.6 kilograms in 2011, according to the U.S. Department of Agriculture.

50. Jägerskog and Clausen, *Feeding a Thirsty World*.

51. By comparison, all the world's cars, trains, planes, and boats account for a combined 13 percent of greenhouse-gas emissions. Steinfeld et al., *Livestock's Long Shadow*, xxi.

52. Steinfeld et al., *Livestock's Long Shadow*, xxi–xxii.

53. Steinfeld et al., *Livestock's Long Shadow*, xxiii.

54. World Health Organization, "Obesity and Overweight," Fact-Sheet No. 311 (Paris: WHO, May 2012).

55. Goodarz Danaei, Mariel M. Finucane, Yuan Lu, Gitanjali M. Singh, Melanie J. Cowan, Christopher J. Paciorek, John K. Lin, Farshad Farzadfar, Young-Ho Khang, Gretchen A. Stevens, Mayuree Rao, Mohammed K. Ali, Leanne M. Riley, Carolyn A. Robinson, and Majid Ezzati, "National, Regional, and Global Trends in Fasting Plasma Glucose and Diabetes Prevalence Since 1980," *Lancet* (June 25, 2011), doi:10.1016/S0140-6736(11)60679-X.

56. The five sheikhdoms are Kuwait, Qatar, Saudi Arabia, Bahrain, and the UAE. International Diabetes Federation, *IDF Diabetes Atlas*, 5th ed. (Brussels: IDF, 2009).

57. Danaei et al., "National, Regional, and Global Trends in Fasting Plasma Glucose."

58. U.S. National Intelligence Council, *Global Trends 2025: A Transformed World* (Washington, DC: National Intelligence Council, November 20, 2008), viii.

59. Sarah C Walpole, David Prieto-Merino, Phil Edwards, John Cleland, Gretchen Stevens, and Ian Roberts, "The Weight of Nations: An Estimation of Adult Human Biomass," *BMC Public Health* 12, no. 439 (2012), doi:10.1186/1471-2458-12-439.

60. Kristen Averyt, Jeremy Fisher, Annette Huber-Lee, Aurana Lewis, Jordan Macknick, Nadia Madden, John Rogers, and Stacy Tellinghuisen, *Freshwater Use by U.S. Power Plants: Electricity's Thirst for a Precious Resource*, report by the Energy and Water in a Warming World Initiative (Cambridge, MA: Union of Concerned Scientists, November 2011); Wendy Wilson, Travis Leipzig, and Bevan Griffiths-Sattenspiel, *Burning Our Rivers: The Water Footprint of Electricity* (Portland, OR: River Network, April 2012).

61. U.S. Department of Energy, *Energy Demands on Water Resources*, Report to Congress on the Interdependency of Energy and Water (Washington, DC: Department of Energy, December 2006), 25.

62. Ronnie Cohen, Gary Wolff, and Barry Nelson, *Energy Down the Drain: The Hidden Costs of California's Water Supply* (Oakland, CA: National Resources Defense Council and Pacific Institute, August 2004), 11–12.

63. Joan F. Kenny, Nancy L. Barber, Susan S. Hutson, Kristin S. Linsey, John K. Lovelace, and Molly A. Maupin, *Estimated Use of Water in the United States in 2005*, U.S. Geological Survey Circular 1344 (Reston, VA: U.S. Geological Survey, 2009), 52; Sharlene Leurig, *The Ripple Effect: Water Risk in the Municipal Bond Market* (Boston: Ceres and New York: Water Asset Management, October 2010), 4; Electric Power Research Institute, "Summary of Presentations: EPRI Advanced Cooling Technology Workshop" (Charlotte, North Carolina, July 8–9, 2008), 2.

64. European Environment Agency, *Water Resources across Europe: Confronting Water Scarcity and Drought* (Copenhagen: EEA, March 2009); Cleo Paskal, "The Vulnerability of Energy Infrastructure to Environmental Change," briefing paper (London: Chatham House and Global EESE, July 2009), 5.

65. International Energy Agency, *World Energy Outlook 2012* (Paris: IEA, November 2012), 7.

66. U.S. Department of Energy, *Energy Demands on Water Resources*, 23.

67. Electric Power Research Institute, *Water & Sustainability*, vol. 3, *U.S. Water Consumption for Power Production—The Next Half Century* (Palo Alto, CA: EPRI, 2002), 1.

68. U.S. Department of Energy, *Energy Demands on Water Resources*, 34.

69. Kenny et al., *Estimated Use of Water in the United States in 2005*, 38.

70. T. J. Feeley, T. J. Skone, G. J. Stiegel, Andrea McNemar, Michael Nemeth, Brian Schimmoller, J. T. Murphy, and Lynn Manfredo, "Water—A Critical Resource in the Thermoelectric Power Industry," *Energy* 33 (2008): 1–11.

71. U.S. Department of Energy, *Energy Demands on Water Resources*, 10.

72. Jordan Macknick, Robin Newmark, Garvin Heath, and K. C. Hallett, *A Review of Operational Water Consumption and Withdrawal Factors for Electricity Generating Technologies*, Report No. NREL/TP-6A20-50900 (Golden, CO: National Energy Technology Laboratory, March 2011), 15.

73. Christine Dell'Amore, "Nuclear Reactors, Dams at Risk Due to Global Warming," *National Geographic News*, February 26, 2010.

74. National Energy Technology Laboratory, *Estimating Freshwater Needs to Meet Future Thermoelectric Generation Requirements: 2010 Update* (Pittsburg, PA: National Energy Technology Laboratory, September 2010), 58–62.

75. Macknick et al., *A Review of Operational Water Consumption and Withdrawal Factors*, 15.

76. U.S. Department of Energy, *Energy Demands on Water Resources*, 44.

77. Electric Power Research Institute, "Summary of Presentations," 2.

78. International Energy Agency, *World Energy Outlook 2012*, 1.

79. National Energy Technology Laboratory, "Cost Effective Recovery of Low-TDS Frac Flowback Water for Re-Use," May 31, 2011, http://goo.gl/zlj7k.

80. See James T. Bartis, Rom LaTourrette, Lloyd Dixon, et al., *Oil Shale Development in the United States: Prospects and Policy Issues* (Santa Monica, CA: Rand Corporation, 2005).

81. Steven Mufson, "Can the Shale Gas Boom Save Ohio?" *Washington Post*, March 4, 2012.

82. Clean Water Action, "Environmental Violations at Marcellus Shale Drilling Sites January 1, 2010 to December 31, 2010," http://bit.ly/eESn5P.

83. National Energy Technology Laboratory, "Cost Effective Recovery of Low-TDS Frac Flowback."

84. "Ohio: Gas-Drilling Injection Well Led to Quakes," *Wall Street Journal*, March 9, 2012.

85. John R. Dyni, *Geology and Resources of Some World Oil-Shale Deposits*, Scientific Investigations Report 2005–5294 (Reston, VA: U.S. Geological Survey, 2006), 17.

86. National Energy Technology Laboratory, "Cost Effective Recovery of Low-TDS Frac Flowback."

87. Erik Mielke, Laura Diaz Anadon, and Venkatesh Narayanamurti, "Water Consumption of Energy Resource Extraction, Processing, and Conversion," Discussion Paper 2010-15 (Belfer Center for Science and International Affairs, Harvard Kennedy School, October 2010), 19.

88. Ralph E. H. Sims, "Hydropower, Geothermal, and Ocean Energy," *MRS Bulletin* 33 (April 2008): 390.

89. A widely accepted definition of large dams is the one put forward by the International Commission on Large Dams (ICOLD)—"those having a height of 15 meters from the foundation or, if the height is between 5 to 15 meters, having a reservoir capacity of more than 3 million cubic meters." This definition has been adopted by the World Commission on Dams.

90. See Jiyu Chen, "Dams, Effect on Coasts," in *Encyclopedia of Coastal Science*, ed. Maurice L. Schwartz, vol. 24 of Encyclopedia of Earth Sciences Series (Berlin: Springer, 2005).

91. See International Commission on Large Dams, "Intranet," online data; World Commission on Dams, "Dams and Water: Global Statistics," online data; Peter Bosshard, "China Dams the World," *World Policy Journal* 26, no. 4 (Winter 2009/10): 43–51; and UN Environment Program (*UNEP*), "Freshwater and Industry: Facts and Figures," *UNEP Industry and Environment* 27, no. 1 (January–March 2004).

92. U.S. Energy Information Administration, http://www.eia.gov/emeu/international/electricitygeneration.html.

93. Wilson et al., *Burning Our Rivers*, 24; P. H. Gleick, "Water and Energy," *Annual Review of Energy and Environment* 19 (1994): 267–299.

94. Ivan B. T. Lima, Fernando M. Ramos, Luis A. W. Bambace, and Reinaldo R. Rosa, "Methane Emissions from Large Dams as Renewable Energy Resources: A Developing Nation Perspective," *Mitigation and Adaptation Strategies for Climate Change* 13, no. 2 (2007): 193–206; Nathan Barros, Jonathan J. Cole, Lars J. Tranvik, Yves T. Prairie, David Bastviken, Vera L. M. Huszar, Paul del Giorgio, and Fábio Roland, "Carbon Emission from Hydroelectric Reservoirs Linked to Reservoir Age and Latitude," *Nature Geoscience* 4 (2011): 593–596.

95. Intergovernmental Panel on Climate Change (IPCC), *Climate Change 2007: Impacts, Adaptation and Vulnerability*, Working Group II Contribution to the Fourth Assessment Report of the IPCC (New York: Cambridge University Press, 2007), 49.

96. Electric Power Research Institute, *Water & Sustainability*, vol. 3, 1–2.

97. Averyt et al., *Freshwater Use by U.S. Power Plants*, 17.

98. Rob Edwards, "UK Nuclear Sites at Risk of Flooding, Report Shows," *Guardian*, March 7, 2012.

99. James Kanter, "Climate Change Puts Nuclear Energy into Hot Water," *New York Times*, May 20, 2007.

100. Susan Sachs, "Nuclear Power's Green Promise Dulled by Rising Temps," *Christian Science Monitor*, August 10, 2006.

101. National Energy Technology Laboratory, *Impact of Drought on U.S. Steam Electric Power Plant Cooling Water Intakes and Related Water Resource Management Issues*, Report No. DOE/NETL-2009/1364 (Pittsburgh, PA: National Energy Technology Laboratory, 2009).

102. Joel E. Cohen, Christopher Small, Andrew Mellinger, John Gallup, and Jeffrey Sachs, "Estimates of Coastal Populations," *Science* 278, no. 5341 (November 14, 1997): 1209–1213.

103. Kanter, "Climate Change Puts Nuclear Energy into Hot Water."

104. U.S. Atomic Energy Agency chairman Lewis L. Strauss, in a speech to the National Association of Science Writers in New York on September 16, 1954, famously said, "Our children will enjoy in their homes electrical energy too cheap to meter." *New York Times*, September 17, 1954. "Too Cheap to Meter" indeed became a key phrase at the 1964 Atoms for Peace Conference.

105. Benjamin Sovacool and Scott Victor Valentine, *The National Politics of Nuclear Power: Economics, Security, and Governance* (London: Routledge, 2012), 46–99. Also see Charles D. Ferguson, *Nuclear Energy: What Everyone Needs to Know* (New York: Oxford University Press, 2011).

106. The share of nuclear power in the worldwide electric power supply stagnated at between 16 and 17 percent from 1986 to 2000. By 2011, it fell to about 13 percent of global electricity generation and 5.5 percent of commercial primary energy. This share is set to decline further over the next decade.

107. Mycle Schneider, Antony Froggatt, and Steve Thomas, *The World Nuclear Industry Status Report 2010–2011* (Washington, DC: Worldwatch Institute, 2011), 4.

108. Jeffrey Friedman, "Congressional Committee with Oversight on Nuke Safety Takes in Big Dollars from Nuclear Power Industry," MAPLight.org, March 18, 2011, http://maplight.org/content/72579.

109. The nuclear industry's spending on U.S. lobbying actually doubled to almost $54 million between 2005 and 2010. Paul Blumenthal, "Nuclear Industry Lobbyists Battle Fallout from Japan Reactor Crisis," Sunlight Foundation, March 23, 2011, http://goo.gl/mvrCz.

110. International Energy Agency, *Global Energy Outlook 2012*, 6.

111. Richard Dobbs, Jeremy Oppenheim, Fraser Thompson, Marcel Brinkman, and Marc Zornes, *Resource Revolution: Meeting the World's Energy, Materials, Food, and Water Needs* (McKinsey Global Institute and McKinsey Sustainability & Resource Productivity Practice, November 2011), 51.

112. International Energy Agency, *Energy for All: Financing Access for the Poor*, World Energy Outlook: Special Report (Paris: IEA, 2011).

113. International Energy Agency, *Global Energy Outlook 2012*, 7.

114. Electric Power Research Institute, *Water & Sustainability*, vol. 3, vii.

115. National Energy Technology Laboratory, *Estimating Freshwater Needs to Meet Future Thermoelectric Generation Requirements*; U.S. Department of Energy, *Energy Demands on Water Resources*.

116. Kenny et al., *Estimated Use of Water in the United States in 2005*, 38–39.

117. National Energy Technology Laboratory, "Cost Effective Recovery of Low-TDS Frac Flowback."

118. For details, see Tony Allan, *Virtual Water: Tackling the Threat to our Planet's Most Precious Resource* (London: I. B. Tauris, 2011). It was Allan who, from the 1990s, promoted the term "embedded water" and subsequently the notion of "virtual water," for which he was awarded the 2008 World Water Prize.

119. Naota Hanasaki, Toshiyuki Inuzuka, Shinjiro Kanae, and Taikan Oki, "An Estimation of Global Virtual Water Flow and Sources of Water Withdrawal for Major Crops and Livestock Products Using a Global Hydrological Model," *Journal of Hydrology* 384, nos. 3–4 (2010): 232–244, doi:10.1016/j.jhydrol.2009.09.028.

120. M. Fader, D. Gerten, M. Thammer, J. Heinke, H. Lotze-Campen, W. Lucht, and W. Cramer, "Internal and External Green-Blue Agricultural Water Footprints of Nations, and Related Water and Land Savings through Trade," *Hydrology and Earth System Science Discussions* 8 (2011): 502–503, doi:10.5194/hessd-8-483-2011.

121. Fader et al., "Internal and External Green-Blue Agricultural Water Footprints," 483–527; J. A. Allan, "Virtual Water—the Water, Food and Trade Nexus: Useful Concept or Misleading Metaphor?" *Water International* 28, no. 1 (2003): 106–113.

122. Hong Yang and Alexander Zehnder, "Virtual Water: An Unfolding Concept in Integrated Water Resources Management," *Water Resources Research* 43 (2007): 1–10, doi:10.1029/2007WR006048.

123. UN-Water, online statistics, http://www.unwater.org/statistics.html. Levi Strauss & Company, however, estimates that a typical pair of blue jeans it manufactures consumes 3,480 liters of water during its entire life cycle, including in cotton-crop irrigation, stitching, and washing.

124. See Williams E. Rees, Mathis Wackernagel, and Phil Testemale, *Our Ecological Footprint: Reducing Human Impact on the Earth* (Gabriola Island, Canada: New Society Publishers, 2008). This book popularized the notion of "ecological footprint."

125. M. M. Mekonnen and A. Y. Hoekstra, *National Water Footprint Accounts: The Green, Blue and Grey Water Footprint of Production and Consumption*, Value of Water Research Report Series No. 50 (Delft, Netherlands: UNESCO-IHE, 2011), 9.

126. Country-specific studies include K. Feng, K. Hubacek, J. Minx, Y. L. Siu, A. Chapagain, Y. Yu, D. Guan, and J. Barrett, "Spatially Explicit Analysis of Water Footprints in the UK," *Water* 3, no. 1 (2011): 47–63; X. Zhao, B. Chen, and Z. F. Yang, "National Water Footprint in an Input-Output Framework—a Case Study of China 2002," *Ecological Modelling* 220, no. 2 (2009): 245–253; D. B. Guan and K. Hubacek, "Assessment of Regional Trade and Virtual Water Flows in China," *Ecological Economics* 61, no. 1 (2007): 159–170; D. A. Kampman, A. Y. Hoekstra, and M. S. Krol, *The Water Footprint of India*, Value of Water Research Report Series No. 32 (Delft, Netherlands: UNESCO-IHE, 2008); D. Vincent, A. K. De Caritat, S. Stijn Bruers, A. Chapagain, P. Weiler, and A. Laurent, *Belgium and Its Water Footprint* (Brussels: WWF-Belgium, 2011); K. Hubacek, D. B. Guan, J. Barrett, and T. Wiedmann, "Environmental Implications of Urbanization and Lifestyle Change in China: Ecological and Water Footprints," *Journal of Cleaner Production* 17, no. 14 (2009): 1241–1248; F. Bulsink, A. Y. Hoekstra, and M. J. Booij, "The Water Footprint of Indonesian Provinces Related to the Consumption of Crop Products," *Hydrology and Earth System Sciences* 14, no. 1 (2010): 119–128; M. M. Aldaya, A. Garrido, M. R. Llamas, C. Varelo-Ortega, P. Novo, and R. R. Casado, "Water Footprint and Virtual Water Trade in Spain," in *Water Policy in Spain*, ed. A. Garrido and M. R. Llamas (Leiden, Netherlands: CRC Press), 49–59; P. R. Van Oel, M. M. Mekonnen, and A. Y. Hoekstra, "The External Water Footprint of the Netherlands: Geographically-Explicit Quantification and Impact Assessment," *Ecological Economics* 69, no. 1 (2009): 82–92; and F. Z. Zhao, W. H. Liu, and H. B. Deng, "The Potential Role of Virtual Water in Solving Water Scarcity and Food Security Problems in China," *International Journal of Sustainable Development and World Ecology* 12, no. 4, (2005): 419–428.

127. Yang and Zehnder, "Virtual Water: An Unfolding Concept," 1.

128. Dennis Wichelns, "An Economic Analysis of the Virtual Water Concept in Relation to the Agri-food Sector," background paper (Paris: Organisation for Economic Co-operation and Development, 2010), 6.

129. One analyst, in fact, has described "virtual water" as a misnomer, saying virtual-water trade is nothing but a plain and simple food trade. Stephen Merrett, "Virtual Water and the Kyoto Consensus: A Water Forum Contribution," *Water International* 28, no. 4 (2003): 540–542.

130. This includes the estimates in Fader et al., "Internal and External Green-Blue Agricultural Water Footprints of Nations," and Mekonnen and Hoekstra, *National Water Footprint Accounts*.

131. Yang and Zehnder, "Virtual Water: An Unfolding Concept," 5.

132. H. Yang, L. Wang, K. C. Abbaspour, and A. J. B. Zehnder, "Virtual Water Highway: An Assessment of Water Use Efficiency in the International Food

Trade," *Hydrology and Earth System Sciences* 10 (2006): 443–454, doi:10.5194/hess-10-443-2006.

133. J. Ramirez-Vallejo and P. Rogers, "Virtual Water Flows and Trade Liberalization," *Water Science & Technology* 49, no. 7 (2004): 25–32.

134. Arjen Y. Hoekstra and Mesfin M. Mekonnen, "The Water Footprint of Humanity," *Proceedings of the National Academy of Sciences, Early Edition* (February 13, 2012): 2.

135. See A. K. Chapagain, A. Y. Hoekstra, H. H. G. Savenije, and R. Gautam, "The Water Footprint of Cotton Consumption: An Assessment of the Impact of Worldwide Consumption of Cotton Products on the Water Resources in the Cotton-Producing Countries," *Ecological Economics* 60, no. 1 (2006): 186–203.

136. U.S. Census Bureau, *The 2011 Statistical Abstract*, table 1374.

137. Fader et al., "Internal and External Green-Blue Agricultural Water Footprints," table 1.

138. Hoekstra and Mekonnen, "The Water Footprint of Humanity," 3. The Nile's average annual yield is nearly eighty-four cubic kilometers. The Amazon is by far the world's largest river by volume. But in length, it is believed to be slightly shorter than Africa's Nile. In 2007, however, a team of Brazilian scientists claimed to have traced the Amazon's actual source to a snow-capped Peruvian mountain and demanded that 284 kilometers be added to the river's length. The rivalry between the Nile and the Amazon for "the world's longest river" title has been legendary.

139. T. Oki and S. Kanae, "Virtual Water Trade and World Water Resources," *Water Science & Technology* 49, no. 7 (2004): 203–209; Yang and Zehnder, "Virtual Water," 1–10.

140. Galloway et al., "International Trade in Meat," 622–629.

141. FAO, Aquastat online data; A. Y. Hoekstra and A. K. Chapagain, *Globalization of Water: Sharing the Planet's Freshwater Resources* (Oxford, UK: Blackwell Publishing, 2008), appendix 20, "Water Footprints of Nations."

142. A. K. Chapagain, A. Y. Hoekstra, and H. H. G. Savenije, "Water Saving through International Trade of Agricultural Products," *Hydrology and Earth System Sciences* 10 (2006): 455–468, doi:10.5194/hess-10-455-2006.

143. Mekonnen and Hoekstra, *National Water Footprint Accounts*.

144. Charles Fishman, *The Big Thirst: The Secret Life and Turbulent Future of Water* (New York: Free Press, 2011); Maude Barlow, "Life, Liberty, Water," *Yes! Magazine* (Summer 2008).

145. Mekonnen and Hoekstra, *National Water Footprint Accounts*.

146. Hoekstra and Chapagain, *Globalization of Water*, appendix 20. Also see A. Y. Hoekstra and A. K. Chapagain, "Water Footprints of Nations: Water Use by People as a Function of Their Consumption Pattern," *Water Resources Management* 21, no. 1 (2007): 35–48.

147. Hoekstra and Chapagain, *Globalization of Water*, appendix 20.

148. Wichelns, *An Economic Analysis of the Virtual Water Concept*, 8.

149. U.S. National Security Council, *National Security Study Memorandum 200: Implications of Worldwide Population Growth for U.S. Security and Overseas*

Interests (Washington, DC: NSC, December 10, 1974), http://pdf.usaid.gov/pdf _docs/PCAAB500.pdf.

150. A. A. R. Ioris, "Virtual Water in an Empty Glass: The Geographical Complexities behind Water Scarcity," *Water International* 29, no. 1 (2004): 119–121.

151. Royal Embassy of Saudi Arabia, "Agricultural Achievements," Washington, DC, http://goo.gl/Qc27A.

152. See United Nations Development Programme, *Human Development Report 2007–2008: Fighting Climate Change—Human Solidarity in a Divided World* (New York: UNDP, 2008).

153. Michael Blackhurst, Chris Hendrickson, and Jordi Vidal, "Direct and Indirect Water Withdrawals for U.S. Industrial Sectors," *Environmental Science & Technology* 44, no. 6 (2010): 2126–2130, doi:10.1021/es903147k.

154. See Brahma Chellaney, *Water: Asia's New Battleground* (Washington, DC; Georgetown University Press), 134–140; Edward Wong, "Plan for China's Water Crisis Spurs Concern," *New York Times*, June 1, 2011; and Jing Ma, Arjen Y. Hoekstra, Hao Wang, Ashok K. Chapagain, and Dangxian Wang, "Virtual versus Real Water Transfers within China," *Philosophical Transactions of the Royal Society*, no. 361 (2006): 835–842.

155. U.S. Department of Agriculture, *USDA Wheat Baseline, 2012–21* (Washington, DC: USDA, May 2012).

156. Figure for 2011–2012. U.S. Department of Agriculture, "Table 23—World Supply & Utilization of Major Crops, Livestock, & Products," http://goo.gl/O2PFS.

157. United Nations Economic and Social Commission for Asia and the Pacific, *The State of the Environment in Asia and the Pacific 2005* (Bangkok: UNESCAP, 2006), 65.

158. M. Falkenmark, J. Rockström, and L. Karlberg, "Present and Future Water Requirements for Feeding Humanity," *Food Security* 1 (2009): 59–69, doi:10.1007 /s12571-008-0003-x.

159. SAB Miller and World Wildlife Fund-UK, *Water Footprinting: Identifying and Addressing Water Risks in the Value Chain* (London: SAB Miller and World Wildlife Fund-UK, 2010).

CHAPTER 3: THE FUTURE OF WATER

1. Samuel Taylor Coleridge, *The Rime of the Ancient Mariner and Other Poems* (North Chelmsford, MA: Courier Dover Publications, 1992).

2. See United Nations, *Water in a Changing World*, United Nations World Water Development Report (Colombella, Italy: United Nations World Water Assessment Program, 2009); Daniel Wild, Carl-Johan Francke, Pierin Menzli, and Urs Schön, *Water: A Market of the Future* (Zurich, Switzerland: Sustainable Asset Management, 2007); and Jill Boberg, *Liquid Assets: How Demographic Changes and Water Development Policies Affect Freshwater Resources* (Santa Monica, CA: Rand Corporation, 2005).

3. Yoshihide Wada, Ludovicus P. H. van Beek, Cheryl M. van Kempen, Josef W. T. M. Reckman, Slavek Vasak, and Marc F. P. Bierkens, "Global

Depletion of Groundwater Resources," *Geophysical Research Letters* 37 (2010), doi:10.1029/2010GL044571.

4. International Boundary and Water Commission of the United States and Mexico, "Interim International Cooperative Measures in the Colorado River Basin through 2017 and Extension of Minute 318 Cooperative Measures to Address the Continued Effects of the April 2010 Earthquake in the Mexicali Valley, Baja California," Minute No. 319, November 20, 2012, http://goo.gl/0DB5r.

5. Francesc Gallart and Pilar Llorens, "Water Resources and Environmental Challenge in Spain: A Key Issue for Sustainable Integrated Catchment Management," *Cuadernos de Investigación Geográfica*, no. 27 (2001): 7–16.

6. Commonwealth Scientific and Industrial Research Organization, *Climate Variability and Change in South-Eastern Australia: A Synthesis of Findings from Phase 1 of the South-Eastern Australian Climate Initiative* (Canberra: CSIRO, May 2010).

7. Food and Agriculture Organization (FAO) of the United Nations, Aquastat online data.

8. Canada's per capita freshwater availability, by comparison, is 84,483 cubic meters per year. FAO, Aquastat online data.

9. Maude Barlow and Tony Clarke, *Blue Gold: The Fight to Stop the Corporate Theft of the World's Water* (New York: New Press, 2003); Anthony DePalma, "Free Trade in Freshwater? Canada Says No and Halts Exports," *New York Times*, March 8, 1999; Council of Canadians, *Canadian Water Under Pressure: Five Reasons to Oppose Bulk Water Exports* (Ottawa: Council of Canadians, October 2007).

10. David Johansen, *Bulk Water Removals, Water Exports and the NAFTA* (Ottawa: Law and Government Division, Government of Canada, January 31, 2002); Gregory F. Szydlowski, "The Commoditization of Water: A Look at Canadian Bulk Water Exports, the Texas Water Dispute, and the Ongoing Battle under NAFTA for Control of Water Resources," *Colorado Journal of International Environmental Law and Policy* 18 (Summer 2007).

11. The editorial claimed Bush "wants Canada to export its freshwater in bulk to help the dry southwestern states water their impossible lawns," *Globe and Mail*, July 19, 2001.

12. See Cynthia Baumann, "Water Wars: Canada's Upstream Battle to Ban Bulk Water Export," *Minnesota Journal of Global Trade* 10 (Winter 2001).

13. The carbon intensity of the dredging process is linked both with the stripping of peat and fen—which naturally store carbon dioxide—and with the enormous energy employed for extraction. Energy equivalent to one barrel of oil is required to extract four to eight barrels from the tar sands. Because the oil sands are mixed with warm water by injecting steam into the earth so that the sand sinks to the bottom and the oil rises to the top, this "steam-assisted gravity drainage" process is water intensive. In addition, the oil-sands development in Alberta is leading to the loss of wetlands and to huge "tailings ponds" where mining waste laced with toxic napthic acid and solvents accumulates. The steam-injection process and the tailings pose a potential contamination threat to aquifers feeding the rivers.

14. Ezra Levant, *Ethical Oil: The Case for Canada's Oil Sands* (Toronto: McClelland & Stewart, 2011).

15. Kate Galbraith, "Amid Texas Drought, High-Stakes Battle over Water," *New York Times*, June 18, 2011.

16. M. N. Baker and M. J. Taras, *The Quest for Pure Water: The History of the Twentieth Century*, 2 vols. (Denver: AWWA, 1981). The authors contend that water-treatment methods in that era were designed more for improving the taste of drinking water. Also see Alice Outwater, *Water: A Natural History* (New York: Basic Books, 1996).

17. See Elizabeth Royte, *Bottlemania: How Water Went on Sale and Why We Bought It* (New York: Bloomsbury, 2008).

18. Shankar Vedantam, "What's Colorless and Tasteless and Smells Like . . . Money?" *Washington Post*, June 30, 2008. One study of twenty-five brands of bottled water collected randomly from three Alabama cities in the United States found that several samples exceeded the contaminant levels set by the U.S. Environmental Protection Agency and the European Union. Abua Ikem, Seyi Odueyungbo, Nosa O. Egiebor, and Kafui Nyavor, "Chemical Quality of Bottled Waters from Three Cities in Eastern Alabama," *Science of the Total Environment* 285, nos. 1–3 (February 21, 2002): 165–175, doi:10.1016/S0048-9697(01)00915-9.

19. John Stossel, "Is Bottled Water Better Than Tap?" ABC News, May 6, 2005.

20. Charles Fishman, *The Big Thirst: The Secret Life and Turbulent Future of Water* (New York: Free Press, 2011).

21. Wild et al., *Water: A Market of the Future*, 16.

22. The U.S. Environmental Protection Agency has set an enforceable drinking-water standard of four milligrams of fluoride per liter for adults and not more than two milligrams per liter for children under nine. Despite a vocal minority expressing skepticism about fluoridation benefits, the U.S. Department of Health and Human Services strongly supports water fluoridation. See Centers for Disease Control and Prevention, "Fluoridation Basics" (Atlanta, GA: CDC, January 14, 2011), http://goo .gl/056tl. Too much fluoride in water, however, causes fluorosis, characterized by mottled teeth and an increased risk of kidney failure.

23. See, for example, Barbara Pinto and Daniela Realia, "Screening of Estrogen-Like Activity of Mineral Water Stored in PET Bottles," *International Journal of Hygiene and Environmental Health* 212, no. 2 (March 2009): 228–232, doi:10.1016 /j.ijheh.2008.06.004; Martin Wagner and Jörg Oehlmann, "Endocrine Disruptors in Bottled Mineral Water: Total Estrogenic Burden and Migration from Plastic Bottles," *Environmental Science and Pollution Research* 16 (2009): 278–286, doi:10.1007 /s11356-009-0107-7.

24. See Andrea C. Gore, ed., *Endocrine-Disrupting Chemicals: From Basic Research to Clinical Practice* (Totowa, NJ: Humana Press, 2010).

25. Hoa H. Le, Emily M. Carlson, Jason P. Chua, and Scott M. Belcher, "Bisphenol A is Released from Polycarbonate Drinking Bottles and Mimics the Neurotoxic Actions of Estrogen in Developing Cerebellar Neurons," *Toxicology Letters* 176, no. 2 (January 30, 2008): 149–156, doi:10.1016/j.toxlet.2007.11.001.

26. National Institute of Environmental Health Sciences, "Bisphenol A (BPA)," http://goo.gl/8p2Om.

27. U.S. Food and Drug Administration, "Bisphenol A (BPA): Use in Food Contact Application," March 30, 2012, http://goo.gl/Oh8Mz; National Toxicology Program, U.S. Department of Health and Human Services, *NTP-CERHR Monograph on the Potential Human Reproductive and Developmental Effects of Bisphenol A* (Research Triangle Park, NC: National Toxicology Program, September 2008), http://goo.gl/v6sA5.

28. See World Health Organization and Food and Agriculture Organization, *Understanding the Codex Alimentarius* (Rome: Codex Secretariat, 2006).

29. List of current official food standards contained in Codex Alimentarius: http://goo.gl/t687I.

30. In a separate case of manufacturing-related contamination, Coca-Cola products were banned for ten days in Belgium in 1999 after thirty schoolchildren became sick by drinking Coca-Cola laced with a chemical used in the cleaning of transportation pallets.

31. Peter Gleick, *Bottled and Sold: The Story behind Our Obsession with Bottled Water* (Washington, DC: Island Press, 2010), ix–x.

32. See David H. Getches, *Water Law in a Nutshell* (St. Paul, MN: West Publishing, 2008); Douglas L. Grant and Gregory S. Weber, *Cases and Materials on Water Law* (St. Paul, MN: West Publishing, 2010).

33. R. Quentin Grafton, Gary D. Libecap, Eric C. Edwards, R. J. O'Brien, and Clay Landry, *A Comparative Assessment of Water Markets: Insights from the Murray-Darling Basin of Australia and the Western U.S.*, Working Paper No.8/2011 (Torino, Italy: International Centre for Economic Research, June 2011).

34. Alexei Barrionuevo, "Chilean Water Rights System Leaves Town Parched," *International Herald Tribune*, March 15, 2009.

35. Rob Taylor (Reuters), "Australia, the Frontier of a Global Rush to Commercialize Water," *International Herald Tribune*, August 31, 2008.

36. See Oscar Olivera and Tom Lewis, *¡Cochabamba! Water War in Bolivia* (Cambridge, MA: South End Press, 2008).

37. Full text of the 1993 Joint Statement by the Governments of Canada, the United States, and Mexico in Edith Brown Weiss, Laurence Boisson de Chazournes, and Nathalie Bernasconi-Osterwalder, *Freshwater and International Economic Law* (New York: Oxford University Press, 2005), appendix D.

38. Szydlowski, "The Commoditization of Water"; Baumann, "Water Wars."

39. International Boundary and Water Commission, Minute No. 319, sec. III.6.

40. Tony Perry and Richard Marosi, "U.S., Mexico Reach Pact on Colorado River Water Sale," *Los Angeles Times*, November 20, 2012; Michael Gardner and Sandra Dibble, "Mexico-U.S. Sign Historic Colorado River Deal," *San Diego Union-Tribune*, November 21, 2012.

41. Cited in Scott Peterson, "Turkey's Plan for Mideast Peace," *Christian Science Monitor*, April 18, 2000.

42. Vandana Shiva, *Water Wars: Privatization, Pollution, and Profit* (Cambridge, MA: South End Press, 2002), 87–106.

43. Maude Barlow, *Blue Covenant: The Global War Crisis and the Coming Battle for the Right to Water* (New York: New Press, 2007), xii.

44. Robin Clarke and Jannet King, *The Water Atlas* (New York: New Press, 2004); Wild et al., *Water: A Market of the Future*, 4.

45. Peter M. Vitousek, Harold A. Mooney, Jane Lubchenco, and Jerry M. Melillo, "Human Domination of Earth's Ecosystems," *Science* 277, no. 5325 (July 25, 1997): 494–499.

46. Text of the 1982 World Charter for Nature, in Scott S. Olson, *International Environmental Standards Handbook* (Boca Raton, FL: CRC Press, 1999), 39.

47. Jan Zalasiewicz, Mark Williams, Will Steffen, and Paul Crutzen, "The New World of the Anthropocene," *Environmental Science & Technology* 44, no. 7 (2010): 2228–2231.

48. Intergovernmental Panel on Climate Change, *Climate Change 2007: The Physical Science Basis—Contribution of Working Group I to the Fourth Assessment Report of the Intergovernmental Panel on Climate Change*, ed. S. Solomon, D. Qin, M. Manning, Z. Chen, M. Marquis, K. B. Averyt, M. Tignor, and H. L. Miller (Cambridge, UK: Cambridge University Press, 2007), sec. 2.5.1; Vitousek et al., "Human Domination of Earth's Ecosystems," 496–497.

49. Breaking that link was emphasized by one of the four principles adopted by various stakeholders at the Stockholm Water Symposium, held in Stockholm, August 11–17, 2002. The symposium declaration stated, "*We must break now the link between economic growth and water degradation.* Activities generating wealth often contaminate water, resulting in pollution of rivers and groundwater throughout the world. If this continues unabated, available water becomes too polluted to use, and the world has less water available. Positive, proactive national and local action toward water-pollution abatement and restoration is essential today to avoid even more severe problems in coming decades" (emphasis added).

50. *Plan of Implementation of the World Summit on Sustainable Development*, World Summit on Sustainable Development, Johannesburg, August 26–September 4, 2002, sec. 3.

51. Al Gore, "We Can't Wish Away Climate Change," *New York Times*, February 27, 2010.

52. Intergovernmental Panel on Climate Change (IPCC), *Climate Change 2007: Impacts, Adaptation and Vulnerability, Contribution of Working Group II to the Fourth Assessment Report of the Intergovernmental Panel on Climate Change*, ed. M. L. Parry, O. F. Canziani, J. P. Palutikof, P. J. van der Linden, and C. E. Hanson (Cambridge, UK: Cambridge University Press, 2007).

53. S. Kanae, T. Oki, and K. Musiake, "Impact of Deforestation on Regional Precipitation over the Indochina Peninsula," *Journal of Hydrometeorology* 2 (2001): 51–70; Vitousek et al., "Human Domination of Earth's Ecosystems," 496–497.

54. U.S. Geological Survey, "Glaciers and Sea Level," http://goo.gl/jXTVY.

55. Joel E. Cohen, Christopher Small, Andrew Mellinger, John Gallup, and Jeffrey Sachs, "Estimates of Coastal Populations," *Science* 278, no. 5341 (November 14, 1997): 1209–1213.

56. John Tibbetts, "Coastal Cities: Living on the Edge," *Environmental Health Perspectives* 110, no. 11 (November 2002), doi:10.1289/ehp.110-a674.

57. Richard S. Lindzen, "The Climate Science Isn't Settled," *Wall Street Journal*, November 30, 2009.

58. Institute for Security Studies, *The New Global Puzzle: What World for the EU in 2025?* (Paris: Institute for Security Studies, 2006), 76.

59. Douglas J. Kennett, Sebastian F. M. Breitenbach, Valorie V. Aquino, Yemane Asmerom, Jaime Awe, James U. L. Baldini, Patrick Bartlein, Brendan J. Culleton, Claire Ebert, Christopher Jazwa, Martha J. Macri, Norbert Marwan, Victor Polyak, Keith M. Prufer, Harriet E. Ridley, Harald Sodemann, Bruce Winterhalder, and Gerald H. Haug, "Development and Disintegration of Maya Political Systems in Response to Climate Change," *Science* 338, no. 6108 (November 9, 2012): 788–791, doi:10.1126/science.1226299.

60. The British government report by Nicholas Stern, released in November 2006, contended that a temperature increase in the range of five degrees Celsius would over time cause a sea-level rise enough to threaten the world's top cities like London, Shanghai, New York, Tokyo, and Hong Kong.

61. Investigations revealed that the Climate Research Unit at the University of East Anglia flouted Britain's Freedom of Information regulations in its handling of requests for data from climate skeptics. James Randerson, "University in Hacked Climate Change Emails Row Broke FOI Rules," *Guardian*, January 27, 2010.

62. The IPCC report flatly stated, "Glaciers in the Himalaya are receding faster than in any other part of the world (see Table 10.9) and, if the present rate continues, the likelihood of them disappearing by the year 2035 and perhaps sooner is very high if the earth keeps warming at the current rate." IPCC, *Climate Change 2007: Impacts, Adaptation and Vulnerability*, chap. 10, sec. 10.6.2, "The Himalayan Glaciers."

63. The magazine quoted glaciologist Syed Hasnain as saying, "All the glaciers in the middle Himalayas are retreating." It added that Hasnain's study "indicates that all the glaciers in the central and eastern Himalayas could disappear by 2035 at their present rate of decline." Fred Pearce, "Flooded Out," *New Scientist*, June 15, 1999.

64. In another account of how the false claim crept into the IPCC report, J. Graham Cogley, a professor at the Ontario Trent University, contended that the IPCC got the date wrong by more than three hundred years by misreading the 2350 date mentioned in the work of hydrologist V. M. Kotlyakov as 2035. Kotlyakov had stated in his 1996 report, "The extrapolar glaciation of Earth will be decaying at rapid, catastrophic rates—its total area will shrink from 500,000 to 100,000 square kilometers by the year 2350."

65. David Rose, "Glacier Scientist: I Knew Data Hadn't Been Verified," *Daily Mail* (London), January 24, 2010.

66. Intergovernmental Panel on Climate Change (IPCC), "IPCC Statement on the Melting of Himalayan Glaciers" (Geneva: IPCC, January 20, 2010).

67. Closer scrutiny of the IPCC's report, *Climate Change 2007: Impacts, Adaptation and Vulnerability*, showed that some of its sources ranged from a mountaineering magazine to a student paper. The report also had relied on World Wide Fund for Nature (WWF), an environmental pressure group, for claims regarding "transformation of natural coastal areas," "destruction of more mangroves," "glacial lake

outbursts causing mudflows and avalanches," and changes in the ecosystem of the "Mesoamerican reef."

68. IPCC, *Climate Change 2007: Impacts, Adaptation and Vulnerability*, chap. 13, sec. 13.4.1, "Natural Ecosystems."

69. The IPCC report stated, "The Netherlands is an example of a country highly susceptible to both sea level rise and river flooding because 55 percent of its territory is below sea level." IPCC, *Climate Change 2007: Impacts, Adaptation and Vulnerability*, 547. As the Netherlands Environmental Assessment Agency pointed out, the correct position is that "55 percent of the Netherlands is at risk of flooding; 26 percent of the country is below sea level; and 29 percent is susceptible to river flooding." Netherlands Environmental Assessment Agency, "Correction Wording Flood Risks for the Netherlands in IPCC Report," http://goo.gl/xjzgY.

70. Mark Siddall, Thomas F. Stocker, and Peter U. Clark, "Retraction: Constraints on Future Sea-Level Rise from Past Sea-Level Change," *Nature Geoscience* 3 (2010): 217; Mark Siddall, Thomas F. Stocker, and Peter U. Clark, "Constraints on Future Sea-Level Rise from Past Sea-Level Change," *Nature Geoscience* 2 (2009): 571–575.

71. See Thomas Homer-Dixon, *Environment, Scarcity, and Violence* (Princeton, NJ: Princeton University Press, 2001); Peter H. Gleick, "Water and Conflict: Fresh Water Resources and International Security," *International Security* 18, no. 1 (1993): 79–112; Ashok Swain, *Managing Water Conflict: Asia, Africa, and the Middle East* (London: Routledge, 2004); David Michel, "A River Runs Through It: Climate Change, Security Challenges, and Shared Water Resources," in *Troubled Waters: Climate Change, Hydropolitics, and Transboundary Resources*, ed. David Michel and Amit Pandya (Washington, DC: Stimson Center, 2009).

72. Noah D. Hall, Bret B. Stuntz, and Robert H. Abrams, "Climate Change and Freshwater Resources," *Natural Resources and Environment* 22, no. 3 (Winter 2008): 35.

73. Some environmentalists, as one media commentator noted, have undermined their cause with "claims bordering on the outlandish: They've blamed global warming for shrinking sheep in Scotland, more shark and cougar attacks, genetic changes in squirrels, an increase in kidney stones, and even the crash of Air France Flight 447." Dana Milbank, "Global Warming's Snowball Fight," *Washington Post*, February 14, 2010. Also see Joel Achenbach, "Global Warming Did It! Well, Maybe Not," *Washington Post*, August 3, 2009.

74. Daniel W. Moulton and John S. Jacob, *Texas Coastal Wetlands Guidebook* (Texas Sea Grant College Program, 2000).

75. Glenn De'ath, Katharina E. Fabricius, Hugh Sweatman, and Marji Puotinen, "The 27-Year Decline of Coral Cover on the Great Barrier Reef and Its Causes," *Proceedings of the National Academy of Sciences, Early Edition*, October 1, 2012, doi:10.1073/pnas.1208909109.

76. Food and Agriculture Organization of the United Nations, *State of the World's Forests 2005* (Rome: FAO, 2005), viii.

77. Vitousek et al., "Human Domination of Earth's Ecosystems," 495.

78. The U.S. National Intelligence Council lists Russia and Canada as the two most likely climate-change winners. While noting that climate change could open up

millions of square kilometers to development in Canada, besides lengthening agricultural growing seasons and spreading forests to the tundra, the council said, "Russia has the potential to gain the most from increasingly temperate weather. Russia has vast untapped reserves of natural gas and oil in Siberia and also offshore in the Arctic, and warmer temperatures should make the reserves considerably more accessible." U.S. National Intelligence Council, *Global Trends 2025: A Transformed World* (Washington, DC: National Intelligence Council, November 20, 2008), 52.

79. The U.S. Geological Survey believes a quarter of the world's undiscovered oil and gas reserves may lie beneath the Arctic Ocean. Global warming is expected to make drilling and shipping more feasible by thinning the Arctic ice. Under the 1982 U.N. Convention on the Law of the Sea, each Arctic nation was given ten years, after its ratification of the convention, to map out the Arctic seabed. The maps, sediment samples, and other scientific data can be used to claim parts of the seabed as a physical extension of the continental shelf of the claimant state. The claim is to apply to the buried resources, not to the waters above, which would remain international. But to be valid, the claim has to prove that the sought seabed portion is a physical extension of the claimant's continental shelf.

80. Vaclav Klaus, *Blue Planet in Green Shackles—What Is Endangered: Climate or Freedom?* (Washington, DC: Competitive Enterprise Institute, 2007). In his book, Klaus insists that global warming is attributable to natural phenomena and that human activity plays a minor role.

81. Arctic Climate Impact Assessment (ACIA), *Arctic Climate Impact Assessment* (Cambridge, UK: Cambridge University Press, 2005), 3.

82. IPCC, *Climate Change 2007: Impacts, Adaptation and Vulnerability.*

83. President Anote Tong of Kiribati in February 2009 called for an international fund to help his nation's citizens relocate to higher ground elsewhere, while Maldivian president Mohamed Nasheed, even before assuming office in late 2008, began advocating a national investment fund, with some earnings from tourism, so that the Maldives could buy an overseas haven for its people. Nasheed was ousted at gunpoint in early 2012.

84. Emily Wax, "Food Costs Push Bangladesh to Brink of Unrest," *Washington Post*, May 24, 2008.

85. The number of migrants from Bangladesh illegally resident in India jumped from 279,878 in the 1991 Indian census to 3,084,826 in the 2001 census. India's 2001 census, while reporting continuing "large-scale migration" from Bangladesh, noted that many illegal migrants had likely not been counted in the various censuses. By many estimates, more than 15 million Bangladeshis are now resident in India. "The Land That Maps Forgot," *Economist*, February 15, 2011.

86. The Mexican-born population in the United States—legal and illegal—fell to 12 million in 2011 from the 2007 peak of 12.6 million, according to an analysis of census data from American and Mexican governments.

87. Richard Monastersky, "Geography Conspires against Bangladesh," *Science News*, May 11, 1991.

88. See, for example, Intergovernmental Panel on Climate Change, *Climate Change 2001: The Scientific Basis*, edited by J. T. Houghton, Y. Ding, D. J. Griggs,

M. Noguer, P. J. van der Linden, X. Dai, K. Maskell, and C. A. Johnson (Cambridge, UK: Cambridge University Press, 2001); P. Whetton, A. B. Pittock, and R. Suppiah, "Implications of Climate Change for Water Resources in South and Southeast Asia," in Asian Development Bank, *Climate Change in Asia: Thematic Overview* (Manila: ADB, 1994); Intergovernmental Panel on Climate Change, *Special Report on the Regional Impacts of Climate Change: An Assessment of Vulnerability*, edited by Robert T. Watson, Marufu C. Zinyowera, and Richard H. Moss (Cambridge, UK: Cambridge University Press, 1997); R. Suppiah, "The Asian Monsoons: Simulations from Four GCMs and Likely Changes under Enhanced Greenhouse Conditions," in *Climate Impact Assessment Methods for Asia and the Pacific*, ed. A. J. Jakeman and B. Pittock, proceedings of a regional symposium, March 10–12, 1993, Canberra, Australia (1994); Climate Impact Group, *Climate Change Scenarios for South and Southeast Asia* (Aspendale, Australia: Commonwealth Scientific and Industrial Research Organization, 1992).

89. IPCC, *Climate Change 2007: The Physical Science Basis*, 235–336.

90. Intergovernmental Panel on Climate Change, Working Group I: The Physical Science Basis, "Summary for Policymakers" (Paris: IPCC, February 2, 2007).

91. Hall et al., "Climate Change and Freshwater Resources."

92. Z. W. Kundzewicz, L. J. Mata, N. W. Arnell, P. Döll, P. Kabat, B. Jiménez, K. A. Miller, T. Oki, Z. Sen, and I. A. Shiklomanov, "Freshwater Resources and Their Management," in IPCC, *Climate Change 2007: Impacts, Adaptation and Vulnerability*.

93. Claudia Tebaldi, Katharine Hayhoe, Julie M. Arblaster, and Gerald A. Meehl, "Going to the Extremes: An Inter-Comparison of Model-Simulated Historical and Future Changes in Extreme Events," *Climatic Change* 76 (December 2006): 185–211.

94. Aiguo Dai, Taotao Qian, Kevin E. Trenberth, and John D. Milliman, "Changes in Continental Freshwater Discharge from 1948 to 2004," *Journal of Climate* 22, no. 10 (May 15, 2009).

95. James Hansen, Larissa Nazarenko, Reto Ruedy, Makiko Sato, Josh Willis, Anthony Del Genio, Dorothy Koch, Andrew Lacis, Ken Lo, Surabi Menon, Tica Novakov, Judith Perlwitz, Gary Russell, Gavin A. Schmidt, and Nicholas Tausnev, "Earth's Energy Imbalance: Confirmation and Implications," *Science* 308, no. 5727 (June 3, 2005): 1431–1435.

96. Kundzewicz et al., "Freshwater Resources and Their Management," 173–210.

97. T. P. Barnett, J. C. Adam, and D. P. Lettenmaier, "Potential Impacts of a Warming Climate on Water Availability in Snow-Dominated Regions," *Nature* 438 (November 17, 2005): 303–309.

98. Noah S. Diffenbaugh, Martin Scherer, and Moetasim Ashfaq, "Response of Snow-Dependent Hydrologic Extremes to Continued Global Warming," *Nature Climate Change*, November 11, 2012, doi:10.1038/nclimate1732.

99. Gregory T. Pederson, Stephen T. Gray, Connie A. Woodhouse, Julio L. Betancourt, Daniel B. Fagre, Jeremy S. Littell, Emma Watson, Brian H. Luckman, and Lisa J. Graumlich, "The Unusual Nature of Recent Snowpack Declines in the North American Cordillera," *Science* 333, no. 6040 (June 9, 2011), doi:10.1126/science.1201570.

100. L. G. Thompson, H. H. Brecher, E. Mosley-Thompson, D. R. Hardy, and B. G. Mark, "Glacier Loss on Kilimanjaro Continues Unabated," *Proceedings of the National Academy of Sciences* 106 (2010): 19770–19775, doi:10.1073/pnas.0906029106.

101. A. M. Duan and G. X. Wu, "Role of the Tibetan Plateau Thermal Forcing in the Summer Climate Patterns over Subtropical Asia," *Climate Dynamics* 24, no. 7/8 (June 2005): 793–807.

102. National Academy of Sciences, *Himalayan Glaciers: Climate Change, Water Resources, and Water Security* (Washington, DC: National Academies Press, September 2012).

103. T. Bolch, A. Kulkarni, A. Kääb, C. Huggel, F. Paul, J. G. Cogley, H. Frey, J. S. Kargel, K. Fujita, M. Scheel, S. Bajracharya, and M. Stoffel, "The State and Fate of Himalayan Glaciers," *Science* 336, no. 6079 (April 20, 2012): 310–314.

104. Chinese Academy of Sciences, *China Glacier Inventory* (Lanzhou, China: World Data Center for Glaciology and Geocryology, Lanzhou Institute of Glaciology and Geocryology, 2004).

105. Yao Tandong, Wang Youqing, Liu Shiying, Pu Jianchen, Shen Yongping, and Lu Anxin, "Recent Glacial Retreat in High Asia in China and its Impact on Water Resource in Northwest China," *Science in China, Series D: Earth Sciences* 47, no. 12 (December 2004): 1065.

106. Veerabhadran Ramanathan, Muvva V. Ramana, Gregory Roberts, Dohyeong Kim, Craig Corrigan, Chul Chung, and David Winker, "Warming Trends in Asia Amplified by Brown Cloud Solar Absorption," *Nature* 448, no. 7153 (2007): 575–578.

107. Yao, "Recent Glacial Retreat in High Asia in China," 1065–1075.

108. United Nations Environment Program, *Global Outlook for Ice and Snow* (Nairobi, Kenya: UNEP, 2007), 131.

109. Jeffrey Mazo, *Climate Conflict: How Global Warming Threatens Security and What to Do about It* (London: Routledge and the International Institute for Strategic Studies, 2010), 138–141.

110. See, for example, Joshua Busby, *Climate Change and National Security: An Agenda for Action* (Washington, DC: Council on Foreign Relations, 2009).

111. Defense Science Board, *Trends and Implications of Climate Change for National and International Security* (Washington, DC: Department of Defense, October 2011), 62.

112. Kundzewicz et al., "Freshwater Resources and Their Management."

113. James Stuhltrager, "Global Climate Change and National Security," *Natural Resources and Environment* 22, no. 3 (Winter 2008), 38.

114. German Advisory Council on Global Change (WBGU), *Climate Change as a Security Risk* (Berlin: WBGU, June 2007), 18.

115. Elinor Ostrom, Joanna Burger, Christopher B. Field, Richard B. Norgaard, and David Policansky, "Revisiting the Commons: Local Lessons, Global Challenges," *Science* 284, no. 5412 (1999), 278–282.

116. For a discussion of geoengineering options, see the Royal Society, *Geoengineering the Climate: Science, Governance, and Uncertainty* (London: Royal Society,

2009); U.S. Government Accountability Office, *Climate Change: A Coordinated Strategy Could Focus Federal Geoengineering Research and Inform Governance Efforts*, Report GAO-10-903 (Washington, DC: GAO, September 2010); National Research Council of the National Academies, *Advancing the Science of Climate Change* (Washington, DC: National Academies Press, 2010).

117. Paul Harrison and Fred Pearce, *AAAS Atlas of Population & Environment* (Washington, DC: American Association for the Advancement of Science, 2007), "Part 2—Natural Resources: Freshwater."

118. Food and Agriculture Organization, *General Summary: Latin America and the Caribbean—Water Resources* (Rome: FAO, 2012).

119. FAO, Aquastat online data.

120. China's dependency ratio is just 0.96 percent, and Turkey's is 1.007 percent. FAO, Aquastat online data.

121. World Resources Institute and United Nations Development Programme, *World Resources 2005: The Wealth of the Poor—Managing Ecosystems to Fight Poverty* (Washington, DC: World Resources Institute, United Nations Development Programme, United Nations Environment Programme, and World Bank, 2005); Diane Bates, "Environmental Refugees? Classifying Human Migrations Caused by Environmental Change," *Population and Environment* 23, no. 5 (May 2002): 469.

122. Aaron T. Wolf, "Conflict and Cooperation along International Waterways," *Water Policy* 1, no. 2 (1998): 254–255. Also see Ismail Serageldin, "Water: Conflicts Set to Arise within as Well as between States," *Nature* 459, no. 7244 (May 14, 2009).

123. Center for Strategic and International Studies and Sandia National Laboratories, *Global Water Future: Addressing Our Global Water Future* (Washington, DC: CSIS and Sandia National Laboratories, September 30, 2005), 7.

124. United Nations Environment Programme, *Environmental Degradation Triggering Tensions and Conflict in Sudan* (Nairobi: UNEP, June 2007).

125. For details, see United Nations Office for the Coordination of Humanitarian Affairs, *Sudan: UN and Partners Work Plan 2012—Mid-Year Review* (New York: United Nations, 2012).

126. Edward Miguel, Shanker Satyanath, and Ernest Sergenti, "Economic Shocks and Civil Conflict: An Instrumental Variables Approach," *Journal of Political Economy* 112, no. 4 (2004): 725–753.

127. See Philip Fredkin, *A River No More: The Colorado River and the West* (New York: Knopf, 1981); and Remi A. Nadeau, *The Water Seekers* (Garden City, NY: Doubleday, 1950).

128. Andrew Wice, *To the Last Drop: A Novel of Water, Oppression, and Rebellion* (Boulder, CO: Bauu Institute, 2008).

129. Robert Glennon, "Our Water Supply, Down the Drain," *Washington Post*, August 23, 2009.

130. See Bonnie G. Colby and Katherine L. Jacobs, eds., *Arizona Water Policy* (Washington, DC: Resources for the Future, 2007), chap. 14.

131. "Indians' Water Rights Give Hope for Better Health," *New York Times*, August 30, 2008.

132. Fred Pearce, *When the Rivers Run Dry: What Happens When Our Water Runs Out?* (Cornwall, UK: Eden Project Books, 2006); U.S. National Intelligence, *Global Water Security*, Intelligence Community Assessment ICA 2012-08, February 2, 2012 (Washington, DC: Office of the Director of National Intelligence, March 22, 2012).

133. Philip Micklin, "The Aral Sea Disaster," *Annual Review of Earth and Planetary Sciences* 35 (May 2007): 47–72.

134. For details, see Iwao Kobori and Michael H. Glantz, eds., *Central Eurasian Water Crisis: Caspian, Aral, and Dead Seas* (Tokyo: United Nations University Press, 1998); Philip Micklin and Nikolay V. Aladin, "Reclaiming the Aral Sea," *Scientific American* (April 2008): 64–71; P. Wouters and V. Dukhovny, eds., *Implementing Integrated Water Resources Management in Central Asia* (Dordrecht, Netherlands: Springer, 2007).

135. Judith Miller, "Cold War Leaves a Deadly Anthrax Legacy," *New York Times*, June 2, 1999.

136. William T. Vollmann, *Imperial* (New York: Viking, 2009), 782–786; Jennifer Medina, "For Desolate, Shrinking Salton Sea, another Dream," *New York Times*, July 29, 2012.

137. Sylvia K. Sikes, *Lake Chad versus the Sahara Desert: A Great African Lake in Crisis* (Tyneside, UK: Mirage Newbury, 2003); J. Lemoalle, "Lake Chad: A Challenging Environment," in *Dying and Dead Seas: Climatic versus Anthropic Causes*, ed. Jacques C. J. Nihoul, Peter O. Zavialov, and Philip P. Micklin, NATO Science Series: IV (Dordrecht, Netherlands: Kluwer, 2004), 321–340.

138. The Lake Chad Basin extends to parts of four more countries—the Central African Republic, Sudan, Algeria, and Libya.

139. Michael T. Coe and Jonathan A. Foley, "Human and Natural Impacts on the Water Resources of the Lake Chad Basin," *Journal of Geophysical Research*, no. 106 (2001): 3349–3356. The study used an integrated biosphere model to help track changes in the basin since 1953.

140. The Lake Chad Basin Commission is run jointly by Chad, Niger, Cameroon, Nigeria, the Central African Republic, and Libya.

141. For a geological analysis of the Dead Sea, see Aharon Horowitz, *The Jordon Rift Valley* (Lisse, Netherlands: A. A. Balkema, 2001).

142. Food and Agriculture Organization, *Irrigation in the Middle East Region in Figures—Aquastat Survey 2008*, Water Report 34 (Rome: FAO, 2009), sec. "Israel."

143. I. Gavrieli and A. Oren, "The Dead Sea as a Dying Lake," in *Dying and Dead Seas: Climatic versus Anthropic Causes*, ed. Jacques C. J. Nihoul, Peter O. Zavialov, and Philip P. Micklin, NATO Science Series: IV (Dordrecht, Netherlands: Kluwer, 2004), 287–306.

144. Haim Watzman, "Israel's Incredible Sinking Sea," *New York Times*, July 29, 2007.

145. Kevin Peraino, "The Myth of Water," *Newsweek*, June 28, 2008; FAO, Aquastat online data; U.S. Central Intelligence Agency, *The CIA World Factbook 2013* (New York: Skyhorse Publishing, 2012), sec. "Israel: Economy."

146. FAOSTAT data, http://faostat.fao.org.

147. World Wide Fund for Nature, *World's Top 10 Rivers at Risk* (Gland, Switzerland: WWF International, 2007).

148. The Ramsar Convention got its name from Ramsar, Iran, where it was concluded in 1971. The convention's full text is at http://goo.gl/b6Hbs.

149. International Atomic Energy Agency, "In Zaragoza, It's Water, Water Everywhere," http://goo.gl/eJcRk.

150. See National Research Council of the National Academies, *Desalination: A National Perspective* (Washington, DC: National Academies Press, 2008).

CHAPTER 4: CHANGING WATER COOPERATION, COMPETITION, AND CONFLICT

1. Ban Ki-moon, "Quenching a Global Thirst," *International Herald Tribune*, March 21, 2008.

2. United Nations Educational, Scientific and Cultural Organization (UNESCO), *Internationally Shared (Transboundary) Aquifer Resources Management: Their Significance and Sustainable Management*, IHP-VI, IHP Non-Serial Publications in Hydrology (Paris: UNESCO, 2001), 9.

3. Defense Science Board, *Trends and Implications of Climate Change for National and International Security* (Washington, DC: Department of Defense, October 2011), 59.

4. See, for example, Jerome Delli Priscoli and Aaron T. Wolf, *Managing and Transforming Water Conflicts* (New York: Cambridge University Press, 2008).

5. Robert Mandel, "Sources of International River Basin Disputes," *Conflict Quarterly* 12, no. 4 (1992): 25–56.

6. Hiroko Tabuchi, "Chinese Developers Tap into Japanese Insecurity," *New York Times*, September 29, 2010.

7. Hideki Hirano, *Japan's Forests under Siege: How Foreign Capital Threatens Our Water Sources* (Tokyo: Shinchosha, March 2010), in the Japanese language, http://bit.ly/yQhtMB.

8. The Tokyo Foundation, "The Conservation of Land Resources," http://goo.gl/4lewx.

9. Policy recommendations on the crisis facing Japan's forestry and water resources have been published in the Japanese language by the Tokyo Foundation every January since 2009. These 2010 recommendations stress the threat from foreign direct investment in forest real estate: http://goo.gl/PJv7Y.

10. Mark Zeitoun and Jeroen Warner, "Hydro-Hegemony—a Framework for Analysis of Transboundary Water Conflicts," *Water Policy* 8, no 5 (2006): 435–460.

11. Kai Wegerich, "Hydro-Hegemony in the Amu Darya Basin," *Water Policy* 10, supp. 2 (2008): 71–88.

12. Martin Ira Glassner, "The Rio Lauca: Dispute over an International River," *Geographical Review* 60, no. 2 (April 1970): 192–207; Robert D. Tomasek, "The Chilean-Bolivian Lauca River Dispute and the OAS," *Journal of Interamerican Studies* 9, no. 3 (October 1967): 351–366.

13. Hilal Elver, *Peaceful Uses of International Rivers: The Euphrates and Tigris Rivers Dispute* (New York: Transnational Publishers, 2002).

14. U.S. Central Intelligence Agency, *The CIA World Factbook 2013* (New York: Skyhorse Publishing, 2012), sec. "Disputes—International."

15. Pal Tamas, *Water Resource Scarcity and Conflict: Review of Applicable Indicators and Systems of Reference* (Paris: UNESCO, 2003), 1.

16. Worldwatch Institute, *State of the World: Redefining Global Security* (Washington, DC: Worldwatch Institute, 2005), 3.

17. For an account of the arbitrary boundary making after World War I, see Margaret MacMillan, *Paris 1919: Six Months That Changed the World* (New York: Random House, 2003), 347–458.

18. Condoleezza Rice, "Syria Is Central to Holding Together the Mideast," *Washington Post*, November 24, 2012.

19. Pierre Englebert, Stacy Tarango, and Matthew Carter, "Dismemberment and Suffocation: A Contribution to the Debate on African Boundaries," *Comparative Political Studies* 35, no. 10 (2002): 1093–1118.

20. Marten van Harten, "Europe's Troubled Waters—a Role for the OSCE: The Case of the Kura-Araks," *Helsinki Monitor* 13, no. 4 (2002): 338–349; Food and Agriculture Organization, *Irrigation in the Middle East Region in Figures: Aquastat Survey 2008*, ed. Karen Frenken, Water Report 34 (Rome: FAO, 2009); Berrin Basak Vener and Michael E. Campana, "Conflict, Cooperation, and the New 'Great Game' in the Kura-Araks Basin of the South Caucasus," presentation at the 2008 UCOWR Conference, Southern Illinois University, Carbondale, http://goo.gl/YnISD.

21. Azerbaijan, however, has signed an accord with Iran to help protect the Araks River.

22. Frank E. Trout, "Morocco's Boundary in the Guir-Zousfana River Basin," *African Historical Studies* 3, no. 1 (1970): 37–56.

23. Food and Agriculture Organization (FAO) of the United Nations, Aquastat online data.

24. Y. Zarhloule, M. Boughriba, and M. Chanigui, "Water as a Parameter of Cooperation between Morocco and Algeria: The Case of the Angad-Maghnia Transboundary Stressed Aquifers of the Bounaïm-Tafna Basin," in UNESCO-IHP, *ISARM2010 International Conference: Transboundary Aquifers—Challenges and New Directions* (Paris: UNESCO, December 2010).

25. Luc Lambs and Mohamed Labiod, "Climate Change and Water Availability in Northwest Algeria: Investigation by Stable Water Isotopes and Dendrochronology," *Water International* 34, no. 2 (June 2009): 280.

26. John Bulloch and Adel Darwish, *Water Wars: Coming Conflicts in the Middle East* (London: Victor Gollancz, 1993), 69.

27. Figure cited in United Nations Office for the Coordination of Humanitarian Affairs, *Sudan: UN and Partners Work Plan 2012—Mid-Year Review* (New York: United Nations, 2012), 11.

28. Robert O. Collins, "The Ilemi Triangle," *Annales d'Ethiopie* 20 (2005), 5–12.

29. Since those killings, human-rights groups have documented a new pattern of reprisals and abuses against Kyrgyzstan's Uzbek population, rather than an attempt

to identify and prosecute the killers and promote ethnic reconciliation. The ethnic polarization has only hardened, threatening to destabilize the region. Uzbeks make up about 15 percent of Kyrgyzstan's population but are concentrated in the Kyrgyz portion of the Fergana Valley. Human Rights Watch, *Distorted Justice: Kyrgyzstan's Flawed Investigations and Trials of the June 2010 Violence* (New York: Human Rights Watch, June 2011).

30. The state government in 2011 actually hired an international consultant to help assess the cumulative economic losses suffered by Jammu and Kashmir owing to the Indus treaty.

31. Pakistan Ministry of Water and Power, *Pakistan Water Sector Strategy*, Detailed Strategy Formulation, vol. 4 (Islamabad: Ministry of Water and Power, October 2002), 7.

32. See, for example, S. Paul Kapur and Sumit Ganguly, "The Jihad Paradox: Pakistan and Islamist Militancy in South Asia," *International Security* 37, no. 1 (Summer 2012): 111–141; C. Christine Fair, "What to Do about Pakistan," *Foreign Policy*, June 21, 2012; Ahmed Rashid, *Pakistan on the Brink: The Future of America, Pakistan, and Afghanistan* (New York: Viking, 2012).

33. Addressing a meeting of former senior civil servants in Islamabad, President Asif Zardari said the terrorist groups were established as a matter of state policy. "Let us be truthful to ourselves and make a candid admission of the realities," he said. "The terrorists of today were the heroes of yesteryears until 9/11 occurred and they began to haunt us as well." Dean Nelson, "Pakistan President Asif Zardari Admits Creating Terrorist Groups," *Telegraph* (UK), July 8, 2009, http://bit.ly/iN7Qh.

34. Asif Ali Zardari, "Partnering with Pakistan," *Washington Post*, January 28, 2009.

35. Cited in Anand Mohan, "The Historical Roots of the Kashmir Conflict," *Studies in Conflict & Terrorism* 15, no. 4 (1992): 283–308.

36. *Dawn* (Pakistan), September 5, 1960. Also see Sisir Gupta, "The Nehru-Ayub Meeting," *Foreign Affairs Reports* 9, no. 10 (October 1960): 122–133.

37. A 1957 State Department telegram to the American Embassy in New Delhi, for example, suggested an Indus settlement "by India assuring Pakistan . . . as originally suggested by Lilienthal." Department of State, *Foreign Relations of the United States, 1955–57*, vol. 8, South Asia, Document 63.

38. David E. Lilienthal, "Another Korea in the Making?" *Collier's*, August 4, 1951.

39. These key provisions included Article 5 (titled, "Financial Provisions") and Article 10 ("Emergency Provision"), as well as annexes relating to transitional arrangements. With many of the original purposes completed, the World Bank's role is now confined to the appointment of a neutral expert or a seven-member court of arbitration to settle Pakistan-India differences or disputes. Treaty's full text at http://goo.gl/abmzX.

40. Set up in 1893 as the border between British-led India and Afghanistan and named after the then British foreign secretary for India, the Durand Line has long been despised and rejected by Afghanistan as a colonial imposition. After Pakistan was carved out of India in 1947, Afghanistan was the only country to openly oppose Pakistan's admission to the United Nations.

41. Leila M. Harris, "Water and Conflict Geographies of the Southeastern Anatolia Project," *Society and Natural Resources* 15 (2002): 743–759.

42. Nilgun Harmancioğlu, Necdet Alpaslan, and Eline Boele, "Irrigation, Health, and the Environment: A Review of Literature from Turkey," Working Paper 6 (Colombo, Sri Lanka: International Water Management Institute, 2001), 11.

43. Ali Akanda, Sarah Freeman, and Maria Placht, "The Tigris-Euphrates River Basin: Mediating a Path towards Regional Water Stability," *Al Nakhlah*, Spring 2007, 1–12.

44. Özden Zeynep Oktav, "Water Dispute and Kurdish Separatism in Turkish-Syrian Relations," in *The Turkish Yearbook of International Relations* (Ankara: University of Ankara Press, 2003), 102.

45. Mostafa Dolatyar and Tim S. Gray, *Water Politics in the Middle East: A Context for Conflict or Cooperation?* (New York: Palgrave Macmillan, 1999), 66.

46. Stephen Kinzer, "Accord Set for Syria and Turkey," *New York Times*, October 22, 1998.

47. Ilter Turan and Gün Kut, "Political-Ideological Constraints on Intra-Basin Cooperation on Transboundary Waters," *Natural Resources Forum* 21, no. 2 (May 1997), 140.

48. See Aliza Marcus, *Blood and Belief: The PKK and the Kurdish Fight for Independence* (New York: New York University Press, 2009); David McDowall, *A Modern History of the Kurds* (London: I. B. Tauris, 2004); John Bulloch and Harvey Morris, *No Friends but the Mountains: The Tragic History of the Kurds* (New York: Oxford University Press, 1993).

49. In 2012, Turkey had 789 dams, including some under construction.

50. Steven Solomon, *Water: The Epic Struggle for Wealth, Power, and Civilization* (New York: HarperCollins, 2010), ii.

51. Many of the domesticated grains that helped in the rise of Western civilization originated in the Tigris-Euphrates belt in Iraq, according to Jared Diamond, whose thesis is that "history followed different courses for different peoples because of differences among peoples' environments, not because of biological differences among peoples themselves." Jared Diamond, *Guns, Germs and Steel: The Fates of Human Societies* (New York: Norton, 1999).

52. Karl August Wittfogel, *Oriental Despotism: A Comparative Study of Total Power* (New York: Vintage, 1981).

53. See Jean Fairley, *The Lion River: The Indus* (London: Allen Lane, 1975).

54. Steven P. Erie, *Beyond Chinatown: The Metropolitan Water District, Growth and the Environment in Southern California* (Stanford, CA: Stanford University Press, 2006); Jeff Putman and Genny Smith, eds., *Deepest Valley—a Guide to Owens Valley: Its Roadsides and Mountain Trails* (Cupertino, CA: Genny Smith, 1979); Robert A. Sauder, *The Lost Frontier: Water Diversion in the Growth and Destruction of Owens Valley Agriculture* (Tucson, AZ: University of Arizona Press, 1994); Marc Reisner, *Cadillac Desert: The American West and Its Disappearing Water* (New York: Penguin, 1993).

55. See Reisner, *Cadillac Desert*.

56. John Hart, *Storm over Mono: The Mono Lake Battle and the California Water Future* (Berkeley, CA: University of California Press, 1996); Norris Hundley, *The Great Thirst: Californians and Water—a History* (Berkeley, CA: University of California Press, 2001).

57. J. H. W. Verzijl, *International Law in Historical Perspective*, vol. 3 (Leyden, Netherlands: A. W. Sijthoff, 1970), 113.

58. Albert Lepawsky, "International Development of River Resources," *International Affairs* 39, no. 4 (October 1963): 533.

59. Signed by the five original members of the commission—Belgium, France, Germany, the Netherlands, and Switzerland—and Britain, the revised Mannheim Convention came into force in 1963.

60. Bela Vitanyi, *The International Regime of River Navigation* (Berlin: Springer, 1980), 98; Richard Baxter, *The Law of International Waterways* (Cambridge, MA: Harvard University Press, 1964), 149.

61. For details, see the official website of the Joint Commission of Canada and the United States, http://goo.gl/9ZdRE, and the website of the U.S. section of the International Boundary and Water Commission of the United States and Mexico, http://goo.gl/wB006.

62. FAO, Aquastat online data.

63. Attorney General Judson Hudson wrote that "the fundamental principle of international law is the absolute sovereignty of every nation, as against all others, within its own territory." U.S. Attorney General Opinions, 21 Op. Att'y Gen. 274 (1895) ("Harmon Opinion").

64. The dam, known as the Elephant Butte Dam and located in New Mexico, was completed in 1916, along with the canals to move the apportioned water to the international boundary.

65. Article 5 of the 1906 Convention between the United States and Mexico Concerning the Equitable Distribution of Waters of the Rio Grande, U.S.-Mex., May 21, 1906, also found at 34 Stat. 2953.

66. Article 10(b) of the Treaty Relating to the Utilization of Waters of the Colorado and Tijuana Rivers and of the Rio Grande, February 3, 1944, U.S.-Mex., 3 U.N.T.S. 313 ("U.S.-Mexico Water Treaty").

67. See John E. Carroll, *Environmental Diplomacy: An Examination and a Prospective of Canadian-U.S. Transboundary Environmental Relations* (Ann Arbor, MI: University of Michigan Press, 1983).

68. Article 2 of the 1909 Treaty Relating to Boundary Waters between the United States and Canada, http://goo.gl/Ohw9s.

69. Aaron T. Wolf and Joshua T. Newton, "Case Study—the International Joint Commission: Canada and the United States of America," Institute for Water and Watersheds, Oregon State University, http://goo.gl/6cTI2.

70. Stephen C. McCaffrey, "The Harmon Doctrine One Hundred Years Later: Buried, Not Praised," *Natural Resources Journal* 35 (Summer 2006): 583.

71. Lepawsky, "International Development of River Resources," 538.

72. See Blaine Harden, *A River Lost: The Life and Death of the Columbia* (New York: Norton, 1996).

73. The comprehensive 1987 protocol called for the implementation of an eco-system approach to pollution control through the development of "lakewide management plans" to combat critical pollutants. The protocol also laid emphasis on tackling source pollution, groundwater contamination, contaminated sediment, and airborne toxic chemicals. In 1991, the two countries further expanded their cooperation through an "Agreement of Air Quality," with the commission given limited authority over joint air resources.

74. See Norris Hundley, *Dividing the Waters: A Century of Controversy between the United States and Mexico* (Berkeley and Los Angeles, CA: University of California Press, 1966).

75. To enable Mexico to divert its share of the allotted waters, the treaty provided for the Mexican construction of a main diversion structure. Known as the Morales Diversion Dam, it is located just below the point where the California–Baja California land boundary line intersects the Colorado River.

76. International Boundary and Water Commission of the United States and Mexico, "Interim International Cooperative Measures in the Colorado River Basin through 2017 and Extension of Minute 318 Cooperative Measures to Address the Continued Effects of the April 2010 Earthquake in the Mexicali Valley, Baja California," Minute No. 319, November 20, 2012, http://goo.gl/0DB5r.

77. Section III.6.e.iii of Minute No. 319.

78. Section III.6.d of Minute No. 319.

79. See Patrick McCully, *Silenced Rivers: The Ecology and Politics of Large Dams* (London: Zed Books, 1997); and Reisner, *Cadillac Desert.*

80. Daniel Seligman, *World's Major Rivers* (Las Vegas, NV: Colorado River Commission of Nevada, November 2008), 86.

81. John H. Coghlin, "All-American Canal Project Sparks Test Case for Trans-boundary Groundwater Law," *Boston College International and Comparative Law Review* 14, no. 1 (1991): 194.

82. Herbert Brownell and Samuel D. Eaton, "The Colorado River Salinity Problem with Mexico," *American Journal of International Law* 69, no. 255 (1975): 255–271.

83. International Boundary and Water Commission of the United States and Mexico, "Permanent and Definitive Solution to the International Problem of the Salinity of the Colorado River," Minute No. 242, August 30, 1973; Reisner, *Cadillac Desert,* 7.

84. The chemical spill occurred near Basel on November 1, 1986, after a fire at a warehouse belonging to Sandoz A. G., one of Switzerland's three largest chemical companies.

85. The International Commission for the Protection of the Danube River (ICPDR) came into being in 1998, the year the Danube River Protection Convention entered into force.

86. G. T. Raadgever, *Transboundary River Basin Management Regimes: The Rhine Basin Case Study*, Background Report (Delft, Netherlands: RBA Center, Delft University of Technology, April 2005), 2.

87. "Directive 2000/60/EC of the European Parliament and of the Council of October 23, 2000, Establishing a Framework for Community Action in the Field of Water Policy," http://goo.gl/5pltO.

88. FAO, Aquastat online data.

89. Charles J. Vörösmarty, Pamela Green, Joseph Salisbury, and Richard B. Lammers, "Global Water Resources: Vulnerability from Climate Change and Population Growth," *Science* 289, no. 5477 (July 14, 2000): 288.

90. The Pew Forum on Religion & Public Life, *The Future of the Global Muslim Population: Projections for 2010–2030* (Washington, DC: Pew Research Center, 2011). According to the report, Muslims will number 2.2 billion by 2030 compared to 1.6 billion in 2010, making up 26.4 percent of the world population compared to 23.4 percent in 2010.

91. See United Nations Population Division, *World Population Prospects: The 2010 Revision* (New York: United Nations, 2011).

92. James L. Gelvin, *The Arab Uprisings: What Everyone Needs to Know* (New York: Oxford University Press, 2012), 21–22; John Vidal, "What Does the Arab World Do When Its Water Runs Out?" *Observer*, February 20, 2011.

93. Jon B. Alterman and Michael Dziuban, *Clear Gold: Water as a Strategic Resource in the Middle East* (Washington, DC: Center for Strategic and International Studies, December 2010), v.

94. United Nations Development Programme, *Arab Human Development Report 2009: Challenges to Human Security in the Arab Countries* (New York: UNDP, 2010).

95. Human Rights Watch, *The Iraqi Government Assault on the Marsh Arabs* (Washington, DC: Human Rights Watch, January 2003).

96. United Nations Environment Program, "World Heritage Push for Garden of Eden" (Kyoto/Nairobi: UNEP, September 5, 2008).

97. A. R. Khater, "Intensive Groundwater Use in the Middle East and North Africa," in *Intensive Use of Groundwater: Challenges and Opportunities*, ed. Ramon Llamas and Emilio Custodio (Lisse, Netherlands: A. A. Balkema Publishers, 2002), 355–386; Food and Agriculture Organization, *Irrigation in the Middle East Region in Figures—Aquastat Survey 2008*, Water Report 34 (Rome: FAO, 2009).

98. FAO, *Irrigation in the Middle East in Figures*.

99. United Nations Environment Program, Environmental Data Explorer, online database, http://geodata.grid.unep.ch.

100. Eugenia Ferragina and Francesca Greco, "The Disi Project: An Internal/External Analysis," *Water International* 33, no. 4 (2008): 451–463; Greg Shapland, *Rivers of Discord: International Water Disputes in the Middle East* (London: C. Hurst, 1997), 148–150.

101. See Ayşegül Kibaroğlu, *Building a Regime for the Waters of the Euphrates-Tigris River Basin* (Berlin: Springer, 2002).

102. The economic-cooperation protocol, signed July 17, 1987, carried a section on water. The section stated that Turkey "undertakes to release a yearly average of more than 500 m³/second at the Turkish-Syrian border."

103. Ferhad Ibrahim and Gülistan Gürbey, eds., *The Kurdish Conflict in Turkey* (New York: St. Martin's, 2000), 109.

104. Oktav, "Water Dispute and Kurdish Separatism in Turkish-Syrian Relations," 99–100.

105. Hillel I. Shuval and Hassan Dwiek, *Water Resources in the Middle East: The Israeli-Palestinian Water Issues—from Conflict to Cooperation* (Berlin: Springer, 2007), 158.

106. Stephen Foster and Daniel P. Loucks, eds., *Nonrenewable Groundwater Resources: A Guidebook on Socially Sustainable Management for Water Policymakers* (Paris: UNESCO, 2006), 19–20.

107. UK Environment Agency, *Do We Need Large-Scale Water Transfers for South-East England?* (London: Environment Agency, September 2006).

108. See Peter Perdue, *China Marches West: The Qing Conquest of Central Eurasia* (Cambridge, MA: Belknap, 2005); Tsering Shakya, *The Dragon in the Land of Snows: A History of Modern Tibet since 1947* (New York: Columbia University Press, 1999); Michael C. van Walt van Praag, *The Status of Tibet: History, Rights, and Prospects in International Law* (Boulder, CO: Westview Press, 1987); Warren W. Smith Jr., *Tibetan Nation: A History of Tibetan Nationalism and Sino-Tibetan Relations* (Boulder, CO: Westview Press, 1996).

109. The Chinese Ministry of Water Resources claims that China has "signed water cooperation agreements or memorandums of understanding with over 40 countries," including in Africa and Latin America. It also signed an agreement with Mongolia in 1994 for "protection" of border water resources.

110. Convention on the Law of the Non-Navigational Uses of International Watercourses, adopted by the General Assembly of the United Nations on May 21, 1997, 36 I.L.M. 700 (1997), Article 9.

111. This definition of what constitutes a large dam was made by the nongovernmental International Commission on Large Dams and accepted by the World Commission on Dams, which was established jointly by the World Bank and the World Conservation Union.

112. Up to 24,000 large dams were built in China by the end of the twentieth century. Jiyu Chen, "Dams, Effect on Coasts," in *Encyclopedia of Coastal Science*, ed. Maurice L. Schwartz, vol. 24, Encyclopedia of Earth Sciences Series (Berlin: Springer, 2005).

113. It was not until the political transition in November 2012 that the domination of engineers in the top leadership since the Deng Xiaoping era gave way to a more diverse lineup with backgrounds that include law, economics, and history, although the group is led by Xi Jinping—an engineer by training.

114. White House, "31. Herbert Hoover 1929–1933," http://goo.gl/H3b5D.

115. Food and Agriculture Organization, *Irrigation in Southern and Eastern Asia in Figures—Aquastat Survey 2011*, Water Report 37 (Rome: FAO, 2012), sec. "China."

116. The Chinese Commerce Ministry's August 2006 regulations called upon Chinese enterprises to hire local workers, respect local customs, and adhere to safety norms in overseas projects. The regulations followed a backlash against Chinese businesses in Zambia over the death of fifty-one Zambian workers in an explosion at a Chinese-owned copper mine and the subsequent shooting of six Zambians at a Chinese-run mine.

117. HydroChina Corporation, "Map of Planned Dams," 2010, http://goo.gl/4VBVy.

118. Xinglin Lei, "Possible Roles of the Zipingpu Reservoir in Triggering the 2008 Wenchuan Earthquake," *Journal of Asian Earth Sciences* 40, no. 4 (2011): 844–854; Shemin Ge, Mian Liu, Ning Lu, Jonathan W. Godt, and Gang Luo, "Did the Zipingpu Reservoir Trigger the 2008 Wenchuan Earthquake?" *Geophysical Research Letters* 36 (2009), doi:10.1029/2009GL040349.

119. China's State-Owned Assets Supervision and Administration Commission, in an August 2011 report, described Myitsone as a model of party-led overseas expansion and noted that the dam project was designed to "principally serve our nation's southern power grid."

120. FAO, Aquastat online data.

121. The population sizes of Canada, China, and India cited in this section are based on World Bank figures for 2011, while the sizes of their water resources are based on 2012 Aquastat data.

122. See Xie Jian, with Andres Liebenthal, Jeremy J. Warford, John A. Dixon, Wang Manchuan, Gao Shiji, Wang Shuilin, Jiang Yong, and Ma Zhong, *Addressing China's Water Scarcity: Recommendations for Selected Water Resource Management Issues* (Washington, DC: World Bank, January 2009), 9–11.

123. Mara Hvistendahl, "China's Three Gorges Dam: An Environmental Catastrophe?" *Scientific American*, March 25, 2008.

124. Yi Si, "The World's Most Catastrophic Dam Failures: The August 1975 Collapse of the Banqiao and Shimantan Dams," in *The River Dragon Has Come! The Three Gorges Dam and the Fate of China's Yangtze River and Its People*, by Dai Qing, ed. John G. Thibodeau and Philip B. Williams, trans. Ming Yi (Armonk, NY: M. E. Sharpe, 1998), 25–26; Wayne J. Graham, *A Procedure for Estimating Loss of Life Caused by Dam Failure* (Denver, CO: U.S. Bureau of Reclamation, September 1999), 1.

125. See, for example, Tan Yingzi, "Yellow River Dams on Verge of Collapse," *China Daily*, June 19, 2009, http://goo.gl/fvoK7. Xu Yuanming's comment published in *China Economic Weekly*, a magazine run by the official *People's Daily* newspaper, was cited in Malcolm Moore, "More Than 40,000 Chinese Dams at Risk of Breach," *The Telegraph* (London), August 26, 2011, http://goo.gl/6m4Rs.

126. "Water Wars Feared over Mekong," Radio Free Asia, September 30, 2012, http://goo.gl/NbktQ.

127. See, for example, Richard Stone, "For China and Kazakhstan, No Meeting of the Minds on Water," *Science* 337, no. 6093 (July 27, 2012): 405–407.

128. Xinhua figures cited in Edward Wong, "Plan for China's Water Crisis Spurs Concern," *New York Times*, June 1, 2011.

129. U.S. Corps of Engineers, National Inventory of Dams, 2010 database.

130. Christine E. Gudorf and James Edward Huchingson, *Boundaries: A Casebook in Environmental Ethics* (Washington, DC: Georgetown University Press, 2010), 172.

CHAPTER 5: SHAPING WATER FOR PEACE AND PROFIT

1. Anton Earle, Anders Jagerskog, and Joakim Ojendal, eds., *Transboundary Water Management: Principles and Practice* (New York: Earthscan, 2010); United Nations,

Managing Water under Uncertainty and Risk, vol. 1 of World Water Development Report 4 (Paris: United Nations Educational, Scientific and Cultural Organization and UN-Water, March 2012); Ken Conca, *Governing Water: Contentious Transnational Politics and Global Institution Building* (Cambridge, MA: MIT Press, 2005); Meredith A. Giordano and Aaron T. Wolf, "Sharing Waters: Post-Rio International Transboundary Water Management," *Natural Resources Forum* 27, no. 2 (2004).

2. Lake Tanganyika Biodiversity Project, "Overview," http://goo.gl/VoKrh.

3. Salman M. A. Salman, "The Helsinki Rules, the UN Watercourses Convention and the Berlin Rules: Perspectives on International Water Law," *Water Resources Development* 23, no. 4 (December 2007): 638.

4. UN-Water, *Status Report on the Application of Integrated Approaches to Water Resources Management* (New York: United Nations, 2012).

5. United Nations Environment Program, *Challenges to International Waters: Regional Assessments in a Global Perspective* (Nairobi: UNEP, 2006); United Nations, *Managing Water under Uncertainty and Risk.*

6. United Nations Educational, Scientific and Cultural Organization, *Internationally Shared (Transboundary) Aquifer Resources Management: Their Significance and Sustainable Management*, IHP-VI, IHP Non-Serial Publications in Hydrology (Paris: UNESCO, 2001), 15.

7. United Nations Economic Commission for Europe, UNECE Statistical Database, http://w3.unece.org/pxweb/Dialog.

8. United Nations Educational, Scientific and Cultural Organization, *Atlas of Transboundary Aquifers: Global Maps, Regional Cooperation, and Local Inventories*, ed. S. Puri and A. Aureli (Paris: UNESCO-IHP ISARM Program, 2009).

9. Cited in BBC News, "Thirsting for War," October 5, 2000, http://goo.gl/1BJlB.

10. Statement of Chinese envoy Gao Feng at the United Nations General Assembly, May 21, 1997, in "General Assembly Adopts Convention on the Law of the Non-Navigational Uses of International Watercourses," United Nations Press Release GA/9248, http://goo.gl/12DqY.

11. Daniel Seligman, *World's Major Rivers* (Las Vegas, NV: Colorado River Commission of Nevada, November 2008), 119.

12. *State of Wyoming v. State of Colorado*, 259 U.S. 419, 42 S.Ct. 552, 66 L.Ed. 999 (1922).

13. Article 36(1) reads, "The present Convention shall enter into force on the ninetieth day following the date of deposit of the thirty-fifth instrument of ratification, acceptance, approval or accession with the Secretary-General of the United Nations." Convention on the Law of the Non-Navigational Uses of International Watercourses, adopted by the General Assembly of the United Nations on May 21, 1997, 36 I.L.M. 700 (1997) ("*U.N. Convention*"), http://goo.gl/FXjxJ.

14. The Helsinki Rules on the Uses of the Waters of International Rivers, adopted by the International Law Association at Helsinki in August 1966, http://goo.gl/WmTTU.

15. See Slavko Bogdanovic, *International Law of Water Resources—Contribution of the International Law Association 1954–2000* (The Hague: Kluwer Law International, 2001); C. B. Bourne, "The International Law Association's Contribution to International Water Resources Law," *Natural Resources Journal* 36 (1996): 155–216.

16. Chapters 2 and 3 of the Helsinki Rules.

17. Articles 26–34 of the Helsinki Rules.

18. For details of the ILC's work, see Patricia Wouters, ed., *Codification and Progressive Development of International Water Law: The Work of the International Law Commission of the United Nations* (The Hague: Kluwer, 1998); Sir Arthur Watts, ed., *The International Law Commission 1949–1998*, vols. 1–3 (New York: Oxford University Press, 2000).

19. *The Berlin Rules on Water Resources*, adopted at the International Law Association's Berlin Conference, August 2004, http://goo.gl/UQOa0. Also see Joseph W. Dellapenna, "The International Law Association's Berlin Rules on Water Resources" (Gland, Switzerland: International Union for Conservation of Nature, n.d.).

20. See Articles 6 and 7 of the Berlin Rules.

21. International Legal Association, Berlin Conference 2004, Water Resources Committee Report Dissenting Opinion, http://goo.gl/mAsbq.

22. It was in 1970 that the United Nations General Assembly asked the International Law Commission to help draft the international convention on the basis of the Helsinki Rules. It, however, took more than a quarter century for the draft convention to come before the General Assembly for a vote because the process was hobbled by controversies and political foot-dragging by countries that control major water resources.

23. See Stephen McCaffrey, *The Law of International Watercourse—Non-Navigational Uses* (Oxford, UK: Oxford University Press, 2001); Joseph W. Dellapenna and Joyeeta Gupta, eds., *The Evolution of the Law and Politics of Water* (Berlin: Springer, 2010).

24. Article 6 of the UN Convention.

25. Article 10(1) of the UN Convention.

26. Article 7 of the UN Convention.

27. Salman, "The Helsinki Rules," 634.

28. Article 33 of the UN Convention.

29. Article 33 (paragraphs 3 to 9) of the UN Convention.

30. United Nations, "General Assembly Adopts Convention on the Law of the Non-Navigational Uses of International Watercourses," UN Press Release GA/9248, May 21, 1997.

31. This exclusion happened even though the supplementary 1986 Groundwater Rules to the 1966 Helsinki Rules had included "fossil" aquifers within their scope.

32. Article 2(a) of the UN Convention.

33. Draft Articles on the Law of Transboundary Aquifers, full text, http://goo.gl/Iqo6O. For background on the draft articles, see Chusei Yamada, "Codification of the Law of Transboundary Aquifers (Groundwaters) by the United Nations," *Water International* 36, no. 5 (September 2011): 557–565.

34. Statement by Steven Hill, counselor, United States Mission to the United Nations, "The Law of Transboundary Aquifers," at a General Assembly Sixth (Legal) Committee session, October 18, 2011, http://goo.gl/wbKgt.

35. United Nations General Assembly, Resolution 1803 (XVII) of December 14, 1962, "Permanent Sovereignty over Natural Resources," http://goo.gl/wBt25.

36. Andreas Zimmermann, Karin Oellers-Frahm, Christian Tomuschat, and Christian J. Tams, *The Statute of the International Court of Justice: A Commentary* (New York: Oxford University Press, 2012).

37. International Court of Justice, Judgment on the Gabcíkovo-Nagymaros Project (Hungary/Slovakia), September 25, 1977, http://bit.ly/I4XY5O.

38. The United Nations Economic Commission for Europe, seeking to promote pan-European economic integration, brings together fifty-six countries located in the European Union, non-EU Western and Eastern Europe, Southeast Europe, the Commonwealth of Independent States, and North America.

39. The Convention on the Protection and Use of Transboundary Watercourses and International Lakes, done at Helsinki on March 17, 1992, http://goo.gl/0ZVjx.

40. Amendment to Articles 25 and 26 of the Convention on the Protection and Use of Transboundary Watercourses and International Lakes, done on November 28, 2003.

41. Rio Declaration on Environment and Development, United Nations Conference on Environment and Development, Rio de Janeiro, June 3–14, 1992, http://goo.gl/f0LqW.

42. In the matter of the Indus Waters Kishenganga Arbitration before the Court of Arbitration between the Islamic Republic of Pakistan and the Republic of India, Order on the Interim Measures Application of Pakistan, September 23, 2011, http://goo.gl/Z9Bk8.

43. Defense Science Board, *Trends and Implications of Climate Change for National and International Security* (Washington, DC: Department of Defense, October 2011), 58.

44. The future of the draft articles finalized by the International Law Commission in 2001 under the title Responsibility of States for Internationally Wrongful Acts remains uncertain in the absence of their adoption by the international community. See draft articles in Michael M. Wood and Arnold A. Pronto, *The International Law Commission 1999–2009*, vol. 4, *Treaties, Final Draft Articles and Other Materials* (New York: Oxford University Press, 2011).

45. United Nations World Water Assessment Program, *Water: A Shared Responsibility*, United Nations World Water Development Report 2 (Paris and New York: UNESCO and Berghahn Books, 2006), 186.

46. Hendrik Bruins, "Proactive Contingency Planning vis-à-vis Declining Water Security in the 21st Century," *Journal of Contingencies and Crisis Management* 8, no. 2 (2000): 63–72; Shlomi Dinar, *International Water Treaties: Negotiation and Cooperation along Transnational Rivers* (Abingdon, UK: Routledge, 2008); Joseph W. Dellapenna, *Middle East Water: The Potential and Limits of Law* (The Hague: Kluwer Press, 2002).

47. Article 9(2) of the UN Convention on the Law of the Non-Navigational Uses of International Watercourses.

48. Elena Lopez-Gunn and Manuel Ramón Llamas, "Re-thinking Water Scarcity: Can Science and Technology Solve the Global Water Crisis?" *Natural Resources Forum* 32, no. 3 (2008): 228–238.

49. United Nations, "Water without Borders," Backgrounder, Water for Life 2005–2015 project (New York: United Nations, n.d.), 1.

50. See Electric Power Research Institute and Tetra Tech, *Sustainable Water Resources Management*, vol. 3, *Case Studies on New Water Paradigm* (Palo Alto, CA: EPRI and Tetra Tech, January 2010).

51. U.S. National Intelligence Council, *Global Trends 2025: A Transformed World* (Washington, DC: National Intelligence Council, November 20, 2008), 3.

52. *Israel-Jordan Peace Treaty*, Annex II: Water and Related Matters, http://goo.gl/wcGGC.

53. World Bank, *West Bank and Gaza: Assessment of Restrictions on Palestinian Water Sector Development*, Report No. 47657-GZ (Washington, DC: World Bank, April 2009), v.

54. *Israeli-Palestinian Interim Agreement on the West Bank and the Gaza Strip*, 1995, Protocol Concerning Civic Affairs, Article 40, http://goo.gl/a4VpX.

55. See Anthony Cordesman and Aram Nerguizian, *The Arab-Israeli Military Balance: Conventional Realities and Asymmetric Challenges* (Washington, DC: Center for Strategic & International Studies, June 29, 2010).

56. The figure for Pakistan includes the 167.2-billion-cubic-meter mean yearly flows of the western rivers, which have been reserved for that country under the Indus Waters Treaty, plus the 11.1-billion-cubic-meter bonus waters that flow to Pakistan—according to the Food and Agriculture Organization—from the eastern rivers. That adds up to 85.9 percent of the 207.6 billion cubic meters of total yearly flows of the six Indus-system rivers that enter Pakistan from India. This figure does not include the waters of the Kabul River, which flows from Afghanistan to join the main Indus stream in Pakistan. As for Egypt, it increased its water share to 44 million acre feet (54.3 billion cubic meters) under the 1959 Agreement for the Full Utilization of the Nile Waters that it signed with Sudan.

57. Paul Harrison and Fred Pearce, *AAAS Atlas of Population & Environment* (Washington, DC: American Association for the Advancement of Science, 2007), "Part 2—Natural Resources: Freshwater."

58. Pakistan Ministry of Water and Power, *Pakistan Water Sector Strategy*, Detailed Strategy Formulation, vol. 4 (Islamabad: Ministry of Water and Power, 2002), 65.

59. The Treaty on the River Plate Basin, April 23, 1969, 875 U.N.T.S. 3 ("La Plata River Basin Treaty"). Also see Declaration of Asuncion on the Use of International Rivers, adopted at the Fourth Meeting of the Ministers of Foreign Affairs of the States of the River Plate Basin, June 1–3, 1971, Y.B. INT'L L. COMM'N, Vol. 2, pt. 2:324 (1976); Statute of the River Uruguay between Argentina and Uruguay, February 26, 1975, 1295 U.N.T.S. 340.

60. Transboundary Freshwater Dispute Database, "International Freshwater Treaties," http://ocid.nacse.org/tfdd/treaties.php.

61. The 1969 Vienna Convention on the Law of Treaties codified and progressively developed international treaty law—the customary and other rules governing conclusion, implementation, interpretation, and termination of international agreements. Full text at http://goo.gl/1skZV.

62. International Court of Justice, "Pulp Mills on the River Uruguay (Argentina versus Uruguay)," http://goo.gl/46bTE.

63. Dennis Kux, *Estranged Democracies: India and the United States 1941–1991* (Darby, PA: Diane Publishing Company, 1993), 355.

64. Jerome Delli Priscoli and Aaron T. Wolf, *Managing and Transforming Water Conflicts* (Cambridge, UK: Cambridge University Press, 2009), 108.

65. Jon B. Alterman and Michael Dziuban, *Clear Gold: Water as a Strategic Resource in the Middle East* (Washington, DC: Center for Strategic and International Studies, December 2010); Food and Agriculture Organization, *Irrigation in the Middle East Region in Figures: Aquastat Survey 2008*, ed. Karen Frenken, Water Report 34 (Rome: FAO, 2009), section "Yemen."

66. For an account of the World Bank's policy, see Salman M. A. Salman, *The World Bank Policy for Projects on International Waterways: An Historical and Legal Analysis* (Washington, DC: World Bank, 2009).

67. Malin Falkenmark, "Global Water Issues Confronting Humanity," *Journal of Peace Research* 27, no. 2 (1990): 177–190.

68. Maude Barlow, *Blue Covenant: The Global Water Crisis and the Coming Battle for the Right to Water* (New York: New Press, 2008).

69. China's Water Resources and Hydropower Planning and Design General Institute, Presentation at the ESCAP Ad Hoc Expert Group Meeting on Water-Use Efficiency Planning, Bangkok, October 26–28, 2004; United Nations Economic and Social Commission for Asia and the Pacific, *The State of the Environment in Asia and the Pacific 2005* (Bangkok: UNESCAP, 2006), 63.

70. Food and Agriculture Organization of the United Nations, Aquastat tables, "Freshwater Availability: Precipitation and Internal Renewable Water Resources (IRWR)," and "Area Equipped for Irrigation," December 2012.

71. Patrick J. Mulholland, Ashley M. Helton, Geoffrey C. Poole, Robert O. Hall, Stephen K. Hamilton, Bruce J. Peterson, Jennifer L. Tank, et al., "Stream Denitrification across Biomes and Its Response to Anthropogenic Nitrate Loading," *Nature* 452 (March 13, 2008): 202–205, doi:10.1038/nature06686.

72. Figures in Richard Dobbs, Jeremy Oppenheim, Fraser Thompson, Marcel Brinkman, and Marc Zornes, *Resource Revolution: Meeting the World's Energy, Materials, Food, and Water Needs* (McKinsey Global Institute and McKinsey Sustainability & Resource Productivity Practice, November 2011), 186.

73. Lester R. Brown, *Outgrowing the Earth: The Food Security Challenge in an Age of Falling Water Tables and Rising Temperatures* (New York: Earthscan, 2005), 111–117.

74. UNESCAP, *The State of the Environment in Asia and the Pacific*, 70.

75. International Energy Agency, *World Energy Outlook 2012* (Paris: International Energy Agency, 2012); United Nations, *Managing Water under Uncertainty and Risk*.

76. National Intelligence Council, *Global Trends 2025*, 47.

77. National Research Council of the National Academies, *Desalination: A National Perspective* (Washington, DC: National Academies Press, 2008); Raphael Semiat, "Desalination: Present and Future," *Water International* 25, no. 1 (March 2000): 54–65; N. V. Voutchkov, "The Ocean—A New Resource for Drinking Water," *Public Works*, June 2004, 30–33; Heather Cooley, Peter H. Gleick, and Gary

Wolff, *Desalinization, with a Grain of Salt—a California Perspective* (Oakland, CA: Pacific Institute, 2006); R. Popkin, *Desalination: Water for the World's Future* (New York: Praeger, 1968).

78. For details, see National Research Council, *Desalination: A National Perspective.*

79. Marco Rognoni, M. P. Ramaswamy, and J. Justin Robert Paden, "Energy Cost for Desalination Evaporation versus Reverse Osmosis," *International Journal of Nuclear Desalination* 4, no. 3 (2011): 277–284. Also see Toshio Konishi and B. M. Misra, "Freshwater from the Seas," *IAEA Bulletin* 43, no. 2 (2001): 5–8; International Atomic Energy Agency, *Optimization of the Coupling of Nuclear Reactors and Desalination Systems*, Publication IAEA-TECDOC-1444 (Vienna: IAEA, 2005).

80. National Research Council of the National Academies, *Water Reuse: Potential for Expanding the Nation's Water Supply through Reuse of Municipal Wastewater* (Washington, DC: National Academies Press, 2012), 195.

81. United Nations Educational, Scientific and Cultural Organization (UNESCO), *Water Security and Peace: A Synthesis of Studies Prepared under the PCCP-Water for Peace Process*, comp. William J. Cosgrove (Paris: UNESCO, 2003), 49.

82. Alliance to Save Energy, *Watergy: Energy and Water Efficiency in Water Supply and Wastewater Treatment—Cost-Effective Savings of Water and Energy* (Washington, DC: Alliance to Save Energy, 2007); National Research Council of the National Academies, *Water Reuse.*

83. J. Martínez Beltrán and S. Koo-Oshima, *Water Desalination for Agricultural Applications*, Land and Water Discussion Paper 5 (Rome: Food and Agriculture Organization, 2004), 2.

84. National Research Council of the National Academies, *Water Reuse*, 173.

85. Intergovernmental Panel on Climate Change, *Climate Change 2007: Impacts, Adaptation and Vulnerability*, Working Group II Contribution to the Fourth Assessment Report of the IPCC (New York: Cambridge University Press, 2007), chap. 10, sec. 10.5.2, "Hydrology and Water Resources."

86. In the face of public opposition, San Diego's city council first rejected the wastewater-reuse idea in 1998, only to agree more than a decade later to set up a treatment plant, which began producing about a million gallons of recycled water a day in 2011.

87. Food and Agriculture Organization of the United Nations, Aquastat online data, http://goo.gl/83tfb.

88. FAO, Aquastat online data.

89. Jin Zhu and Liang Chao, "Conservancy Projects to Draw More Investment," Xinhua, October 13, 2011.

90. FAO, Aquastat table, "Freshwater Availability."

91. Intergovernmental Panel on Climate Change, *Climate Change 2007: The Physical Science Basis; Contribution of Working Group I to the Fourth Assessment Report of the Intergovernmental Panel on Climate Change*, ed. S. Solomon, D. Qin, M. Manning, Z. Chen, M. Marquis, K. B. Averyt, M. Tignor, and H. L. Miller (Cambridge, UK: Cambridge University Press, 2007).

92. Figures cited in United Nations, *Managing Water under Uncertainty and Risk.*

93. United Nations Environment Program and Stockholm Environment Institute, *Rainwater Harvesting: A Lifeline for Human Well-Being* (Nairobi: UNEP, 2009); United Nations Environment Program, *Rain and Storm Water Harvesting in Rural Areas* (Dublin, Ireland: Tycooly International Publishing, 1982).

94. Brahma Chellaney, "Asia's Worsening Water Crisis," *Survival* 54, no. 2 (April–May 2012): 143–156.

95. Daniel Wild, Carl-Johan Francke, Pierin Menzli, and Urs Schön, *Water: A Market of the Future* (Zurich, Switzerland: Sustainable Asset Management, December 2007), 13.

96. Ashley Halsey III, "Billions Needed to Upgrade America's Leaky Water Infrastructure," *Washington Post*, January 3, 2012.

97. Willem Buiter, "Essay: Water as Seen by an Economist," Global Themes Strategy, Citigroup Global Markets, July 20, 2011.

98. See, for example, Robert Glennon, *Unquenchable: America's Water Crisis and What to Do about It* (Washington, DC: Island Press, 2009).

99. See, for example, United Nations Development Programme, *Human Development Report 2006* (New York: UNDP, 2006).

100. Organization for Economic Cooperation and Development, "Improving Water Management: Recent OECD Experience," Policy Brief, February 2006, 3.

101. UNESCAP, *The State of the Environment in Asia and the Pacific*, 70.

102. UN-Water, *Status Report on Application of Integrated Approaches*.

103. Because of the length of the Indus treaty's annexes, study after study has glossed over their contents, focusing merely on the text in the main body. The Indus treaty's full text and annexes are available at http://bit.ly/bRL2O6. The Israeli-Jordanian treaty, in its Article 6, mentions the goal of a lasting water settlement. But the agreed principles, quantities, and quality are set out in its Annex 2. Text of Israel-Jordan Peace Treaty at http://goo.gl/UJmhJ.

104. The Anti-Ballistic Missile Treaty, although of unlimited duration, included a provision for either party to withdraw if "extraordinary events related to the subject matter of this Treaty have jeopardized its supreme interests," with the withdrawing party required to give six months' notice of its intention to withdraw, including a statement of the "extraordinary events."

105. According to Article 62 of the Vienna Convention, a fundamental change of circumstances may not be invoked as a ground for terminating or withdrawing from a treaty if that pact involved the territorial realignment of political frontiers or "if the fundamental change is the result of a breach by the party invoking it either of an obligation under the treaty or of any other international obligation owed to any other party to the treaty." Also see Ian Sinclair, *The Vienna Convention on the Law of Treaties* (Oxford, UK: Manchester University Press, 1984), 181–195; Olivier Corten and Pierre Klein, *The Vienna Conventions on the Law of Treaties: A Commentary* (New York: Oxford University Press, 2011), 1411–1436.

106. International Court of Justice, *Fisheries Jurisdiction Case* (Federal Republic of Germany versus Iceland), General List No. 56, Judgment of February 2, 1973, Jurisdiction of the Court, 18.

107. F. Meng, S. Chae, A. Drews, M. Kraume, H. Shin, and F. Yang, "Recent Advances in Membrane Bioreactors (MBRs): Membrane Fouling and Membrane Material," *Water Research* 43, no. 6 (April 2009): 1489–1512.

108. Gordon Conway, *The Doubly Green Revolution: Food for All in the Twenty-First Century* (Ithaca, New York: Cornell University Press, 1999).

109. United Nations, *Managing Water under Uncertainty and Risk*.

Glossary

actual renewable water resources (ARWR): The sum of internally and externally generated water resources in a nation, taking into account the reduction of flows due to upstream withdrawals and the quantity of waters reserved for a downstream country.

algal bloom: A heavy growth of algae in and on a body of water often triggered by environmental conditions such as high nitrate and phosphate concentrations.

aquifer: A water-bearing geologic formation or formations. Aquifer usually refers to water-bearing stratums of permeable rock, sand, or gravel capable of storing, transmitting, and yielding exploitable quantities of water.

artesian spring: When water in a confined aquifer rises above the land surface because of natural pressure, it forms a free-flowing spring.

artificial recharge: Artificially using surface-water supplies to store water underground in aquifers.

average groundwater depletion: The amount of water extracted yearly from aquifers that is not replenished. When such action is continuous, it risks fully depleting the aquifer.

baseflow: The sustained flow in a stream linked to groundwater discharge or seepage. When the groundwater table intersects the land surface, the water escapes as a spring or baseflow of a stream.

basin (or "river basin"): The entire tract of land drained by a river and its tributaries. It covers all the land dissected and drained by many streams and creeks that flow downhill to empty into one central river that goes out to the estuary or sea. Also see "catchment area" and "groundwater basin."

biological diversity (or "biodiversity"): The variability among living organisms and the ecological complexes of which they are a part, including the diversity found within and between species and ecosystems.

blue water: Resources in rivers, lakes, and aquifers used for irrigation, industrial, and other needs. Because blue water is open to human manipulation, irrigation relies on this type of water. See "green water."

brackish water: Less salty water than seawater. It is found in underground aquifers or estuaries, deltas, and mangrove swamps.

catchment area (also known as "drainage basin" or "watershed"): The area of land surface producing runoff. It collects the water originating as precipitation and drains it into a stream, river, lake, reservoir, or other body of water. Other terms used to describe a catchment area include "catchment," "catchment basin," "drainage area," "river basin," and "water basin." Catchment also refers to the land area from which groundwater (in addition to surface water) drains into a stream system.

conservation storage: Storage of water for release later for purposes such as municipal water supply, power, or irrigation. This is distinct from storage capacity used for flood control.

consumptive use: Extracted water that is no longer available for use because it has evaporated or transpired; been incorporated into products and crops, consumed by man or livestock, or discharged directly into the sea; or otherwise removed from freshwater resources.

crop yield: Crop production per unit of harvested area.

dam silting: The accumulation of sediments inside a dam, leading to erosion in its useful capacity.

dead storage: The volume in a dam or reservoir below the lowest controllable level.

dependency ratio: The percentage of the total water supply originating outside a nation's frontiers. This ratio can serve as an important indicator of a country's water security or insecurity.

depletion: The progressive withdrawal of water from surface-water or groundwater bodies at a rate greater than natural replenishment.

desalinated water: Freshwater generated by removing salt from seawater or brackish water.

drainage basin: See "catchment area."

drip irrigation (also known as "trickle irrigation," "dribble irrigation," or "microirrigation"): An irrigation system that saves water and fertilizer by allowing water to drip slowly to the roots of plants (either onto the soil surface or directly onto the root zone) through a network of valves, pipes, tubing, and emitters. Water is applied in small but frequent quantities to maintain the most active part of the soil at quasi-optimum moisture.

ecosystem: A community of interdependent organisms together with the environment they inhabit and with which they interact.

ecosystem services: Services provided by the natural system that benefit people. Some of these services are well known, including food, fiber, and fuel provision and the cultural services that provide benefits to people through recreation and appreciation of nature. Other services are less known, including climate regulation, purification of air and water, flood protection, soil formation, and nutrient cycling.

evapotranspiration: Plant-related water that is subject to the processes of evaporation and transpiration. Evaporation occurs when water changes to vapor on soil and plant surfaces. Transpiration refers to the water lost through the leaves of plants.

external renewable water resources: Inflows of surface water and groundwater from neighboring countries.

flood irrigation: All types of irrigation (such as "flood recession," "spate irrigation," and "wild flooding") that use floodwaters on farmland without the help of major structural works.

floodplain: A strip of relatively smooth land bordering a stream or river, built of sediment carried by that watercourse and subject to recurrent flooding.

fluvial: Of or pertaining to rivers and streams; produced by or found in a river or stream.

fossil water: When geological changes have sealed off a deep aquifer from further recharging, the water locked inside is classified as *fossil water*. The water may have been locked in for hundreds or even thousands of years. The age of the water is what gives it the name "fossil water."

freshwater: Water of sufficient quality to support its intended purpose—human consumption, food production, electric power generation, or industrial processes.

green water: Soil moisture from precipitation that naturally supports rainfed agriculture and other plants and trees.

groundwater: Water beneath the land surface that fills the spaces in rock and sediment. Essentially, it is rainfall and snowmelt lying in underground aquifers.

groundwater basin: A groundwater reservoir together with the overlying land surface and the underlying aquifers that contribute water to the reservoir. It may be separated from adjacent groundwater basins by geologic or hydrologic boundaries.

groundwater recharge: The inflow of surface water to a groundwater reservoir ("zone of saturation"). The addition of water to the groundwater storage can be by natural processes or by artificial methods for subsequent human withdrawals.

groundwater recharge area: The area in which water infiltrates and moves downward into the zone of saturation, thus replenishing groundwater.

hydro: In a broad sense, the term is really a synonym for water and refers to more than just hydropower. Thus, "hydropolitics" connotes water politics, and "hydro-infrastructure" refers to irrigation, dam, hydropower, storage, and canal infrastructure.

hydrogeology: The study of the interrelationships of geological materials and processes with water, especially groundwater.

hydrologic cycle (or "water cycle"): The natural cycle representing the movement of water from the atmosphere to earth and back to the atmosphere, including precipitation, infiltration, percolation, storage, evaporation, transpiration, and condensation.

hydrology: The science that relates to the occurrence, distribution, and chemistry of all waters of earth. In practice, the study of the waters of the oceans and the atmosphere is considered part of the sciences of oceanography and meteorology.

impoundment: A body of water confined by a dam, dike, reservoir, floodgate, or other barrier or structure.

integrated water resources management (IWRM): A policy approach aimed at promoting the holistic development and management of water, land, and related resources to maximize economic and social benefits without compromising the sustainability of ecosystems.

interbasin transfer: The physical transfer of water from one watershed to another through a project.

internal renewable water resources: The average annual flow of rivers and recharge of groundwater that is generated from endogenous precipitation.

irrigated area: The farm areas upon which water is artificially applied for the production of crops, including areas equipped for full- or partial-control irrigation but excluding cultivated wetlands, inland valley bottoms, and flood-recession cropping areas.

irrigation: The controlled application of water for agricultural purposes through human-made systems to supply crops' water requirements not satisfied by rainfall.

irrigation efficiency: The ratio or percentage of the water diverted from a source that is consumed by crops on an irrigated farm.

natural inflow: The amount of water that would naturally flow into a country without human interference. This term is distinct from "actual inflow," or the incoming surface-water and groundwater flow resulting from human influence.

nonrenewable water sources: Water resources that are not replenished by nature or that have a negligible rate of recharge, as in "fossil" aquifers.

outflow: The annual outflow of surface water and groundwater from a country to a neighboring nation (either natural outflow or actual outflow).

overlap between surface water and groundwater: The part of internal water resources that is common to both surface water and groundwater. This is the portion of river runoff that originates from groundwater ("baseflow") as well as the portion of groundwater recharge that comes from surface-water seepage.

permafrost: Permanently frozen ground that remains at or below zero degrees Celsius and is found in cold landscapes of the higher latitudes or at high elevations in the lower latitudes. Most of the world's permafrost has been frozen for millennia.

pondage: Small-scale storage at a run-of-river hydropower plant to counterbalance daily or weekly fluctuations in river flow.

potable: Water safe for human consumption.

precipitation: The deposition of rain, snow, sleet, dew, frost, fog, or hail. It is the process by which atmospheric water becomes surface or underground water.

rainfed agriculture: Land cultivated with rainfall, thus obviating the need for irrigation.

rainwater harvesting area: The area where rainwater is collected and either directly applied to the cropping area or stored in a reservoir for productive use later.

recession agriculture: The practice whereby crops are planted in the receding floodwaters of rivers.

renewable resources: Natural resources that despite human exploitation can return to their previous stock levels by natural processes of growth or replenishment. Overexploitation, however, may upset those natural processes.

reservoir: A pond, lake, or basin—natural or artificial—for the storage, regulation, and control of water. In some regions, the terms "reservoir" and "dam" are used interchangeably.

return water (or "return flow"): The part of irrigation water that is not consumed by evapotranspiration and that returns to its source or another body of water. The term is also applied to the water discharged from industrial plants.

riparian states: States that share the resources of a transboundary watercourse, such as a river, lake, or aquifer.

riparian zones: While the term "riparian" relates to something situated or taking place along or near a natural watercourse, riparian zones are vegetated buffer zones that are ecologically diverse and contribute to the health of other aquatic ecosystems by filtering out pollutants and preventing erosion.

river depletion: The overexploitation of water resources leading to a river's downstream stretch drying up or the river discharging little water and nutrient-rich silt into the sea.

river fragmentation: The interruption of natural flows due to dams, interbasin transfers, or other water diversions. A chain of reservoirs, for example, can fragment a river.

runoff: That part of precipitation appearing in surface streams. Total runoff is the sum of surface runoff (overland flow), baseflow, and interflow (subsurface stormflow) entering a stream or other body of water. "Stormflow" (also called "direct runoff" or "storm runoff") is the runoff entering stream channels soon after rainfall or snowmelt. Also see "streamflow."

run-of-river power plant: The use of a river's natural flow energy and elevation drop to produce electricity without the aid of a large reservoir and dam. Electricity is generated by allowing water to flow through low-head turbines housed in a weir (or headpond), with the outflow virtually the same as the inflow and the water returning directly to the river. Such plants are distinct from the storage-type hydropower plants, which impound large volumes of water.

salinization: The accumulation of soluble salts in farmlands to levels that may adversely affect soil conditions and crop yields. Salt-laden lands often result from poor irrigation practices or the capillary rise of saline groundwater.

sediment: As a river plunges from high elevations into the plains, it carries with it fragmented material from rock erosion. Sediment discharge (also loosely known as "sediment load") is the rate at which eroded rock debris passes a section of a river or the quantity of sediment discharged in a given time interval. The processes of erosion, transport, and deposition of nutrient-rich sediments (e.g. silt, gravel, and sand) create and nurture deltas.

siltation: The deposit of silt in a water body by sedimentation.

snowline: The general altitude to which the continuous snow cover of high mountains retreats in summer.

snowpack: The combined layers of ice and snow on the ground at any one time. The snowpack stores massive quantities of water through the winter months to provide supply in the warmer months.

sprinkler irrigation: A method of pressure-induced irrigation in which the water is sprinkled in the style of artificial rain.

streamflow: The volume of water, or rate of flow, in a stream or river. The term "streamflow" is more general than "runoff" because streamflow refers to discharge whether or not it is affected by human regulation or diversion.

subbasin: Part of a river basin drained by a tributary or with significantly different characteristics than the other areas of the basin.

supplementary irrigation: The practice of providing additional water to rainfed crops to make up for rainfall deficits and to boost yields.

surface drainage system: A system of drainage measures, such as channels, to prevent waterlogging on croplands.

surface irrigation: A system of irrigation centered on the gravity flow of water over the soil surface. This is the oldest irrigation approach known to humanity.

surface water: Water that flows in streams, rivers, ponds, lakes, wetlands, and reservoirs.

sustainability: The balancing of opportunities for economic growth with the need to protect the environment. In essence, a capacity to meet the needs of the present without compromising the ability of future generations to meet their own needs.

tailings: The chemically laced waste material remaining after metal is extracted from ore.

transpiration: The process by which plants take up water through their roots and then ooze water vapor through their leaves.

unconventional (or nontraditional) water resources: Water obtained through new technologies, including wastewater treatment for reuse, desalination of seawater and brackish water, and commercial-scale rainwater harvesting. Such resources are more expensive that conventional surface water and groundwater.

virtual water (or "embedded water"): The amount of freshwater used in the production processes and embedded in traded goods and services.

wastewater treatment: Any of the mechanical or chemical processes used to modify the quality of wastewater in order to make it fit for human reuse or for discharge into the environment.

water cycle: See "hydrologic cycle."

water gap: The shortfall between water supply and demand.

water infrastructure: The physical and organizational water-related structures to meet a society's functional needs, including reservoirs, treatment plants, and piped collection and distribution systems.

water productivity: Water productivity, or efficiency, is the ratio of product output (goods and services) over water input.

water scarcity: A term that generally refers to less than 1,000 cubic meters of annual water availability per inhabitant in a country or region. However, UN-Water—the mechanism set up by the United Nations—offers a simple definition of water scarcity that obviates the distinction between stress and scarcity: "The point at which the aggregate impact of all users impinges on the supply or quality of water under prevailing institutional arrangements to the extent that the demand by all sectors, including the environment, cannot be satisfied fully."

water security: A concept advanced by the United Nations Development Programme as part of its human-security agenda: "In broad terms, water security is about ensuring that every person has reliable access to enough safe water at an affordable price to lead a healthy, dignified, and productive life, while maintaining the ecological systems that provide water and also depend on water."

water stress: The symptoms of water shortages, which may include growing competition for water, declining standards of water supply, and rising food-insecurity risks. Water stress is widely defined as the availability of less than 1,700 cubic meters of freshwater per person per year.

water table (or "groundwater table"): The upper level of an underground surface in which the soil or rocks are permanently saturated with water.

water withdrawal (or "water abstraction"): The water removed from the ground or diverted from a stream or lake for human use.

watercourse: A system of surface waters and groundwater constituting by virtue of their physical relationship a unitary whole and normally flowing into a common terminus. The term includes rivers, streams, lakes, and aquifers. A broader definition of watercourse also includes deep "fossil" aquifers, although they do not contribute water to, or receive water from, a surface-water basin.

waterlogging: The condition of land in which the water table is located at or near the surface, resulting in declining crop yields. Irrigation can raise the level of aquifers.

watershed: The term "watershed" traditionally meant the divide separating one drainage basin from another. But now the term has become a synonym for drainage basin or catchment area. See "catchment area."

watershed management (or "river-basin management"): The planned used of watersheds to achieve clearly determined objectives.

wetlands: Areas, such as swamps, marshes, and bogs, that are inundated or saturated by surface water or groundwater and support vegetation and biological activity that are adapted to a wet environment. Wetlands are transitional areas between terrestrial and aquatic systems. The water table is usually at or near the surface, or shallow waters cover the land.

zone of saturation: The zone in which the functional permeable rocks are saturated with water.

Index

Figures, glossary, maps, notes, and tables are indicated with f, g, m, n, and t following the page number

Abbotsford-Sumas Aquifer, 245
absolute territorial integrity, 249–50; and doctrine of restricted sovereignty, 250
absolute territorial sovereignty, 208, 209, 249–50; U.S.-Canada treaty as example of, 208–9
Abyei, Sudan-South Sudan dispute over, 183, 189–90
Act of the Congress of Vienna (1815), 206
Act of Mannheim (1868), 206
actual renewable water resources (ARWR), 172, 357g
Administrative Commission of the River Uruguay, 274
Afghanistan: demographic trend in, 13, 30; and Durand Line, 196–97, 342n40; and Helmand and Kabul rivers, 196–97; relationship with Pakistan, 185, 192, 196–97, 342n40; renewable water resources in, 13, 155t; Tibetan rivers as water source for, 155; water-storage capacity in, 197; water stress in, 183, 219
AfPak, 13, 31, 219, 221

Africa: agricultural water withdrawals in, 74, 319n25; artificial political borders in, 184–185; changes in hydrological map of 184; China as new colonial power in, 25–26, 229, 235; Darfur conflict in, xii, 163; demographic trends in, 31; and farmland leasing, 70–72, 113–14; and food, 83, 84, 110, 113; and irrigation, 62, 74, 75; meat consumption in, 81; revanchist and identity struggles in, 184; water availability in, 10, 37, 62, 123, 160, 167, 219t, 230, 247; water cooperation in, 40, 272; water conflict in, vii, 1, 37, 38, 48, 72, 162, 163, 190–191. *See also specific basins*
African Development Bank, 168
aging populations as a driver of consumption growth, 33
Agreement Concerning the International Commission for the Protection of the Rhine against Pollution (1963), 215
agriculture: carbon intensity of, 73; and climate change, 150, 152–54;

as contributor to water-resource depletion, 5, 13, 68–69, 73–75; crop-yield growth, xv, 76, 281; and farmland leases overseas, 70–72; as largest freshwater user, 13, 73, 105, 113; global pressures on, 68; investments in, 67; main grain exporters, 67; and new crop varieties, xv, 66, 84, 146; and per capita cropland decrease, 69; and poverty reduction, 113; rainfed agriculture, xx, 62–63, 104; water efficiency in, 68, 77, 102, 277, 280–81; and water crisis, 68–69, 282. *See also* food prices; irrigation; *specific crops or practices*

Aksai Chin Plateau, 242

Alaskan water, 124, 128, 129, 137

al-Assad, Bashar and Hafez, 199, 200

albedo, 142

Al-Asi/Orontes River, 45, 244

Alberta (Canada), 6, 124, 125, 210, 329*n*13

al-Disi (Saq) aquifer, 176, 224

algal blooms, xv, 9, 76, 82, 85, 148, 281, 357*g*

Algeria, 8, 339*n*138; territorial conflict with Morocco, 187–188; water resources in, 8, 188

All-American Canal, 212, 213, 245

Alps, 129, 155, 216

Al Qaeda, 13

Al Qaeda in the Arabian Peninsula, 13

Amazon River, 37, 38, 145, 149, 160, 244, 271, 327*n*138; and 1978 cooperation treaty, 244

Amazonian forests, 145

American West, xxi, 126, 133, 135, 154, 155, 156*t*, 251

Amu Darya River, xiii, 53, 156*t*, 158*t*, 166, 178, 186, 190, 275

Amur River, 179, 231, 232*t*, 233

Andes, 129, 155, 156*t*

Anglo-German Heligoland Treaty (1890), 184

Angola, 47, 161

Anguilla, 166

Annan, Kofi, 7

Ansar al-Sharia terrorist group (Yemen), 13

Antarctica, 37, 60, 155, 231

Anthropocene, 141

antibiotic-resistant bacteria, 144

Antiplano, 160

Aquafina, 128, 129

aquifers: in arc of Islam, 224–25; artificial recharge of, 35, 173, 285, 357*g*; as bottled-water source, 129; depletion of, 161, 166, 170; "fossil" type, xii, xiv, 6, 12, 161, 256–57; and fracking, 91; and international law, 256–57, 259; ocean rise due to depletion of, 120; rate of recharge of, 75; saltwater intrusion into coastal, xvi, 119–20, 146, 148, 165; transnational, xxiii, 2, 28, 37, 42, 44, 175, 176, 181, 183, 247–48, 256–57, 261. *See also specific aquifers and countries*

Arab Headwater Diversion Project, 50

Arab Spring uprisings, xiv, 4, 27, 66, 222; link with water crisis, xiv, 4, 27, 222

Arab world, xiv, 13, 27, 109, 222; population explosion in, 27; rejection of Turkey's water pipeline proposal by, 226–27; subsidies in, 109; threat from water scarcity to, xiv, 27, 222

Arabian Peninsula: depletion of water resources in, 112, 166, 224, 247; groundwater in, 224; irrigation in, 112, 114; water scarcity in, 12, 166, 219, 226, 228–229. *See also specific countries*

Aral Sea, 119, 166–67, 186, 239; and anthrax burial, 167; ecological consequences of shriveling of, 166–67

Arava/Wadi Araba aquifer (Jordan), 53, 169

arbitral tribunals, 249, 253
arc of Islam: aquifers in, 224–25; artificial states in, xxii; demographic trends in, 220–21, 222–23; dependency ratio on external water inflows, 224; as hub of water-poor states, xxii, 219–20; nexus between water scarcity, overpopulation, and terrorism, xxii, 219–25
Arctic, 6, 123, 129, 149–50, 154, 155, 160, 231, 334n78, 335n79
Argentina, 67, 68, 104, 115, 116t, 156t, 169, 224, 248, 272; political row with Uruguay, 274; and water agreement with Uruguay, 274
Arizona, 103, 115, 135, 138, 164–65, 211, 214–15
Armenia, 132, 186–87
Arunachal Pradesh (India), 55, 177, 181, 235, 262; China's resurrected claim to, 177, 235; as a water-rich region, 177
Arun River, 179, 232t, 233
Asia: dam racing in, 2, 261–62; demographic trends in, 31, 33; dietary preferences in, 80–81; food production in, 73, 115, 281, 299; as global irrigation hub, 73, 74, 280; as international center of gravity, 23; as leanest continent in average adult body mass, 33; one country's chokehold on riverheads, 229–31, 242; per capita annual water supply in, 10, 74, 123, 219t, 230; resource challenges in, 101; vulnerability to climate change, 143, 155–58; water crisis in, 10, 100; water disputes in, xii, 232. *See also specific basins*
Asi-Orontes River, 187, 188–89, 225; "friendship dam" on, 189
Aswan Dam (Egypt), 179
Atatürk Dam (Turkey), 199, 249
Atatürk, Kemal, 200
Atlanta, 4

Atlantic Ocean, 6, 160, 210
atmosphere, 30, 85, 95, 99, 141, 142, 144, 146, 147, 148, 152, 154, 157, 299
"Atoms for Peace" program, 98, 324n104
Atrek River, 244, 304
Australia: and biodiversity, 148; buyback of water rights in, 121–22; and climate change, 121, 151; drought in, 121; "Every Drop Counts" program in, 282; food production in, 67, 104, 115; fossil aquifers in, 12; greenhouse-gas emissions of, 123; importance of Murray-Darling Basin for, 121–23, 134, 136; intrastate water issues in, xiii, 123, 164, 288; irrigation in, 121–23; per capita water availability in, 123, 219, 230; water market in, xxi, 133, 134–35, 136, 277; water pipeline in Western Australia, 227; water withdrawals for agriculture, 74, 123
Australian Competition and Consumer Commission, 134
Austria, 150, 205
Azerbaijan, 186–87, 341n21

Baghdad's water-supply network, 56
Bahrain: water dependency ratio of, 111, 224; water scarcity in, 5, 220
Baikal Lake, 210
balance of power, 23, 205
Balkhash Lake, 239
Baluchistan, 183
Bangladesh: dependency on external water inflows, 37, 160; food-related issues in, 113, 116t; and Ganges Water Treaty, 45, 275; impact of Chinese hydro projects on, 239; likely impact of global warming on, 151–52; monsoonal flooding in, 147; population density in, 151; refugee

movement from, 151–52, 162; secession from Pakistan, 185; water footprint of, 110

Ban Ki-moon, 7, 175

Banqiao Dam (China), 238

Banyas River, 50, 52

Barcelona, water crisis in, 120

Barlow, Maude, 140

Barnaby, Wendy, 41

basin agreements. *See* transboundary basins

beef production, water intensity of, 78–79

Bekaa Valley, 53, 199

Belarus, 37

Belgium, 8, 10, 111, 206, 331*n*30, 349*n*59

Bellagio draft treaty (1989), 253

Ben-Gurion, David, 50

Berlin Rules on Water Resources (2004), 252–53; Website link to, 301

Bermuda, 166

Bhutan, 192

Bible, 61, 169, 170

biodiesel. *See* biofuels

biodiversity: loss of, 8, 21, 23, 83, 173; water and, 60, 129, 216, 244, 265

biofuel production: as contributing to deforestation, 67; impact on food prices, xx, 67–68, 79; and irrigation, 84; subsidies for, xx, 67–68; water intensity of, 67, 90

biosphere, 20, 59, 141, 173

birth control, 221, 222

birth rates. *See* demographic trends

bisphenol A (BPA), 131

Black Irtysh River, 232*t*

Black Sea, 4, 205

Blair, Tony, 221

blue/green water equation, xx, 62–64, 111, 358*g*; and agriculture, 62–63, 74; ways to enhance green-water utilization, 63, 77; and virtual-water crop exports, 104–5, 107, 108, 111

Blue Nile River, 179

Blue Planet in Green Shackle—What Is Endangered: Climate or Freedom? (Vaclav Klaus), 150

BMC Public Health journal, 33

body mass index (BMI), 32–33. *See also* obesity epidemic

Bolivia, 136, 156*t*, 160, 179, 244, 272; water conflict with Chile over Rio Lauca, 179

boreal forests, 6, 125, 149

Bosporus, 230

bottled water: as aggravating rich-poor divide, 128; as costlier than crude oil, xii, 5, 128; in Europe, 133; as generating enormous garbage, 130, 132–33; in Germany, 129; international trade in, 128, 137; leaching of estrogenic compounds from bottles, 131–32; under NAFTA, 123–24; as new wine, 128; and overexploitation of water resources, 129; price of, 125–26, 290, 291; recalls of contaminated, 132; rise of bottled-water industry, 5, 128–33, 290, 291; and soft drinks, 128, 130; and tap water, 128–29; in the U.S., 129. *See also* commoditization of water

Botswana, 37

Bounaïm-Tafna Basin, 188

brackish water, xvii, 6, 85, 102, 104, 140, 166, 269, 279, 282–84, 298, 358*g*

Brahmaputra Canyon, 235

Brahmaputra River: hydroengineering projects on, 93, 177, 179, 233, 235; intercountry water disputes over, 53; transboundary flows from, 37, 151, 239; water resources of, 156*t*, 158*t*, 232*t*

Brazil: agriculture in, 68, 116; groundwater in, 248; as king of freshwater world, 14; and La Plata Basin, 244, 272; meat consumption in, 80; natural endowments of, 11,

14, 70, 231, 237, 291; as producer of hydropower, 93; water conflict with Paraguay, 53–54; water-storage capacity of, 287; water stress in northeast, 160

Britain, 24, 71, 132, 144, 206, 214; seaside nuclear plants in, 97; water situation in, 123

Budapest Treaty (1977), 258

Bulgaria, 37, 205

bulk-water trade: Canadian opposition to, 124–25, 137, 138, 226; constraints on, 108, 127, 291; internationally, xxi, 8, 125, 127, 137; between Malaysia and Singapore, 138; and NAFTA, 124, 138; potential windfall for water-rich states, 291

Burkina Faso, 13, 111

Burma (Myanmar), 25, 79, 162, 177, 192; and Chinese hydropower projects, 26, 47, 233, 236, 241, 262; water resources in, 156*t*, 192, 232*t*; water footprint of, 111; Western sanctions against, 237

Burundi, 13, 42, 111, 244, 254

Bush, George W., 124, 295

California, xxiv, 3, 4, 25, 74, 85, 102, 115, 124, 133, 135, 138, 156*t*, 164, 167, 204, 212, 215, 290

Cambodia, 114, 116*t*, 241

Cameroon, 168

Canada: abundant water resources in, 1, 14, 124, 160, 237, 291; and climate change, 149; dependency on transboundary water inflows, 161, 207, 237; food exports of, 104; greenhouse-gas emissions of, 125, 329*n*13; hydropower in, 93; and Keystone pipeline, 125; opposition to bulk-water exports, 124–25, 137, 138, 226; per capita freshwater availability in, 14, 123, 219, 329*n*8; tar sands in, 6, 125; water relations

with U.S., 46, 123–25, 137, 207–14, 217, 244, 245, 250, 259, 271, 304

cancer risk from water contamination, 132, 167

carbon capture and sequestration (CCS), 87*f*, 90

Carbon Disclosure Project, 11

Caribbean, 83, 141, 219*t*

Carter, Jimmy, 275

Cascade Mountains, 129, 156*t*

Caspian Sea Basin, 101, 186

Catalonia (Spain), xxi, 120–21

Catawba River, 164

Caucasus, 186, 190, 219*t*; Great Game in, 186

Central Africa, 149, 160

Central African Republic, xiv, 111, 339*n*138

Central America, 5, 144, 160, 218; water resources in, 219*t*

Central Arizona Project, 214–15

Central Arizona Water Conservation District, 138

Central Asia: artificial borders in, 190; disputes in, xiii, 2, 55, 178, 183, 186, 190, 191, 231; irrigation inefficiency in, 75; water situation in, 75, 119, 123, 155, 156*t*, 162, 166–67, 219*t*

Central Commission for the Navigation of the Rhine (CCNR), 205

cereals. *See* grains

Ceyhan River, 161, 226

Chad, 111, 163, 168, 248

Chad, Lake, 119, 167–69, 339*n*138; ecological disaster relating to, 167–69

Chari/Logone river system, 168

Chile: water conflict with Bolivia over Rio Lauca, 179; water footprint of, 110; water situation in, 11, 156*t*, 160; water as a tradable commodity in, xxi, 135–36, 277

China: as Asia's hydro-hegemon, xxiii, 46, 47, 161, 177, 179, 229–42, 248; as biggest global lender, 275; dam collapses in, 238;

engineers in leadership of, 233–34; environmental impacts, 31, 129, 237–38, 239; food production, 64, 70, 74, 104, 113, 114–15, 116*t*; forced relocation of citizens, 238; hydroengineering projects, 26, 39, 42, 47, 85, 115, 177, 179, 180, 181, 193, 195, 212–13, 228, 234–35, 241–42, 261; hydrological divide in, 237; intercountry water disputes, 26, 232–33, 236, 238–39, 241–42; intrastate water conflict, 38; irrigation systems, 9, 74, 237, 280; investments in Japanese forest real estate, 176–77; as largest emitter of greenhouse gases, 31; as leading source of cross-border flows, xxiii, 229–42; meat consumption, 77, 81; natural endowments, 70, 237; one-child policy, 31; opposition to water-sharing cooperation, 231–33, 241–42, 250, 266, 275; as potential master of Asia's water taps, 230, 242; strategy to control resources, 21, 25–26, 229–33; and territorial conflicts, 23, 48–49, 55, 177, 182, 183, 192; Tibet as water source for, 155, 161, 201, 231; use of military force, 49, 229, 230–31, 242; water situation, 3, 11, 12, 62, 70, 108, 110, 153, 237, 278, 286, 310*n*16; water-storage capacity of, 287; weather-modification experiments, 159; as world's most dammed nation, xxiii, 93, 201, 233–34, 238. *See also specific issues and hydroengineering projects*

civil society, 120, 267

clean-water technologies, xvii, 85, 140, 260, 269–70, 279, 282–86, 290–91, 298; and "adoption lag," 270; as key to future of international peace, 298; as means to create sustainable and diverse sources of water supply, 298; promise of new technological breakthroughs, 270, 298; and renewable energy, 260, 284. *Also see* technology

climate change, effects of: anthropogenic connections, 21, 22, 143–47, 300; on Arctic, Antarctica, and Tibet, 150, 155–57; on atmosphere, 146; on coastal communities, 143; as contributor to conflict, 158, 159, 277; distinction between climate and environmental change, xxi, 147–49; on glaciers and snowpack, 143, 154–58; on hydropower and nuclear power, xix, 95, 96–98, 100; on oceans, xxi, 96, 120; on polar sea icecap, 143; on precipitation patterns, 152–53, 288; as a threat multiplier, 150, 277; on water resources, xix, 43, 86, 141–60, 146, 149, 150, 151–59

climate-change science, 144–46, 159

"Climategate," 144

cloud-seeding program. *See* geoengineering

Club of Rome, xv–xvi

coal, 19, 21, 27, 35, 85, 86, 91, 92, 102, 104, 143; plants burning, 88, 89, 90, 93

coal-to-liquids (CTL) technology, 90

Coastal Aquifer (Israel/Gaza), 52

coastal zones, 143

Codex Alimentarius Commission, 132

Coleridge, Samuel Taylor, 119

Colombia, 42, 156*t*, 160

"Colorado Doctrine," 126, 251. *Also see* doctrine of prior appropriation

Colorado River, 3, 44, 63, 91, 115, 121, 138, 154, 156*t*, 164, 167, 173, 208, 209, 211–13, 214–15, 217, 244

Colorado River (Texas), 126

Columbia River, 153, 156*t*, 210

combined cycle gas turbine (CCGT) plants, 87*f*, 89, 90, 102

commoditization of water, xxi, 125–41; business opportunities from scarcity driving, xix, 125–27. *See also* bottled water *and* water markets

concentrating solar power (CSP), 86, 88, 89, 90

Congo, Democratic Republic of, 110, 111, 168, 244

Congo River, 37, 145, 153, 160, 169; and free navigation among colonial powers, 206, 246

consumption growth, xiii, xvi, 3, 21, 28–35, 61, 68, 79, 80, 82, 83, 85, 115, 143, 268; factors contributing to, 31–35; linkage with obesity epidemic, 32–33; link with aging populations, 33; link with rising incomes, 33–34

Convention on Biological Diversity (1992), 148–49

Convention for the Protection of the Rhine against Pollution by Chlorides (1976), 215

Convention on the Protection and Use of Transboundary Watercourses and International Lakes ("Helsinki I Convention," 1992), 260; Website link to, 301

Convention of Strasbourg (1963), 206

Convention on the Transboundary Effects of Industrial Accidents ("Helsinki II Convention," 1992), 260

Convention on Wetlands of International Importance ("Ramsar Convention"), 173, 340*n*148

convict laborers, 235

coral reefs, 21, 82, 148, 149

corn production, 65, 67, 71, 72, 77, 90, 102, 104, 109, 115, 116; and biofuels, 67, 90; in U.S., 67, 90, 104

cotton production, 64, 68, 72, 108, 114, 116, 123, 166, 171, 186; as an environmental curse, 166, 186; and grains, 68

Crete, 139, 203

Croatia, 205

crop yields: achieving higher, xv, 68, 76, 77, 281, 299; in Egypt and Pakistan comparatively, 272; and irrigation, 64, 74, 281

customary international water law, 40, 56, 213, 248, 249–52, 253, 258, 295; basic principles of, 248, 249–52; and ICJ, 257–58; and UN Convention, 249, 252; weakness of, 40, 249, 251

Cyprus, water situation in, 8, 10, 166, 228; Turkish water pipeline to, 139, 228

Czech Republic, 150, 205

Dalai Lama, 48

dam racing, xxiii, 40, 261–63

dams: building spree, 93, 233–34, 241, 243, 262; collapses of, 238; constraints on finding new sites for, 93–94, 234–35, 241; as contributors to conflict, 93–94, 176, 180, 236, 245, 261, 263, 270; effect on ecosystems, 92–95, 176, 235–36, 239, 241; as emitters of methane and carbon dioxide, 94; engineer-politicians behind building, 233–34; evaporation losses from, 94; and forcible eviction of residents, 94, 234–35, 238; and hydropower needs, 92–95; as a political weapon, 177–78, 241–42, 256; and reservoir-triggered seismicity (RTS), 235–36; world's most dammed countries, 121, 233, 234, 241. *See also specific projects or rivers by name*

Danjiangkou Dam (China), 233

Dan River, 52

Danube Commission, 205, 206, 215

Danube River, 37, 205, 215–16, 244, 246; and ICJ, 258–59; as world's most international basin, 205

Dead Sea, 169–71; extraction of potash and magnesium from, 170; linked to Abraham, Moses, and Jesus, 170; as lowest point on earth, 169

Dead Sea–Red Sea canal link plan, 170–71

deforestation, hydrological effects of, 5, 67, 82, 147, 148, 149, 265

demand management, 223, 270

Demirel, Süleyman, 227, 249–50

democracy, 150, 201

demographic trends: in arc of Islam, xxii, 54, 220–23; in developing world, 30–31; globally, xi, 29–31; slum population, 30, 290; in sub-Saharan Africa, 31; as ticking time bomb, 27, 221; linkage with water crisis, 29–31, 34, 220–23

Deng Xiaoping, 347*n*113

Denmark, 149

dependency ratio, 160–61, 207, 217, 237, 338*n*120, 358*g*

desalination, xiv, 103, 127, 139, 171, 226, 228–29, 241, 269, 278, 282, 283–84, 285, 286, 358*g*; of brackish water, 284; cost of, 174, 283, 285; energy intensity of, 101; nuclear evaporative, 284; potential to fundamentally change water geopolitics, 184; toxic residues from, 174, 283

desertification, 167, 168

diabetes prevalence, 83, 165, 321*n*56

diets, changing: and diabetes prevalence, 83; greater intake of meat, xx, 77–79, 80–87; link with prosperity, 34, 70, 77, 79, 83, 144; meat-based vs. vegetarian diets, 78–79, 82; share of animal-based protein in human diets, 82; shift toward Western dietary preferences, 68, 80; water-consumption linkage with, 77–84. *See also* meat production

Djibouti, 109, 220

doctrine of prior appropriation, xxiii, 40, 126, 251, 261–62; as driver of dam racing, xxiii, 40, 261–63

doctrine of restricted sovereignty, 250. *See* absolute territorial sovereignty

drinking water: access to, xxi, 5–6, 10, 11, 12, 17, 19, 128, 137, 189, 264; availability as a human right, 17; arsenic threat, 132; contamination of, 75, 91, 127, 160, 281; cross-border supplies of, 47, 74; fluoride content in, 130; and mobile phones, 10; UN fudges millennium goal on, 19; unconventional sources of, 174, 282–86. *See also* bottled water

drip irrigation, 76, 102, 171, 174, 280–81, 358*g*; global acreage under, 76, 280; and paddy rice, 281; upfront capital costs of, 281; yield improvement from, 281

droughts, 8, 35, 37, 61, 85, 89, 95, 109, 119, 133, 143, 146, 147, 148, 149, 152, 153, 158, 161, 165, 166, 190, 212; crop varieties resistant to, 66, 146, 281; impact on food prices, 66. *See also specific regions*

Durand Line, 196–97, 342*n*40

earth: agricultural production capacity of, xv; as a blue planet, 60; drinkable water portion on, 60; extent of croplands, 70; extent of river basins, 37; human alterations of, 21, 141–42, 143–44, 149, 154; water-renewable capacity of, xii, 30, 60, 79, 219*t*, 287

earthquakes: dams as a trigger of, 235–36; disposal of "fracking" wastewater as trigger of, 91

East Africa, 156*t*, 162, 171

East China Sea, 177, 182

East Ghor Canal (King Abdullah Canal), 52, 172, 225

East Timor, 185

Ebro River, 97, 120

Echeverria, Luis, 214

ecological footprint, 106, 326*n*124

ecological restoration, 149

economic growth: as driver of consumption growth, 31–32, 35; geopolitical impact of, 21–22, 46, 123, 269–70, 275; link with depletion of natural resources, 31–35, 142–43; 299–300, 332*n*49; and water resources, xxi, 2–3, 11, 16, 30, 73, 86, 123; water scarcity as constraint on, xiii, xv, 11, 16, 232, 269

economic risks related to water stress: compared with other risks, xxiv; risks and opportunities for investors, xii, xiii, xix, xxiv, 8, 11, 12, 71, 86, 125, 127, 129, 134–35, 140, 177; need for fundamental shifts in policies and practices, 269; as a spur to overseas farmland acquisition, 71–72; new technologies and practices to mitigate, 12, 269–70, 282; ways to address, 277–91

ecosystems: anthropogenic impacts on, xxi, 9, 18, 20, 21, 68, 76, 85, 98, 114, 129, 141–42; 143–47, 148, 165, 176, 238, 299; balances affected by overexploitation, 9, 18, 20–21, 23, 60, 62, 63, 76, 106, 119, 142, 147–49, 260, 268; freshwater and marine ecosystems, xxi, 16, 17, 20–21, 35, 62, 70, 82, 88, 92, 147, 172, 239, 259, 261, 265; poverty adversely affects, 86, 164

Ecuador, 136, 156*t*, 160, 161

Egypt: agricultural trade of, 108, 109, 116; as an ancient entity, 184, 203; demographic trend in, 54; dependency on external water inflows, 11, 160, 224; and Nile accord with Sudan (1959), 72, 178; and the principle of absolute territorial integrity, 250; riparian ascendancy of, 178–79, 189, 231, 271, 352*n*56; reprisal threat against upstream diversions, 178–79; and

Sinai Peninsula, 50; and Switzerland, 272; water stress in, 11, 108, 224

electricity. *See* energy

electrodialysis reversal (EDR) technology, 283

El Jucar River, 121

El-Kabir River, 45, 306

El Niño, 121

El Segora River, 121

energy: biofuels and, xx, 67–68, 84, 90; and electricity-for-water-pump subsidies, 277; fuel-cell vehicles, 6; as contributing to higher food prices, xxi, 84, 101; electric power production, 84–92; geopolitical rivalries over, xvii, 1, 2, 22, 23, 101, 149–50, 225; grassroots protests over water-intensive plants for, xvi–xvii; link between cooling technology and level of water use, 88–90, 103–4; link with growing prosperity and obesity, 28, 32–33, 34, 85; misguided policies on, 67; nexus with water, 14, 84–104; oil-generated electricity, 6; oil as mainly transportation fuel, 6; rising prices of, xvi; thirsty nature of power plants, 84, 86, 87, 103; and water availability, xxi, 101–4; water constraints on energy choices, 2, 11, 84–85, 86, 101; water consumption by hydrocarbon sector, 86, 90–91. *See also* hydropower; renewable-energy technologies; shale energy; thermoelectric power plants

energy intensity of water supply growing, 35, 84, 85, 101–2, 173, 174, 229, 283–86, 289, 290–91

environment: human impacts on the, xiii, 13, 18, 31, 32, 33, 34, 35, 75, 77, 82, 83, 92, 94, 108, 123, 125, 130, 144, 166–73, 198, 212, 223, 234, 235, 241, 261, 264, 278, 300; poverty as enemy of the, 86, 164; protection of the, 8, 57, 135, 251,

260, 263, 273, 289; water flows
  essential to the, 12, 62, 264
environmental change, xxi, 141–60
environmental degradation, xvi, xxiv, 2,
  5, 21, 28, 31, 71, 94, 136, 147, 157,
  164, 223, 277, 299
environmental security, xxiv, 160, 260,
  263
environmental sustainability, 1, 5, 17,
  19, 35, 79, 83, 102, 117, 140, 247,
  276, 299
equitable distribution principle, 264
Erdogan, Recep Tayyip, 200
Eritrea, 42, 185
Eshkol, Levi, 50, 317n124
Espoo Convention on Environmental
  Impact Assessment in a
  Transboundary Context (1991), 260
Estonia, 91
ethanol, 67, 90
Ethiopia, xiv, 13, 14, 31, 42, 72, 111,
  161, 166, 178, 179, 183, 190–91,
  272; access to Nile resources,
  72, 178, 179, 272; ambition to be
  Africa's leading power exporter, 179;
  and Lake Turkana, 172, 190–91;
  leasing farmland to foreigners, 166;
  water footprint of, 111
ethnic-minority homelands, 2, 39, 197,
  230, 231, 234
Euphrates River, 2, 5, 37, 53, 139, 161,
  179–80, 189, 190, 197, 198–99, 203,
  223, 225, 226–27, 244, 249, 271
Europe: and biofuels, 67, 68; bottled-
  water consumption in, 133; and
  colonialism, 24, 184, 206, 222,
  229; diet in, 81, 82; development
  of freedom of fluvial navigation
  in, 204–7, 215, 216, 244, 246; few
  intercountry water-sharing disputes
  in, 216, 261; greater river runoff in,
  153; water situation in, 4, 10, 110,
  171, 219t, 247, 259–60; water use by
  energy sector in, 86, 87, 97, 108. *See
  also specific basins and countries*

European Commission of the Danube
  (ECD), 205, 215
European Union, 52, 67, 104, 115, 131,
  137; and international food trade, 108,
  115, 116; Water Framework Directive
  of, 216; water sector in, 137

Failed States Index, 13
Falkenmark, Malin, 310n16
family planning. *See* birth control
Farakka Barrage (India), 39
farmland acquisitions by foreigners, xx,
  66, 70–72, 113–14, 166, 179; extent
  of, 71; and food security, 113–14;
  impact on sub-Saharan Africa's food
  supply, 71; potential downstream
  river-flow impacts, 72, 179; role of
  Arab states, 71, 113–14, 166, 179;
  role of East Asian firms, 66, 71–72,
  179; as water grabs, 72
Fatehpur Sikri, 3
Fergana Valley, 2, 55, 183, 191
Ferguson, Niall, 24
fertigation, 102
"Fertile Crescent" region, 203
fertility rates, 30, 31, 33, 220, 221, 222
fertilizers: and microirrigation systems,
  76, 102, 280–81; role in boosting
  food production, xv, 105, 109,
  144; role in water pollution and
  eutrophication, 70, 72, 75, 82, 144,
  239, 280–81, 299
Fiji, 128
Finland, 14
flooding: as an annual feature, 59, 147,
  237; as bringing nutrient-rich silt,
  203, 239; caused by attacks on dams,
  55–56; and climate change, 143, 144,
  146, 147, 152, 153, 158, 161; flood
  management, 43, 63, 85, 94, 149,
  209, 216, 239, 241, 258, 262, 264,
  266, 270, 272; flood-resistant crop
  varieties, 66, 146, 281; and paddy
  rice cultivation, 281; triggered by a
  rival state, 180–81, 256

Florida, 4, 9–10, 97, 164
fluoride in water, 130, 330*n*22
fluvial navigation, 204–7, 215, 216, 244, 246
Food and Agriculture Organization (FAO), xv, 45, 73, 75, 132, 234
food demand: and changing dietary preferences, 68, 70, 77–84; and fall in buffer stocks, 65–66; and genetically engineered food, 72–73, 117; and hunger and malnutrition, 64, 66, 77, 299; irrigation role in meeting, 73–77, 280–81; and "Malthusian catastrophe" thesis, xv; a new green revolution to meet, 298–99; and obesity epidemic, xiii, 32–33, 83; rising prosperity boosting, 34, 68, 70; and water intensity of main food products, 78; water-stressed economies as food exporters, 7, 108–9, 114
food prices, xiv, 4–5, 64, 66, 67, 68, 115, 222; and food riots, xiv, 4–5, 66, 222; impact of biofuel production on, 67–68; impact on world's poor, 68, 84; link with imbalance in supply/demand, 66, 69; link with increasing protein and calorie intake, 64, 79; and already-stretched international markets, 70, 113, 116–17; and "oil shock," 67–68; relationship with energy prices, xxi, 101, 222
food production. *See* agriculture
food security, xv, 17, 18, 19, 27, 73, 107, 111–12, 115, 123, 277; as a driver of farmland acquisition overseas, xx, 70–72, 113–14, 166; food as a weapon, 112; and food consumption, 68, 299; misguided energy policies eroding, 67; and natural endowments, 70–71, 74; role of virtual-water trade in, 104–18; and slowing crop-yield growth, 69; water constraints to building, xv, 17, 64,

66, 67, 68, 70, 74, 76, 83, 84, 115, 164
food wastage, 66, 68, 79; in U.S., 66
forced relocation of residents, 94, 234–35, 238
Ford Randall Dam (U.S.), 233
forest real estate, 176–77
forests: Amazonian, 145, 149; ecological role of, 20, 147–49, 265; oil extraction from tar sands under, 6, 125. *See also* deforestation
fossil fuels, 19, 20, 70, 85, 95, 96, 99, 100, 142, 144, 148, 149, 157, 235, 243
fossil water, xiv, 3, 4, 6, 12, 62, 104, 112, 161, 176, 224, 228, 256, 359*g*; and UN Convention, 256–57, 350*n*31; as virtually nonrenewable resource, 6, 12, 104, 256. *Also see* aquifers
France, xxi, 24, 97–98, 100, 120, 121, 123, 187, 188; as colonial power, 187, 188; nuclear power's water use in, 97–98; water resources in, 97
freshwater. *See various entries under* water
Fukushima disaster, xxi, 59, 95, 96, 98, 99, 100

Gabcíkovo-Nagymaros barrage system (Hungary/Slovakia), 258–59
Gabon, 14, 219, 311*n*34
Ganges River, 37, 45, 151, 153, 156*t*, 158*t*, 203, 217, 244, 271, 275; and Jimmy Carter, 275; special bacteriophagic conditions in, 45; water-sharing agreement over, 45, 217, 244, 271, 275
GAP (Güneydoğu Anadolu Projesi, Turkey), 39, 197–202, 226–27; Arab campaign against, 226–27; and environmental and public-health issues, 198; and Kurds, 197, 200–202; launch of, 200
Garrison Dam (U.S.), 233

Gaza Strip, 13, 50, 52, 220*t*, 271; and
 Israel, 50, 52, 271; water scarcity in,
 13, 52, 220*t*
General Agreement on Tariffs and
 Trade, 138
Generation IV International Forum, 100
genetically engineered (GE) food,
 72–73, 117
Geneva Conventions (1949), 56
geoengineering technology, 61, 159
geographic information system (GIS),
 267
geopolitics: and geoeconomics, 22, 292;
 new Great Game, 1, 23, 186, 225;
 larger picture relating to, xxiii, 1, 6,
 20, 23–24, 53, 146, 150, 292, 294,
 296; of water, xi–xii, 1, 6, 23–24, 53,
 225, 284. *Also see* hydropolitics
Georgia (Republic), 161, 186–87
Germany, 24, 56, 96, 123, 129, 186,
 205, 206, 207, 269*f*, 292, 295;
 bottled water in, 129; and Mohne
 Dam, 56; and nuclear power, 96;
 per capita water availability in, 123,
 269*f*; reunification of, 186; and
 Rhine Commission, 206, 207
Gibe III Dam (Ethiopia), 191
Gibraltar, 166
Gilgit-Baltistan region, 193–95;
 Chinese military presence in, 193;
 hydropower projects in, 193, 194
"Glaciergate," 144–45
glaciers: accelerated thawing of, 95,
 143, 154–57, 288; IPCC scandal
 over Himalayan, 144–45; role in the
 hydrologic cycle, 60; as rich stores
 of freshwater, 60, 143; as source of
 bottled water, 129, 137; as source of
 rivers and lakes, 143, 154, 156*t*
Gleick, Peter, 132
Glen Canyon Dam (U.S.), 212
"global commons," 141
global warming. *See* climate change,
 effects of
Global Water Partnership, 19, 312*n*52

Globe and Mail on U.S. interest in
 Canadian water, 124
Gobi Desert, 237
Golan Heights, 2, 41, 50, 51*m*, 183, 188;
 capture by Israel, 41, 50; as Jordan
 River headwaters, 50, 51*m*, 188
Gorbachev, Mikhail, 59–60
Grand Canal (China), 240
Grand Coulee Dam (U.S.), 93, 233
grassroots protests over dams or water-
 guzzling industries, xvi, 11, 94, 98,
 195, 235, 262, 308*n*14
Great Barrier Reef, 148
Great Himalayan Watershed, 155–58;
 militarized frontiers in, 192; river
 systems of, 155, 158*t*; world's
 highest mountains in, 155, 192
Great Lakes and Great Lakes-St.
 Lawrence system, 4, 123, 124, 207,
 209, 210, 211, 244; and U.S.-Canada
 relations, 124, 207, 209, 210, 211
Great Manmade River Project (Libya),
 3, 228
Great Plains (U.S.), 12, 124
Great South–North Water Diversion
 Project (China): costs of, 115, 239–
 41; as drawing inspiration from U.S.
 projects, 115, 234; environmental
 risks of, 115, 228, 239–41; as means
 to perpetuate a paradox, 114–15,
 239–41; quantum of waters to be
 drawn from Tibetan Plateau, 115
Great Wall of China, 212
Great Western Route (China), 239–41
Greece, 4, 195, 228
green/blue water. *See* blue/green water
greenhouse-gas emissions, xiii, 19, 21,
 31, 32, 60, 68, 73, 81, 82, 85, 94, 95,
 121, 123, 125, 142, 144, 147, 149,
 154, 157
green revolution, xv, 66, 73, 298–99;
 need for a second green revolution,
 298–99; role of irrigation in, 73, 298
groundwater resources: brackish, 102,
 282, 284, 357*g*; contamination

of, 91, 132, 173–74; as drought insurance, 8; energy use in extraction of, 85, 101; extent of, 248; factors contributing to overexploitation of, xvi, 9, 68, 126, 129, 166; fossil water, xiv, 3, 4, 6, 12, 62, 104, 112, 161, 176, 224, 228, 256; hidden nature of, xvi, 37, 166, 175; impacts of unbridled extraction of, xvi, 9, 75, 84, 165, 166, 278, 290; as important source of irrigation, 75; and sea-level rise, 120; and UNESCO's International Hydrological Program, 247; as world's most extracted reserves, xx, 8; weak international arrangements on, 37, 44, 45, 175–76, 244, 247–48, 256–57. *See also specific countries and regions*
Group of Eight (G8), 123
Guaíra Falls, 53–54
Guaraní Aquifer, 248
Guinea, 217
Guir-Zousfana River Basin, 187
Gulf of Mexico, 63, 125, 214
Güneydoğu Anadolu Projesi. *See* GAP
Guyana, 14, 311*n*34

Hague Regulations (1907), 56
Haiti, 14, 66, 111
Harappa, 203
"Harmon Doctrine" (1895), 208, 213, 249, 344*n*63
Harmon, Judson, 208, 344*n*63
Hasbani River, 50, 52
Hatay, Syrian-Turkish dispute over, 188–89
Hayek, Friedrich von, 150
Helsinki Rules on Uses of Waters of International Rivers (1966), 252–53, 350*n*22; Website link to, 301
Hezbollah, 52
Himalayan Asia, 155, 157
Himalayas, 48, 129, 156*t*, 157, 333*n*63
Hindu, 3, 60
Hippocrates, 127

Hirano, Hideki, 177
Homer-Dixon, Thomas, 42
Hong Kong, 138, 333*n*60
Hoover Dam (U.S.), 209, 212, 233, 234; influence on China, 234
Hoover, Herbert, 209, 234
Horn of Africa, 190, 191, 219
Hu Jintao, 234
Hudson River, 96
Human Development Index, 13, 16
Hungary, 37, 160, 205, 160, 258–59; dependency ratio on external water inflows, 160; water dispute with Slovakia, 258–59
Huntington, Samuel, 26
Hurricane Andrew (1992), 97
Hurricane Katrina (2005), 147
Hurricane Sandy (2012), 97, 147
Hussein, Saddam, 223–24
hydraulic civilizations, 203
hydroconflicts. *See* water wars
hydroelectricity. *See* hydropower
hydroengineering projects. *See specific projects*
hydro-hegemony, xxiii, 161, 177–80, 226, 229, 230–31, 236, 242, 248; of lower riparians, 178–79; of upper riparians, xxiii, 161, 226, 248; with no modern historical parallel, 229–242. *See also specific countries*
hydrological data, 43, 56, 180–81, 233, 266–67, 272; potential manipulation of, 56, 180–81, 267; quality like rice, 181; and UN Convention, 233, 266–67
hydropolitics: cooperative politics matters, 291–93; and desalination, 284; and economic interdependence, 292–93; examples of sharpening, 176–202, 217–28; politics vs. law, 293–96; reconciling geopolitical interests to improve, 294
hydronationalism, 2, 238, 289
hydropower: distinction between plants with and without reservoir

storage, 92–93; economics of, 92; environmental costs of, 92, 94, 172, 191, 235–36; and reservoir-triggered seismicity (RTS), 235–36; vulnerability to climate change, 94–95, 154, 155; water losses attributable to, 94; world's leading producers of, 93, 241. *Also see* run-of-river plants

hydrosphere, 141, 142

hydrosupremacy. *See* hydro-hegemony

Iceland, 14, 149, 295, 311*n*34

Ilemi Triangle, 190

Illy River, 179, 231, 232*t*, 233, 239

income levels, as related to consumption levels, 30, 31, 32–33, 35

India: agricultural exports of, 74, 104, 108, 116*t*; and Chinese hydroengineering projects, 47, 181, 235, 242; division of Indus Basin waters, 44–45, 185, 192–97, 250, 293; grassroots protests as damper to new plants, 94, 98, 262; irrigation in, 9, 74; and Jammu and Kashmir, 48–49, 55, 56, 183, 192–97; natural endowments of, 70; refugee flows into, 151–52, 162; run-of-river projects in, 94, 262; storage capacity in, 287*f*; transboundary water flows into, 232*t*; unique dietary profile of, 81; water footprint, 110, 111; water as a new divide in ties with China, 26, 177, 181, 233, 235, 242; water resources in, 237; water-sharing arrangements by, 39, 44–45, 49, 192–96, 250, 262, 275, 293; water stress in, 8, 11, 108. *See also specific issues or projects*

Indian Point nuclear power plant (U.S.), 96

Indonesia: biofuels in, 68; and climate change, 152; rice imports of, 116; water footprint of, 110

Indus Basin Development Fund (IBDF), 275

Indus River system: 44–45, 49, 54, 185, 192–97, 250, 262, 275, 293, 304

Indus Waters Treaty (1960): background on, 44–45, 49, 185, 192–97, 250, 262, 275, 293, 304; as embodying doctrine of restricted sovereignty, 250; as incorporating world's largest water allocations for lower riparian, 44–45, 195; meat of treaty in its annexes, 293; opposition in Indian Kashmir to, 193; virtual river-partitioning line drawn in, 192–93, 194*m*

industrialization, 205, 216

Inner Mongolia, 91, 235, 236

Institute of Medicine, 32

Institute for Security Studies, 28

integrated gasification combined cycle (IGCC), 87*f*, 90

integrated water resources management (IWRM), 19, 103, 118, 268, 269, 278–79, 312*n*52, 360*g*; lip service to, 19

interbasin water transfer projects (IBWT), xiv, 46, 85, 124, 176, 183, 238, 251, 360*g*

Intergovernmental Panel on Climate Change (IPCC): on glaciers, 95, 144–45; on hydropower, 95; on potential conflicts over water, 143; on precipitation trends, 152; on sea-level rise, 145–46, 151

International Atomic Energy Agency (IAEA), 173, 284

International Boundary and Water Commission (IBWC), 207, 209, 201, 211, 212, 214, 271, 344*n*61

International Commission for the Hydrology of the Rhine Basin, 215

International Commission on Large Dams (ICOLD), 323*n*89

International Commission for the Protection of the Danube River (ICPDR), 216, 345*n*85

International Commission for the Protection of the Rhine (ICPR), 215

International Court of Justice (ICJ), 206, 214, 249, 253, 255, 257–59, 265, 274, 295–96; judgment in Gabcíkovo-Nagymaros case, 258–59; judgment in West Germany-Iceland case, 295–96; judgment in Uruguay-Argentina case, 274; and water rules, 257–59

International Danube Commission, 215

International Energy Agency, 87, 100, 101

International Hydrological Program Intergovernmental Council (IC-IHP), 38, 247

International Joint Commission (IJC)

International Journal of Epidemiology, 32

International Law Association (ILA), 252, 253, 301

International Law Commission (ILC), 253, 254, 257, 350*n*22, 351*n*44

international water law, 39–40, 244–45, 246–259; ambiguities and open issues, 39–40, 43, 44, 244–45, 248–49, 254, 257, 259, 263; basic customary principles, 249–52, 252–53; determining "reasonable and equitable share," 250–51, 252, 254, 255; law vs. politics, 293–96; no binding global water law in force, 44, 248, 249; obligation "not to cause significant harm," 250, 252, 254, 255, 259–60. *See also specific laws and treaties*

international water norms and principles. *See* international water law

interriparian relations, xiii, xiv, xvii, 16, 39, 94, 123, 176, 179, 181, 217, 224, 225, 250, 253, 292; centrality of power equations in, xxiii, 179–80, 249, 273

investor risks. *See* economic risks

IPCC. *See* Intergovernmental Panel on Climate Change

Iran: as ancient entity, 184; fertility rate and family planning in, 220–21, 222; feud with Iraq over Shatt al-Arab, 190; as food importer, 116; and Kurds, 201; and Marsh Arabs, 223; meat consumption in, 80; relations with China, 25; transnational water resources in, 186, 197, 341*n*21; water pipeline in, 227–28; water stress in, 14, 123, 219

Iraq: feud with Iran over Shatt al-Arab, 190; largely autonomous Kurdistan region in, 201; and "Marsh Arabs," 223–24; military asymmetry with Turkey, 139, 179–80; as rice importer, 116; transboundary water flows into, 160; and Turkey's appropriation of Tigris-Euphrates waters, 139, 179, 198, 199, 225, 226, 227, 249–50; U.S. military actions in, 24, 56; water stress in, 219

Iron Curtain, 207

iron ore, 1, 21, 26, 28, 35, 308*n*14

Irrawaddy River, 156*t*, 158*t*, 236

irrigation: and atmospheric humidity, 141–42, 143; and biofuel production, 90; Asia as global hub of, 73; countries with largest areas under, 74; as contributing to enhanced crop yields, 74; decline of per capita irrigated acreage, 75; global extent of, 73; groundwater as source of, 9, 75; irrigation subsidies, 109, 114, 172, 290; as largest grain producer, 73–74; and paddy rice cultivation, 117; relationship with rainfed agriculture, 62; role in water depletion, xx, 9, 64, 72, 73–77, 121, 166, 167, 168, 169, 171, 237; role in water pollution, 75, 76, 214, 265, 270, 280–81; salinization linked to, 64, 75, 76, 108, 122, 271–72, 280; underdeveloped in sub-Saharan Africa, 62, 74; water lost in, 75, 76*f*. *Also see* microirrigation; drip irrigation

Irtysh River, 179, 231, 232*t*, 233, 239

Islamic architectural style, 3

Islamic Development Bank, 114
Islamic terrorism, 161. *See also* jihad
Islamic world, 180, 220, 221, 224. *See also* arc of Islam
isotope hydrology, 173
Israel: agriculture in, 108, 171–72; collapsed water deal with Turkey, 139, 227; control over subregional water resources, 48, 49, 50–52, 53, 187, 230, 271; and Dead Sea, 169–71; hydrotechnology in, 171–72; irrigation subsidies in, 172; military accord with Turkey, 199; military actions in Lebanon, 52–53; and National Water Carrier, 49, 50, 74, 169, 228; regional preeminence of, 271; recycling of wastewater in, 1972; water stress in, 11; water-sharing arrangement with Jordan, xii–xiii, 44–45, 46, 53, 225, 271, 293, 305. *Also see* Golan Heights *and* West Bank
Israeli-Palestinian water issues, 40, 45, 247, 271, 305
Itaipú Dam (Brazil and Paraguay), 54
Italy, 30, 123, 207

Jammu and Kashmir. *See* Kashmir
Japan: changing diets in, 109; Chinese investments in forest real estate in, 176–77; dam building in, 94; demographic trends in, 109; and Fukushima nuclear disaster, 99; global power of, 292; meat and other food imports, 109, 116*f*; and natural resources, 24–25, 176–77; and Pearl Harbor attack, 24–25; rare-earth imports from China, 25, 242; territorial conflicts, 177, 182; water footprint of, 110, 111; water quality in, 286; water resources in, 109, 123, 286
Japan's Forests under Siege: How Foreign Capital Threatens Our Water Sources (Hideki Hirano), 177

Jerusalem, 50, 271
Jhelum River, 185, 192, 195
Jiang Zemin, 234
jihad, xxii, 13, 112, 222
Jinsha River, 93
Johnson, Lyndon, 214
Johnson Reef, 242
Jordan: agricultural water withdrawals in, 171, 172; and al-Disi aquifer, 175–76, 224; and East Ghor Canal, 52, 172, 225; groundwater depletion in, 224; per capita water supply in, 223; and Red Sea-to-Dead Sea canal, 170–71; and Six-Day War (1967), 48–50, 52; water scarcity in, 8, 139, 169, 172, 220, 223; water-sharing arrangement with Israel, 44–45, 46, 53, 217, 225, 244, 274, 275, 293, 305
Jordan River, 50, 52, 53, 169–72, 225, 244, 271, 305
Journal of Climate, 153
Justice and Development Party (Turkey), 200

Kabul River, 196–97, 352*n*56
Karakaya Dam (Turkey), 197
Karakoram Range, 155, 156*t*, 157
Kashmir, 2, 48–49, 55, 56, 156*t*, 183, 192–96, 236, 242
Kazakhstan: as cotton exporter, 108, 186; energy resources of, 139, 178, 191; river diversions by China, 26, 47, 230, 231, 233, 239, 262; water footprint of, 110, 111
Kenya, xiv, 8, 172, 190–91; territorial and water disputes, xiv, 190–91
kerosene, 143–44
Khan, Ayub Mohammad, 196, 316*n*119
Khyber Pass, 197
King Abdullah Canal (Jordan). *See* East Ghor Canal
King Abdullah Initiative for Agricultural Investment Overseas (Saudi Arabia), 114
King Hussein (Jordan), 53

Kinneret Lake. *See* Tiberias Lake
Kiribati, 151, 166, 335*n*83
Kishenganga hydropower plant, 262
Kissinger, Henry, 112
Klare, Michael, 27
Klaus, Vaclav, 150
Komati River, 45, 244, 305
Korea. *See* North Korea; South Korea
Kura-Araks river system, 186–87
Kurdistan People's Congress (KGK), 199
Kurdistan Workers' Party (PKK), 199
Kurds, 2, 161, 180, 183, 186, 191, 197, 199, 200, 202*m*; and Abdullah Ocalan's capture, 199; as challenge to Turkey's monoethnic identity, 200, 201; and greater Kurdistan, 201; in Iraq, 201, 202*m*; in Syria, 199, 201, 202*m*; their water-rich traditional homeland, 161, 180, 191, 202*m*
Kuwait, 5, 56, 109, 111, 160, 220, 223, 226, 247; dependency on external water inflows, 160, 224; as one of world's poorest states in water, 5
Kyoto Protocol (1997), 142, 256
Kyrgyzstan: border and ethnic conflict, 190, 191; China as source of water to, 231, 232*t*; dam-building plans of, 178; and Fergana Valley, 55, 191; as transboundary water supplier, 139, 186

Ladakh, 192
Lake Chad Basin Commission, 168, 339*n*140. *Also see* Chad, Lake
Lal, Murari, 145
Lancet journal, 83
land disputes. *See* territorial disputes
Lanier Lake, 4, 164
La Niña, 121
Laos, 192, 232*t*, 241, 262
La Plata River, 172, 244, 271, 272; and 1969 treaty, 244, 272
Las Vegas, 4, 138, 212

Latin America. *See* South America
law of reason, 251
law of riparian rights, 251
Law of Transboundary Aquifers, draft articles on the, 257, 259; Website link to, 301
League of Nations, 206
Lebanon, 45, 52, 53, 72, 169, 188, 199, 271; water agreements with Syria, 45, 188; water relations with Israel, 52, 53, 169
Lesotho, 47; and water treaty with South Africa, 47, 305
Levant, 188, 228
Libya: fossil aquifers in, 3–4, 12, 228, 247, 248; location of main cities, 4; water poverty in, 109, 220; water transfers from Sahara, 3–4, 228
life expectancy, average, 29, 30, 144
light water reactors (LWRs), 95–96; *Also see* nuclear power
Lilienthal, David, 196, 342*n*37
Li Peng, 234
Litani River, 52–53
lithosphere, 141
Little Ice Age, 144
Liujiaxia Dam (China), 233
livestock sector, environmental impacts of, xx–xxi, 82–83
London, 3, 214, 229, 253, 285; recycled water in municipal supply, 285
Los Angeles, 4, 204, 212; and California Water Wars, 204
Lower Colorado River Authority, 126
Luxembourg, 111

Madagascar: deal with Daewoo, 66, 71
Madras Atomic Power Station (India), 96
Maghreb, 123, 166, 219
maize. *See* corn production
Malawi: as a failing state, 13; territorial conflict with Tanzania, 184–85; water footprint of, 111; water stress in, 13, 42

Malawi, Lake (Lake Nyasa), 184–85

Malaysia, xiii, 116*t,* 137, 138, 227; water-export dispute with Singapore, xiii, 137, 138, 227

Maldives: and Kiribati, 151; vulnerability of, 113, 151, 220, 278, 335*n*83; water scarcity in, 8, 166, 220

Mali, 111, 183

malnutrition, xx, 64, 168

Malta, 109, 111, 139, 247

Malthus, Thomas, xv

"Malthusian catastrophe" thesis, xv

Malthusian logic, 229

Manavgat River, 139

Manchuria, 231, 232*t,* 235

Mangla Dam (Pakistan), 195

mangrove forests, 82, 147, 148, 152, 284, 333*n*67

manufacturing. *See* industrialization

Mao Zedong, 212, 233

Marsh Arabs, 223–24

maternal mortality, 29, 30

Mauritania: dependency on external water inflows, 37, 160, 224; water stress and conflict in, 183

Maya civilization, 5, 144, 309*n*10

Mead, Lake, 211, 212

meat production, xx, xxi, 77–82; antibiotic use on food animals, 77–78; and diet of industrially raised cattle, 77; and livestock population explosion, xx–xxi, 82–83; meat consumption in developed world, 80, 81; meat consumption in the East, 77, 80–81; rising human consumption as driver of, 80, 82; threat to earth's water-renewal capacity, 79, 80*t*; water intensity of, 78–79

Mediterranean, 3, 120, 139, 141, 170–71, 188, 225, 226

Mediterranean-to-Dead Sea canal plan, 170–71

Meghna River, 151

Mekong Agreement (1995), 44, 244, 271; China's nonparticipation, 232–33, 266

Mekong River: Chinese dam building on, 93, 179, 232–33, 235, 239; cross-border discharge of, 232*t*; as endangered river, 172; as international river, 37; potential global-warming impact on, 156*t,* 158*t*

Mekong River Commission, 232

Mekorot, 50, 317*n*124

membrane bioreactor (MBR) technology, 283, 298

membrane technology for water purification, 283, 298

Memphis, Tennessee, 37, 164

mercantilism, 24, 27, 229

mercantilist policies, 22, 26, 27, 229, 246; as pursued by China, 26, 229

Mesopotamia, 62, 183, 203

Mesopotamian Marshlands, 223

Mesta/Nestos River, 244, 306

metals, xvi, 1, 7, 12, 21, 23, 24, 25, 26, 34, 35, 127, 246, 251, 281

Metog ("Motuo") Dam (China), 235

Metropolitan Water District of Southern California, 138

Mexicali Valley, 211

Mexico: Colorado River flow into, 44, 45, 126, 208, 211, 212, 214, 215, 217; and International Boundary and Water Commission, 207, 209, 211, 212, 271; meat consumption in, 80; and Rio Grande, 172, 208, 212; water accords with U.S., 44, 46, 121, 138, 207–17, 245, 304; water footprint of, 110; water stress in, 11, 12, 38, 160

microirrigation systems, 76, 102, 171, 174, 280–81; and business opportunities, 280; for fertilizer and pesticide application, 76; global acreage under, 76, 280. *See also* drip irrigation middle class, impact of growing, xiii, 28, 68, 70, 77, 80, 128

Middle East: artificial political borders in, 184; fragile state structure in, xiv, 184; irrigation in, 70, 75; popular uprisings in, 4, 222; water and conflict in, 10, 28, 40, 48, 53, 62, 123, 218–225, 243, 247, 272, 292. *See also specific countries and basins*

Milk River, 210

Millennium Development Goals, 17, 18, 19, 66, 312*n*49

mineral water, 5, 128, 129, 132. *Also see* bottled water

Ming Dynasty, 212

minorities. *See* ethnic-minority homelands

Mischief Reef, 242

Mississippi River, 145, 153

Mississippi State, 37, 164

Missouri River, 156*t*, 228

Mohenjo-Daro ("Mound of the Dead"), 203

Mohne Dam (Germany), 56

Mojave Desert, 212

Moldova, 205

Monaco, 166

Mongolia, 231, 232*t*, 347*n*109

monsoons, 59, 73, 147, 152, 155, 157, 181, 197, 288; potential global-warming impact on, 152, 288

Morales, Evo, 136

Morocco, 187–88; territorial conflict with Algeria, 187–88; water resources in, 188

Moses, 170

Mount Kenya, 155, 156*t*

Mount Kilimanjaro, 155, 156*t*

Mountain Aquifer (West Bank/Israel), 74

Mozambique, 14

Mughal Empire, 3

Murrumbidgee River, 122

Muslims, 185, 220–21, 346*n*90

multi-effect distillation (MED) technology, 283

multistage flash (MSF) technology, 283

Murray-Darling Basin, 122–23, 134

Myanmar. *See* Burma

Myitsone Dam (Burma), 236–37, 348*n*119

Nagorno-Karabakh, 186–87

nanofiltration (NF), 283

Napoleon, 205

NASA, 168

National Center for Atmospheric Research, 153, 154

National Institute of Environmental Health Sciences, 131

National Intelligence Council (U.S.), 42, 334*n*78

National Intelligence Estimate (U.S.), xiii

nationalism, 2, 28, 222, 238. *See also* hydronationalism

national security and water, xiii, xix, xxiv, 38, 48, 53, 123, 160, 177, 178; rise of nontraditional threats, xiii–xiv, 2–3, 20, 22, 38, 47–48, 163, 195, 199–200

National Water Carrier (Israel), 49, 50, 74, 169, 228

Native American, 12, 164–65

natural resources: attempts to corner natural resources, 26, 40, 41, 42, 177, 229, 262; distinction between abiotic and biotic resources, 20; constraints on availability of, xi, 20, 21–23, 25, 27, 28, 29, 34–35; Great Game over, 1, 23, 25, 27; importance of natural capital, 11, 20–21, 24, 25, 28–29, 70; and changing power equations, 21–23, 28; resource curse, 26–27, 291; resource nationalism, 28; resource scarcity, xv, xvi, 1, 11, 13, 21; resource-related shocks, xvi, xvii, 161; resource wars, 2, 13,

24, 26, 28; role of market forces in availability of, xvi, 20, 34–35
Nature Geoscience journal, 146
Nauru, 166
navigational uses of rivers. *See* fluvial navigation
Neelum-Jhelum Hydropower Plant, 262
Negev Desert, 50, 169, 172, 228
Nehru, Jawaharlal, 196
Nepal, 47, 94, 192, 232t, 233, 262
Netherlands: dependency on external water inflows, 111, 160; new water law in, 136, 137; portion under sea level, 146, 334n69
Nevada, 37, 99, 135, 138, 156t, 204, 211
New Mexico, 115, 133, 164
New York City, 96, 294; competition with Philadelphia, 204
New Zealand, 74, 151, 219, 230; water resources in, 219t, 230
Niger, 160, 168, 339n140; dependency on external water inflows, 160
Niger Basin Authority, 275
Niger River, 38, 153, 206, 244, 246, 271, 275; and free navigation among colonial powers, 246
Nigeria, 31, 110, 111, 116, 168, 339n140; irrigation in, 168; water footprint of, 110, 111
Nile River basin, xii, 37, 38, 43, 54, 72, 109, 160, 172, 178, 179, 189, 203, 222, 225, 244, 250, 271, 272, 275; as endangered river, 160, 172; Egypt-Sudan water accord, 72, 179, 244, 250, 304
Nixon, Richard, 214
nongovernmental organizations (NGOs), xvi, 94
nonrenewable resources, 6, 20–21, 28–29, 30, 34–35, 62; fossil waters as, 6, 62, 360g; major new discoveries of, 34–35
nonstate actors, 56, 57. *Also see* terrorism

nontraditional threats, xiii–xiv, 2–3, 20, 22, 38, 47–48, 163, 195, 199–200; and nontraditional sources of water, xiii–xiv, 6, 101, 118, 138, 279, 283, 284, 289, 297, 363g. *See also* unconventional conflicts; unconventional sources of water
Nordic countries, 1, 137, 149, 219
North Africa, 4, 10, 70, 75, 83, 108, 222, 223, 243, 248
North America, 33, 67, 71, 75, 81, 110, 115, 128, 131, 138, 149, 157, 167, 207, 210, 213, 216, 217, 219, 244, 271, 292; irrigation in, 75; as major grain exporter, 67, 115; water resources in, 219t
North American Free Trade Agreement (NAFTA), 123, 124, 137, 138
North American Water and Power Alliance, 124
North Atlantic Treaty Organization, 20, 179, 187
North Korea, 25, 66, 122, 232t, 236; food aid to, 122
North Sea, 205, 216
Northern Hemisphere, 154, 157
Northwest Passage, 150
Norway, 14, 149, 311n34
Nubian Sandstone Aquifer, 248
nuclear power, 56, 95–101, 103, 284; for desalination, 284; potential effects of global warming on, 97; subsidies for, 98–100; vulnerability to natural disasters, 95, 96–97, 98; water intensity of, xxi, 95, 96
Nuozhadu Dam (China), 235, 239

Ob River, 186, 239
Obama, Barack, 99, 100
obesity epidemic: as contributing to water and environmental stresses, xiii, 10, 32–33, 83; and diabetes, 83; as a global killer, 83
Ocalan, Abdullah, 199
Oceania, 31, 219t, 319n25

Ogallala Aquifer, 12, 125
oilseed production, 64, 67–69, 109, 123, 171
Okhotsh Sea, 6
Oman, 226
Omo River, 190, 191
open cycle gas turbine (OCGT) plants, 88
Operation Gibraltar, 49, 317*n*120
Orange (Senqu) River, 47, 244, 305
Organization of American States (OAS), 179
Organization for Economic Cooperation and Development (OECD), 290
Organization for Security and Cooperation in Europe, 187
Oslo II agreement, 271, 305
Osoyoos River, 210
Ottoman Empire, 190, 200
Oubangui River, 168–69
overirrigation, 75, 77
overweight. *See* obesity epidemic
Owens Valley (California), 204
Özal, Turgut, 226, 227
ozone layer, 149

Pacific Institute, 55, 132
Pacific Ocean, 25, 154, 167, 214
Pakistan: and Afghanistan, 185, 196–97; Chinese dam building in, 193–95, 236; and Failed States Index, 13; growing food for export, 108, 116; and Indus Waters Treaty, 44–45, 185, 192–97, 250, 293, 304; as main consumer of Indus waters, 271–72; and proxy war, 48, 56, 294; territorial and water conflicts, 54, 94, 183, 192–97, 262; water distress linked with population explosion and extremism, 13, 219; water as driver of 1965 war against India, 48–49; water quality vs. quantity, 286
Palestine Liberation Organization (PLO), 49, 52

Palestinian territories, 45, 50, 53, 170, 175, 195, 219*t*, 220*t*, 247, 271; and water, 40, 45, 52, 53, 247, 271, 305. *See also* Gaza Strip; West Bank
Pangong Tso (Lukung Lake), Sino-Indian conflict over, 192
Paracel Islands, 242
Paraguay, 54, 244, 248, 272
Paraná River, 53, 54, 153
Patagonia Mountains, 129, 156*t*
Pashtun nationality, 196
Pax Britannica, 23
Peace Treaty of Versailles (1919), 206
Pearl Harbor attack (1941), 24
Pentagon, 158
People's Liberation Army (PLA), 193
permafrost, 143, 361*g*
Permanent Court of Arbitration, 262
Permanent Court of International Justice, 206, 249, 259
Permanent Indus Commission, 273–74
Pew Research Center, 220
Philadelphia, 204
Philippines, 114, 116
Pinochet, Augusto, 136
pipelines, water, 120, 124, 125, 128, 138, 139, 171, 214, 224, 225–29, 286, 289; business opportunities relating to, 225–29; energy intensity of, 289; for peace, 225–29; transnational, 124, 125, 128, 138, 171, 225–29, 289; world's largest underground network of, 228. *See also specific countries*
Polisario Front, 188
politics of water. *See* hydropolitics
"polluter pays" principle, 260, 265, 297
polycarbonate containers, 131
polyethylene terephthalate (PET), 130–31
population growth. *See* demographic trends
Portugal, 4, 24
Potomac-Raritan-Magothy aquifer system, 120

poverty: agriculture as key to
    combating, 113; as enemy of the
    environment, 86, 164; link with
    water poverty, 11, 16, 17, 68, 264
Poyang Lake, 212–13
precautionary principle, 260, 265
pricing of water: as a basic unresolved
    issue, 127, 289–291; concept of full-
    cost pricing, 290; distorting role of
    subsidies, xxii, 6, 71, 108–9, 125,
    126, 140, 269, 277, 290; grassroots
    backlash over, 136–37; Malaysia-
    Singapore dispute over, xiii, 137, 138,
    227; misallocation of underpriced
    water, 140, 290; and OECD, 290;
    price of bottled water vs. crude oil,
    xii, 5, 128; social concerns act as a
    brake, 127; South Africa's example,
    290; valuing water, 133; water as a
    resource with no price, 126
principle of causing no significant harm,
    250, 252, 254, 255, 259–60
privatization of water. *See*
    commoditization of water
protests. *See* grassroots protests over
    dams or water-guzzling industries
public health, xi, xxiv, 10, 17, 99, 123,
    150, 159, 160, 198, 264, 297, 299
pumped-storage technology, 92
pumping race, 175, 224, 247, 270

Qaddafi, Muammar, 3–4, 228. *See also*
    Libya
Qatar, 200, 220, 223, 226
Qing Dynasty, 231
Quantum of Solace, 136
Quaraí/Cuareim River, 244

Radcliffe, Cyril, 185
rainfed agriculture, 62, 63, 73–74,
    104, 361g; dominant in much of
    developed world, 74, 104; share of
    global food produced by, 73–74
rain forests, 67, 70, 145, 148, 149; as
    earth's "lungs," 148

rainwater harvesting, 63, 77, 153,
    174, 269, 279, 287–88, 361g; as an
    ancient technique, 288; and volume
    of precipitation, 152–53, 287–88
Rainy River, 211
Rajasthan State (India), 228
rare-earth minerals, 25, 35, 229, 242;
    use of exports as a weapon, 25, 229,
    242
recycled ("reclaimed") water, 103, 121,
    137, 138, 173, 279, 284–85, 354n86;
    artificial replenishment of aquifers
    with, 173, 285; means to moderate
    water crisis, 173, 279, 284; for
    potable and nonpotable uses, 173,
    285; and "yuck factor," 285
renewable-energy technologies, 88,
    92, 95, 100, 104; non-freshwater-
    consuming types, 88, 104; thirsty
    types, 88
renewable natural resources, 6, 20–21,
    28–29, 30, 34–35, 61, 62, 361g;
    factors contributing to greater
    vulnerability of, 20–21, 29, 30, 34,
    61, 75
reservoirs, xvi, 3, 26, 35, 61, 63, 87,
    92–94, 153, 154, 176, 234, 238, 256,
    287–88, 361g
reservoir-triggered seismicity (RTS),
    235–36
resource capture, 26, 40, 41, 42, 177,
    229, 262
resource curse, 26–27, 291
resource nationalism, 28
resource-related shocks, xvi, xvii, 161
resource scarcity, xv, xvi, 1, 11, 13, 21;
    and violent conflict, xxiv, 27
resource wars, 2, 13, 24, 26, 28
reverse osmosis (RO), 127, 129, 283
Rhine Commission, 205–7, 215
Rhine River, 37, 38, 205–7, 215, 216,
    244, 246, 261, 274; chemical spill of
    1986 into, 215, 261
rice: already-stretched global rice
    market, 65, 116-17; Asia, world's

largest grower and consumer, 281; genetically engineered varieties, 117; major exporters and importers,104, 108, 114, 116; and microrrigation irrigation, 281; and "ponded" culture, 117, 239

Rice, Condoleezza, 184

Rio Grande, 172, 208, 209, 212, 214; and 1906 U.S.-Mexico treaty, 208

Rio Lauca, 179

riparian ascendancy or power, xviii, xxiii, 139, 178, 180, 229, 231, 242, 292

river basin organizations (RBOs), 270–71

river depletion, xvi, 13, 63, 217, 362g

river fragmentation, 18, 92, 94, 176, 362g

riverless regions, 12, 112, 166

Rocky Mountains, 126, 129, 154, 155, 156t

Romania, 205

run-of-river plants, 88, 92–93, 94, 193, 233, 262, 361g, 362g; drawbacks of, 93; limited environmental impacts of, 93

Russia: and Caucasus, 186–87; as grain exporter, 67, 104, 115; natural endowments of, 70, 115, 145, 149; as nuclear and missile power, 230, 295; rich in hydrocarbons, 90; rich in water, 1, 11, 14, 70, 237; water a new divide in ties with China, 26, 47, 180, 231, 232t, 233, 239, 262, 291. *See formerly* Soviet Union

Rwanda, 8, 13, 14

Sadat, Anwar, 53

Saeed Asad, Muhammad, 195

Sahel belt in Africa, 123, 168, 219

Salazar, Ken, 211

salinization, 5, 64, 76, 108, 213, 280, 299, 362g

Salton Sea, 167

Salween River, 37, 93, 156t, 158t, 179, 232t, 233

San Diego, 285, 354n86

Sanaa (Yemen), 3, 161

sanitation services: impact of lack of access to, 10, 17, 160; and mobile-phone access, xvi, 10; and UN, 17, 18, 19

Sanmenxia Dam (China), 233

Santa Cruz Basin, 245

Saudi Arabia: and food production, 112, 114, 116t, 166; groundwater depletion in, 112, 166, 224, 247; irrigated farming in, 112, 114; as leader of Arab oil embargo (1973–1974), 112; to phase out wheat production, 112; transboundary water dispute with Jordan, 176, 224; water distress in, 114, 166, 223, 226, 228–229

Sea of Galilee. *See* Tiberias Lake

seawater incursion: into Bangladesh, 151–52; factors contributing to, xvi, 119, 148, 165, 278; and global warming, 146, 151; into U.S. coastal region, 120; into Yemen's coastal plains, 278

securitization of water resources, xix, 140

sedimentation, 34, 122, 175, 265, 362g

Senate Foreign Relations Committee (U.S.), 44

Senegal, 137

Senegal Basin Permanent Water Commission, 274

Senegal River, 217, 244, 271, 275

Senkaku/Diaoyu Islands, 177

separatist movements, water as a driver of, xxii, 2, 6, 48, 182, 183, 191, 197. *See also* territorial disputes *and* resource wars

sewage, 17, 103, 173, 274, 286

Seyhan River, 226

shale energy: as a game changer, 6, 90; water intensity of fracking, 90, 91; water intensity of shale gas vs. shale oil, 90–91; water contamination threat from, 85, 86, 90–91

Sharon, Ariel, 50
Shatt al-Arab (Arvand Rud) waterway, 90
Shimantan Dam (China), 238
Siachin Glacier, Pakistani-Indian conflict over, 192
Siberia, 143, 149, 160, 210, 334*n*78
Sierra Blanca nuclear-waste repository proposal, 213
Sierra Leone, 163
Sierra Nevada Mountains, 156*t*, 204
silt, 39, 59, 203, 239; nutrient value of, 59, 203, 239
silting of reservoirs and waterways, 34, 122, 175, 362*g*
Sinai Peninsula, 50
Singapore: hydrotechnology in, 138, 285; rainwater capture in, 288; water dispute with Malaysia, xiii, 137, 138, 227; water stress in, 278
Sinohydro Corporation, 234, 241
Six-Day War (1967), 41, 48–52, 187, 230; subregional water map changed by, 48, 50–52, 187
Slovakia, 205, 258–59; water dispute with Hungary, 258–59
Slovenia, 205
soil salinity. *See* salinization
Somalia, 13, 166, 183, 190, 221, 278; drought and famine in, 190
Souris River, 210
South Africa, 3, 8, 11, 45, 46–47, 93, 108, 110, 116*t*, 281, 290; growing food for export, 108; per capita water footprint of, 110; and Lesotho Highlands Water Project, 46–47; "Work for Water" program in, 290
South America, 31, 70, 74, 143, 145, 149, 155, 156*t*, 218, 219*t*, 230, 244, 247, 248, 272, 319*n*25; aquifers in, 247, 248; arable land in, 70; per capita water availability in, 74, 218, 219, 272; uneven distribution of water, 272; water cooperation in, 244. *See also specific basins*

South Asia, 38, 48, 70, 75, 156*t*, 185, 195, 230, 280; irrigated land in, 280. *See also specific countries*
South Atlantic, 6
South Chad Irrigation Project, 168
South China Sea, 23, 157, 177, 229, 242
South Dakota, 12, 228
South Korea: coastal areas in, 103; farmland leasing in Africa by, 66, 71–72, 79; water quality in, 286; water stress in, xiii, 8, 11, 66, 278
South Pacific Islands, 83, 151
South Sudan, Republic of: and Abyei, 189–90; oil resources in, 189; secession from Republic of Sudan, 185, 189; water and other interstate conflicts of, xiv, 189–90; water resources in, 189
Southeast Asia, 40, 149, 155, 180, 231, 236, 280. *See also specific countries*
Southeastern Anatolia Development Project (Turkey). *See* GAP
Southern Africa, 38, 292
Southern Nevada Water Authority, 138
Southwest (U.S.), 4, 74, 89, 124, 156*t*, 161, 164, 208, 212; preference for closed-loop cooling systems in, 89
Soviet Union: divide-and-rule policy in Central Asia, 190; irrigation systems built by, 75, 166–67, 186, 212; unraveling of Soviet empire, 185–86. *See also* Russia
soybean production, 65, 67, 68, 72, 77, 90
Spain: autonomous regions in, 121; Catalonia's water imports, xxi, 120–21; as most dammed country in Europe, 121; intrastate water conflict in, 120, 121, 164; and nuclear power, 97; water stress in, xiii, 4, 10, 120–21, 123
sprinkler irrigation, 280
St. Croix River, 211
St. Lawrence River, 123, 124, 207, 210; and 1854 treaty, 214
St. Lawrence Seaway, 210

St. Mary River, 210
Stalin, Joseph, 190, 191, 212
State Department (U.S.), xiii, xvii, 39, 213
Stern Report, 144, 161
storage capacity for water, xii, xvii, 35, 38, 43, 63, 92, 93, 94, 149, 153, 176, 178, 218, 233, 243, 265, 277, 286, 287–88, 358; in developing world, 35, 287; global average, 35; nature's water storage, 143, 146, 147, 148
Strauss, Lewis L., 98, 324*n*104
sublimation, 61
sub-Saharan Africa: farmland in, 70–71, 74; hunger in, 71, 84, 110; irrigation underdeveloped in, 62, 74; and population explosion, 31; transnational basins in, 37; water resources in, 219*t*
subterranean water. *See* groundwater
Sudan, Republic of: Abyei dispute with South Sudan, 183, 189–90; close ties with China, 25; Darfur conflict in, xiii, 163; foreigners' farmland acquisition in, 72, 166; and its Nile accord with Egypt, 72, 178, 179, 272, 304; water footprint of, 111; as a weak or failing state, 13, 31
sugar, 54, 67, 72, 109, 123
Sumerian civilization, 5, 40, 223
Sunderbans, 152
Supreme Court (U.S.), 126, 164, 251
surface-water resources, xiv, 9, 37, 45, 61, 63, 76, 84, 86, 101, 126, 147, 154, 165, 181, 283, 363*g*; cleaning up contaminated, 285–86; extent of shared, 37. *See also specific countries and regions*
Suriname, 14, 311*n*34
sustainable development, xvi, 19, 94, 142, 147, 168, 172, 260, 261, 266
Sutlej River, 185, 193, 228, 232*t*
Switzerland, 111, 261, 272, 345*n*84
Syr Darya River basin, 156*t*, 158*t*, 166, 186, 190, 271, 275

Syria: and Asi-Orontes, 45, 188–89; civil war in, 200, 201; and Hatay, 188; and Jordan River, 169; and proxy war, 48, 49, 199–200, 201; and Turkey's appropriation of Tigris and Euphrates waters, 139, 179, 180, 198, 199, 225, 226–27, 249–50; water distress in, 42, 108, 226; water and territorial conflicts, 48, 49–50, 52, 53, 188–89, 199–200, 201, 249–50
Syrian Gates, 188

Tajikistan, 139, 178, 186, 190, 191, 192
Taliban, 197
Tanganyika Lake, 244
Tanzania, 108, 136, 156*t*, 184–85, 189, 244; conflict with Malawi, 184–85; as cotton exporter, 108
tariffs on water. *See* pricing of water
technology: and "adoption lag," 270; for clean water, xvii, 85, 140, 260, 268–70, 279, 282–86, 290–91, 298; weaponization of science, 20
Tennessee River, 98
Tennessee Valley Authority, 196, 233
territorial disputes: as legacy of colonially drawn borders, 183–85; territorial-water nexus, 181–202; over water-rich areas, 2, 41, 52, 55, 177, 191, 197, 201. *See also specific countries and territories*
terrorism: transnational threats from, 31, 158, 293–94; nexus with water scarcity and overpopulation, xxii, 31, 218–25; as threat to water infrastructure, 2, 56, 57. *See also specific terrorist organizations*
Texas, 74, 116, 126, 133, 135, 164, 172, 212, 214
Thailand: biofuel production in, 68; irrigation in, 104, 290; meat consumption in, 81; virtual water exports of, 104, 108, 116; world's

largest rice exporter, 116*t*, water
　　stress in, 116
The Limits of Growth (Club of Rome),
　　xv–xvi
The Rime of the Ancient Mariner
　　(Samuel Taylor Coleridge), 119
The Road to Serfdom (Friedrich von
　　Hayek), 150
theory of rational choice, 41
thermoelectric power plants, xxi, 86–88,
　　101, 102; environmental impacts
　　of, 88; non-freshwater sources for
　　cooling, 102–3, 282; types of cooling
　　systems, 88–90, 102
Three Gorges Dam (China):
　　environmental impacts of, 212, 239;
　　forced eviction of residents due
　　to, 234; as symbol of engineering
　　prowess, 93, 212, 235
Three Mile Island, 95, 100
Tiananmen Square, 159
Tiberias Lake (Lake Kinneret/Sea of
　　Galilee), 50, 52, 74, 169, 225, 228
Tibet: as Asia's "water tower," 49;
　　climate-change effects on, 143, 150,
　　155–57; forcibly annexed by China,
　　xxiii; hydroengineering projects in,
　　6, 115, 181, 192, 193, 235, 240;
　　impacts on ecosystems in, 129,
　　172; resource extraction and water
　　pollution in, 91; and Sino-Indian
　　disputes, 55, 177, 183; separatist
　　unrest in, 2, 191, 236; as source of
　　China's hydro-hegemony in Asia,
　　49, 161, 201, 231; as source of
　　international rivers, 49, 185, 231,
　　232*t*, 239; as "Third Pole," 155;
　　unique hydrological role of, 157; as
　　world's largest freshwater repository,
　　155, 231
Tibetan Plateau. *See* Tibet
Tigris River, 5, 37, 53, 139, 161, 179–
　　80, 189, 190, 198–99, 203, 223, 225,
　　226, 249, 271
Tijuana River, 208

To the Last Drop (Andrew Wice), 164
toilets, lack of, 10, 14; and menstruating
　　girls, 14; water-saving toilet
　　technologies, 174
Tokyo Foundation, 177, 340*n*9
Tonga, 166
total renewable water resources
　　(TRWR), 74, 75, 234
tourism, 94, 141, 148, 335*n*83
trade, 24, 25, 61, 246, 260, 292. *See
　　also* bulk-water trade; water trading;
　　virtual water
transboundary basins: and
　　decolonization, 183–85; extent of,
　　37–38, 183, 247; and international
　　law, 246–59; militarily changed
　　equations in, 46–55, 187, 229–32;
　　role of power equations in, xxiii,
　　179–80, 249, 273
Transboundary Freshwater Dispute
　　Database, 272
Treaty of Berlin (1878), 205, 215
Treaty of Paris (1856), 215
Treaty of Peace between Israel and
　　Jordan (1994), xii–xiii, 44–45, 46,
　　53, 225, 271, 293; and its Annex 2
　　on water, 225, 293, 355*n*103
Treaty Relating to Boundary Waters
　　between the United States and
　　Canada (1909), 123, 207, 208–9,
　　214, 250
Truong Tan Sang, 239
tsunamis, xxi, 59, 95, 96, 99, 148
Tunisia, 8
Turkana, Lake, 172, 190–91; Ethiopian-
　　Kenyan disputes over, 190–91
Turkey: appropriation of Tigris and
　　Euphrates waters, 139; collapsed
　　water deal with Israel, 139; dam
　　building in, 201, 225; GAP program
　　in, 39, 197–201; as a hydro-
　　hegemon, xxiii, 161, 179–80, 231;
　　and Kurdish issue, 161, 183, 191,
　　197, 199–201; as largest source
　　of waters to Tigris-Euphrates

Basin, 180; link between water and territorial issues, 180, 188–89, 199–200, 225, 226–27; military accord with Israel, 199; and opposition to UN Water Convention, 226–27, 248, 249–50, 254; political tussle within, 200–201; and proxy war, 48, 199–200; regional geopolitical clout of, 179–80; water dependency ratio, 161; and water pipelines, 226–27, 228

Turkey Point nuclear power plant (U.S.), 97

Turkmenistan, 37, 108, 139, 160, 178, 186, 191, 224; and the curse of cotton, 108, 186; dependency ratio on external water inflows, 37, 160, 224

Tuvalu, 166

Twain, Mark, xxiv

unconventional conflicts, xiii–xiv, 2–3, 20, 22, 38, 47–48, 163, 195, 199–200; and nontraditional (or unconventional) sources of water, xiii–xiv, 6, 101, 118, 138, 279, 283, 284, 289, 297, 363*g*; use of riparian leverage or terrorist proxies in, xiii–xiv, 2, 48, 49, 53, 139, 163, 195, 199–200, 294; water as source of, xiii, 2–3, 38, 47–48, 49, 53, 199, 294

unconventional (or nontraditional) sources of water, xiii–xiv, 6, 101, 118, 138, 279, 283, 284, 289, 297, 363*g*; as sustainable new supply paths, 6, 298

Uganda, 14, 189

UK Environment Agency, 229

Ukraine, 100, 104, 115, 205

Ullman, Richard H., 38

UN Convention. *See* United Nations Convention on the Law of the Non-Navigational Uses of International Watercourses

UN Human Rights Council, 17

UN-Water, 298

UNESCO World Heritage site, 152

United Arab Emirates, 3, 72, 220, 223, 226; and Abu Dhabi's water crisis, 3

United Kingdom. *See* Britain

United Nations: on interstate water cooperation, 42–43, 244–45, 275; on population growth, 31, 221; on right to safe drinking water, 17; role in facilitating water accords, 274–76; on spreading water stress, xii, 61, 62; on waterborne diseases, 18; on water control within national borders, 267; on water-related risks, 245; and World Water Day, 18

United Nations Convention on the Law of the Non-Navigational Uses of International Watercourses (1997), xxiii, 44, 226–27, 232, 233, 248, 249, 252–57; and Berlin Rules, 253; China and Turkey as leading opponents of, 226–27, 231–32, 248, 249–50, 254; entry into force of, 44, 249, 252, 255, 256; and fossil aquifers, 256–57; gaps in, 254–55; and need for a follow-up protocol, 256; principles enshrined in, 252–57; Website link to, 301

United Nations Development Programme, 13, 14, 113, 223, 274; and Human Development Index, 13; role in water agreements, 274

United Nations Economic Commission for Europe (UNECE), 259–60

United Nations Educational, Scientific and Cultural Organization (UNESCO), 38, 152, 247

United Nations Environment Programme (UNEP), 163

United Nations Framework Convention on Climate Change (1992), 142, 256

United Nations General Assembly, 17, 248

United Nations Population Division, 31

United Nations Security Council, 22, 25, 190, 258

United Nations University, 53

United States: aquifers in, 12; antibiotic use on food animals, 77–78; and the Arctic, 149–50; biofuel production in, 67, 68; climate-change impact on Rockies, 154–55, 156*t*; dams in, 233, 234, 241; diabetes prevalence in, 83; as food exporter, 68, 104, 108, 115, 116; and genetically engineered food, 72; greater river runoff in, 153–54; groundwater withdrawals in, 9–10, 12, 75, 166; hydropower in, 93, 94; intrastate water disputes in, 37, 164–65; irrigation in, 9, 12, 74, 75; losses from leaky pipes, 288; meat consumption in, 79, 81; Mexican immigrants in, 152, 335*n*86; with most number of overweight or obese people, 32–33; natural endowments of, 70; nuclear power in, 97, 98, 99; resource acquisition as geopolitical driver, 24; saline-water use, 102; shale energy boom in, 90–91; share of electricity used in water treatment and supply, 85; subsidies in, 68, 81, 290; thermoelectric water needs in, 87, 103; water accords with Mexico, 44, 46, 121, 138, 207–17, 245, 304; water dependency ratio, 207; water intensity of energy sector, 86; water markets and rights in, 126, 133–34, 135, 277; water relations with Canada, 46, 123–25, 137, 207–14, 217, 244, 245, 250, 259, 271, 304; water resources in, 12, 70, 210; water stress in, xii, 4, 11, 74, 110, 123, 208, 268; as world's most heavily consuming society, 33, 85, 110; as world's most water-profligate state, 14. *See also specific issues or projects*

Universal Declaration of Human Rights (1948), 17

urbanization, xii, xx, 10, 21, 30, 69, 268; contribution to reducing global farmland, 69, 299; increase in number of cities, 30; increased water use by urban households, 18, 30, 172, 268, 281, 285

Uruguay, 116*t*, 136–37, 195, 244, 248, 272, 274

Uruguay River, 153, 274

U.S. Army Corps of Engineers, 241

U.S. Atomic Energy Agency, 98

U.S. Food and Drug Administration (FDA), 131, 320*n*33

U.S. Geological Survey, 14, 335*n*79

U.S.-Mexico Water Treaty (1944), 44, 207–17

U.S. National Academy of Sciences, 157, 285

U.S. National Institutes of Health, 131

U.S. Supreme Court. *See* Supreme Court

Utah, 37

Uzbekistan: and the curse of cotton, 108, 186; intercountry water and territorial disputes of, 55, 178, 190, 191; reliance on transboundary water inflows, 139, 178, 186, 191; riparian ascendancy of, 178, 231; and its threat of military reprisals, 178

Vakhsh River, 178

vegetable oils, 67–68

Venezuela, 125, 156*t*

Vienna Convention on the Law of Treaties, 273, 294–96; definition of "fundamental change of circumstances" under, 295–96; on dissolution of a treaty, 294, 296; and water pacts of indefinite duration, 294–96

Vietnam: attacks on water systems during war, 56; meat consumption in, 80, 81; and Mekong River, 239; rice exports, 104, 116; transboundary water flows into, 160, 232*t*; water footprint of, 110; water a new divide in ties with China, 26, 47, 233, 239, 242, 262; water stress in, 11

virtual database (VDB), 267
virtual water, xxi, 54, 104–18, 244, 289,
    363*g*; as a concept, 105, 107, 111;
    factors limiting benefits from trade
    in, 111–18, 289; and food security,
    113–14; green/blue water sources,
    104–5, 108; importance in agricultural
    trade, 104, 109, 115, 117; and world
    food markets, 113; and national
    water footprints, 106, 109–11;
    nascent research field, 106–14; water
    embedded in trade flows, 109; and
    water stress, 106, 107, 108, 109, 110
Volga River, 186

Washington Treaty Relating to the
    Navigation of St. Lawrence River
    (1871), 214
wastewater, 17, 35, 85, 86, 91, 129,
    133, 211, 215, 228, 261, 282, 284,
    285, 286; as contributor to pollution,
    17, 35, 85, 91, 215, 286; as threat to
    ecosystems, 17, 35, 86, 91
wastewater recycling. *See* water
    recycling
water agreements between countries,
    39–44, 244–45, 272–76, 293–96;
    challenges of implementation, 276;
    deceptively long list of, 40, 272–73;
    gaps and inadequacies in, 42, 43,
    46, 244–45, 272–73, 275, 276; list
    of existing sharing accords, 303–6;
    number of existing, xxiii, 40, 42–43,
    244, 273, 303–6; role of a third party
    in, 274–76; toothless, 43; and use of
    proxy/covert means, 294
water allocations, 43, 44, 49, 134, 172,
    175, 178, 180, 208, 264, 268, 271,
    293
water availability: by continent, 9*f*, 74,
    218, 219*t*; decline in global, 61–62;
    defined, 10; seasonal or spatial
    variability in, xii, xvii, 8, 16, 19, 93,
    114, 153, 154, 155, 160, 219, 264,
    270, 279, 287

waterborne diseases, 10, 18, 59
water competition: increased risks
    arising from, xix, 245–46, 267, 268–
    72; linked to regional politics and
    security, 175–202, 217–42, 268–72;
    "securitization" trend, xix, 140
water conflict: arising from information
    gaps, 266–67; as battle for political
    and social justice, 163, 204; by
    covert or nonmilitary means, xiii, 2,
    48, 176, 293–96; increased risks of,
    xii, xviii, 35–58, 162, 165, 244–45;
    linkage with water scarcity, 162,
    163, 164, 204, 218–225; linkage
    between interstate and intrastate, xii,
    191; low intensity, xviii, 180, 242;
    nexus with territorial disputes, 187–
    202; between pastoralists, xiv–xv,
    190–91; and rejiggering of political
    borders, xxii, 183–87; sources of,
    163–64; 182–83, 197, 254, 273. *See
    also* water wars
water consumption growth, xv, 34
water cooperation: building stable,
    259–277; changing patterns of,
    175–76; 245–46; limited utility of
    historical examples, xviii-xix, xxii,
    42, 245, 268; number of existing
    treaties, xxiii, 40, 42–43, 244, 247,
    273; in the past, 202–217, 246;
    plugging information gaps to build,
    266–67; in the form of hydrological-
    data sharing, 43, 56, 180–81, 233,
    266–67, 272
water crisis, xiii, xiv, xvii, xxiv, 2,
    8, 12, 16, 17, 60, 61, 66, 68, 119,
    121, 123, 140, 175, 218–25, 259,
    268–91, 299–300; and new business
    and innovation opportunities, xxiv,
    277–91; skills and technologies to
    alleviate, 289–91
water dependency ratio, 160–61, 207,
    217, 224, 237, 243, 338*n*120, 358*g*
water distribution: building efficiency
    in, 140, 277, 279, 286–89;

Index

imbalances in, xii, 8, 16, 19, 160, 264, 270; irrigation-related losses, 75, 76; potable-water losses in U.S., 288; preventing water leakages and recontamination, 288–89
water efficiency, xix, 12, 16, 19, 68, 77, 86, 104, 106, 107, 114, 117, 118, 140, 173, 243, 246, 269, 277, 278, 279–82; centrality of agriculture in, 280–81; and industry, 281–82
water and energy intersection, 84–104
Water Exploitation Index (WEI), 10
water-filtration technology, 101, 127–28, 282–83
water footprint, 106, 109–11, 117, 281; of food products, 78; national, 85, 106, 109–11
water-induced shocks, xvii, 161, 163
water infrastructure, 286–91, 363g; cross-border investments in, 296–97; leaks and system inefficiencies, 75, 76, 288–89; public spending on, 286, 291; recontamination of treated water, 288–89; role of pipelines, 228–29, 289; storing rain and melt runoff, 287–88
water gap, 11, 17, 172, 243, 277, 298, 363g
water intensity of global economy, 12, 34
water institutions: at bilateral or basin level, 35–37, 39, 44–46, 243–45, 246, 247, 259–268, 270–77; evolutionary process to build, 213–17; and international norms, 246–59; and peace, 291–300
Water Justice Movement, 140
water laws. *See* international water law
waterlogging, 75, 108, 271, 364g; and soil salinity, 75, 108
water management, xxiv, 7, 14, 17, 46, 77, 108, 113, 118, 147, 173, 185, 203, 225, 243, 268, 274, 276, 277, 278, 279, 282; and better allocation,

distribution, and pricing, 277; integrated, 19, 103, 118, 268, 269, 278–79; technologies and practices in, 282; and water-sector reforms, 291
water markets: in Australia, xxi, 133, 134–35, 136, 277; and bulk-water trade, xxi, 8, 125, 127, 137; Catalonia's water imports, xxi, 120–21; in Chile, xxi, 135–36, 277; international, 135, 137–40, 291; national, 133–36; in U.S., 126, 133–34, 135, 251; water as next big investment opportunity, 125; and water rights, xxi, 38, 45, 53, 121, 122, 125, 126, 129, 133, 134, 135, 136, 138, 140, 164, 165, 204, 212, 248, 251, 252, 264, 270, 271, 277. *Also see* commoditization of water *and* water rights
water nationalism, 2, 238
water peace, 40, 225–28, 296–300; investing in, 296–300; pipelines for peace, 225–29
"water peace" school of thought, 40–41
water pollution, xii, 12, 13, 18, 20, 34, 35, 61, 82, 90, 91, 111, 121, 130, 132, 175, 187, 210, 213, 215, 216, 240, 248, 252, 261, 264, 265, 271, 274, 286, 293, 299; as aggravating water crisis, 274, 286, 293; cleaning up, 285–86; and industrial sector, 11, 261, 265, 281–82; as the greatest killer, 5–6; and "polluter pays" principle, 260, 265, 297; through overuse of nitrogen fertilizers, xv, 82, 299
water poverty, xiv, 4–5, 11, 14, 16, 108, 169, 219–23; defined, 4–5; linkage with economic poverty, 11, 16
Water Poverty Index, 14
water productivity. *See* water efficiency
water quality, 3, 5, 13, 18, 20, 43, 63, 91, 92, 102, 115, 165, 173, 211, 214,

216, 261, 265, 270, 271, 274, 286, 293; importance of 20, 63, 270, 274, 286, 293; as a means to alleviate water distress, 270, 274, 286

water recycling, xiv, 6, 12, 85, 103, 104, 118, 140, 172, 173, 174, 228, 269, 279–85, 286, 289, 298, 363g; new technologies for, 140, 269, 298; safe disposal of residues from treatment, 174

water refugees, xxii, 5, 151–52, 161–65

water resources: as asset class, 1, 291; commercialization of, 125–41; decolonization and, 183–85; and disease control, xi, 10; as flash point, 1, 2, 38, 175, 232, 244, 276; global per capita availability of, 8; holistic management of, 19, 103, 118, 268, 269, 278–79; as instigator of economic, social, and technological change, 270; and national security, xix, 140; nature's unequal distribution of, xii, 8, 16, 19, 160, 264, 270; new era of water stress, xi, xxii, 42, 59; as new oil, 5–8, 126, 291; permits for extraction of, 126; supply/demand gap, 11, 17, 172, 243, 277, 298; threat from environmental and climate change, 141–60

water rights, xxi, 38, 45, 53, 121, 122, 125, 126, 129, 133, 134, 135, 136, 138, 140, 164, 165, 204, 212, 248, 251, 252, 264, 270, 271, 277; as akin to property rights, 126, 133, 134, 135, 251; and water laws, 133–34, 251. *Also see* water markets

water scarcity: adaptation to, xi, 153, 300; business opportunities from, xxiv, 277–91; and changing nature of threats and conflicts, xiv, 38; defined, 8, 363g; as driver of human migration patterns, 5; as an existential threat, xiv, 151, 219; largely concentrated in developing

world, 10, 13; leading to rationing, 3, 4, 120; link with religious extremism and terrorism, xiv, 220–22; nexus with overpopulation, xxii, 13, 220–23; and weak states, xvii, 20, 278; in the West, 4

water-sector reforms, 279

water security and insecurity, xix, xxiv, 48, 57, 79, 83, 114, 243, 245, 268, 294, 297, 298; defined, 364g; and harmony with nature, xv, 300; link with regional security/insecurity, 297; obesity and, 32–33; one nation's water security as another state's water insecurity, 57

water-sharing arrangements and treaties, 39–44, 244–45, 272–76, 293–96, 303–6

water stress: defined, 8, 364g; and rise of nontraditional security challenges, 38–39; spread of, 1, 4; and violent conflict, 4–5, 13, 46, 183, 219–221, 245

water tankers, xxi, 120, 124, 137, 139, 141

water tariffs. *See* pricing of water

water trading between nations, 1, 137, 141, 291; Canadian opposition, 124–25, 137, 138, 226; Malaysia-Singapore case, xiii, 137, 138, 227. *See also* water markets; bulk water

water warlords, xxii, 162

water warriors, 161, 162, 165

water wars: attacks on water infrastructure, 55–58; averting, xviii, 55, 57, 256, 276; as a concept, xviii, 1, 55; fought by another name, 46–55; intrastate, 136, 204; through resort to military force, xviii, 39, 48–53, 211, 230; by nonmilitary means, xviii, 2, 6, 47–48; silent types, 2, 6; use of terrorists or other proxies in, xxiv, 2, 48, 176, 293–96. *See also* water conflict water as a weapon, xiii, xviii, xxii, 2, 50, 55, 56, 57, 176–81, 227, 230, 242, 256

Wazzani Springs, 52

weather-modification technology. *See* geoengineering technology

Wellton-Mohawk irrigation project, 214

Wen Jiabao, 238

West Africa, 153, 160

West Bank: Israel's control over water resources of, 41, 50, 52, 53, 183, 271; and Jordan River, 52, 53, 169; and "Oslo II" accord, 271, 305; Palestinian access to water, 53, 271; water resources in, 52

Western Europe, 97, 110, 205

Western Sahara (Spanish Sahara), 187–88

Western Sahara War (1975–1991), 187–88

wetlands, xvi, 9, 21, 121, 129, 143, 148, 149, 161, 165, 168, 169, 173, 176, 192, 223, 224, 285, 329*n*13, 364*g*; coastal, 148, 173; destruction of, xvi, 21, 143, 165, 173, 223; vegetation of, 148

wheat: as animal feed, 77; Egypt's import of, 109; as a factor in water depletion in Saudi Arabia, 112, 114, 166; in Japan, 109; as key staple, 65; leading exporters of, 104; major importers of, 116; production in China, 115; production in EU, 115; production in Jordan, 172; production in U.S., 116; versus rice, 116

Wice, Andrew, 164

Wolf, Aaron, 41

women, as water providers and users, 14–16

World Bank, 33, 193, 195, 196, 271, 274, 276; aid for dam building, 195; and China's emergence as world's largest lender, 275–76; on Palestinians' water access, 271; role in Indus treaty, 193, 196; role in other water agreements, 274–76

World Charter for Nature (1982), 27, 56, 141, 318*n*142

World Health Organization (WHO), 18, 83, 132, 285; and Codex Alimentarius Commission, 132; standards for recycled water, 285

World Summit on Sustainable Development, 19

World Trade Organization, 25

World War II, 56; international institutions set up after, xi; number of dams built after, 61; redrawing of political borders since, 183–84

World Water Council, 259

World Water Day, 18, 298

World Water Forum, 259, 298

Wullar Lake, 56

Wyoming, 126, 133, 251

Xi Jinping, 234, 347*n*113

Xiaowan Dam (China), 235

Xinjiang: climate-change impact on, 157; as source of transboundary water, 231, 232*t*, 239; unrest in, 236; water pollution in, 91

Xu Yuanming, 238

Yalu River, 232*t*

Yamdrok Tso Lake, 172

Yangtze River: and Great South–North Water Diversion Project, 115, 239–41; Three Gorges Dam's impact on, 93, 234, 239, 241; Tibetan Plateau as source of, 155, 156*t*

Yarlung Tsangpo Great Canyon. *See* Brahmaputra Canyon

Yarmouk River, 44, 52, 53, 169, 172
Yellow River: as endangered river, 3,
    119, 153, 159, 160, 173, 237; and
    Great South–North Water Diversion
    Project, 115, 239–41; Tibetan
    Plateau as source of, 155, 156*t*
Yemen: desalination as prohibitive
    option for, 278; as failing state, 13,
    161, 223; as a hydrological basket
    case, 162; parched future for, 13,
    161–62; Sanaa's bleak future, 3, 161;
    water scarcity and violence, 3, 13,
    31, 161, 183, 221
Yenisey River, 153

Yuan Dynasty, 231
Yucca Mountain, 99
Yugoslavia, 185
Yuma Desalting Plant, 215

Zambezi River, 38, 53
Zambia, 37, 244, 347*n*116
Zardari, Asif Ali, 195, 342*n*33
Zimbabwe, 25, 111
zinc, 21, 34, 36*f*
Zinni, Anthony, 39
Zionist dream, 50, 172
Zipingpu Dam (China), earthquake
    linked to, 236

About the Author

Brahma Chellaney is a geostrategist following major international trends. He is a professor of strategic studies at the Center for Policy Research in New Delhi, a fellow of the Nobel Institute in Oslo, a trustee of the National Book Trust, and an affiliate with the International Centre for the Study of Radicalization at King's College London. He has served as a member of the Policy Advisory Group headed by the foreign minister of India and an advisor to India's National Security Council.

As a specialist on international strategic issues, he held appointments at Harvard University, the Brookings Institution, the Paul H. Nitze School of Advanced International Studies at Johns Hopkins University, and the Australian National University. He has also been a Bosch Public Policy Fellow at the Transatlantic Academy in Washington, D.C.

His scholarly essays have been published in numerous journals, including *International Security*, *Orbis*, *Survival*, *Terrorism*, *Washington Quarterly*, *Nature*, and *Security Studies*. He is also a columnist and commentator, including for Project Syndicate. His opinion articles appear in the *International Herald Tribune*, *Wall Street Journal*, *Financial Times*, *Le Monde*, *The Guardian*, *Economic Times*, *Japan Times*, *La Vanguardia*, *Straits Times*, *South China Morning Post*, and other newspapers. And he has often appeared on CNN and BBC, among others.

He sits on a number of national and international organizational boards, including the Academic Council of the Henry Jackson Society, London. He has addressed the annual conference of German ambassadors in Berlin and international business forums like the Asia-Europe Business Forum, CLSA Investors' Forum, Global ARC, and FutureChina Global Forum. In addition, he has participated in high-powered initiatives such as the World Economic Forum at Davos, Doha Forum, Bergedorf Roundtable, Singapore Global Dialogue, Berlin Foreign Policy Forum, and Robert Bosch Annual Forum.

He is the author of seven books, including *Asian Juggernaut* (2010) and *Water: Asia's New Battleground* (2011). Among his other publications are *Controlling the Taps*, a 125-page "Blue Book" for international institutional investors published in 2012 by CLSA, a wholly owned subsidiary of Credit Lyonnais SA, and a monograph, *From Arms Racing to "Dam Racing,"* published by the Transatlantic Academy in 2012. His book, *Water: Asia's New Battleground*, won the 2012 Bernard Schwartz Award from the New York–based Asia Society.